**IET MATERIALS, CIRCUITS AND DEVICES SERIES 57**

# Cross-Layer Reliability of Computing Systems

## Other volumes in this series:

| | |
|---|---|
| Volume 2 | **Analogue IC Design: The current-mode approach** C. Toumazou, F.J. Lidgey and D.G. Haigh (Editors) |
| Volume 3 | **Analogue–Digital ASICs: Circuit techniques, design tools and applications** R.S. Soin, F. Maloberti and J. France (Editors) |
| Volume 4 | **Algorithmic and Knowledge-Based CAD for VLSI** G.E. Taylor and G. Russell (Editors) |
| Volume 5 | **Switched Currents: An analogue technique for digital technology** C. Toumazou, J.B.C. Hughes and N.C. Battersby (Editors) |
| Volume 6 | **High-Frequency Circuit Engineering** F. Nibler *et al.* |
| Volume 8 | **Low-Power High-Frequency Microelectronics: A unified approach** G. Machado (Editor) |
| Volume 9 | **VLSI Testing: Digital and mixed analogue/digital techniques** S.L. Hurst |
| Volume 10 | **Distributed Feedback Semiconductor Lasers** J.E. Carroll, J.E.A. Whiteaway and R.G.S. Plumb |
| Volume 11 | **Selected Topics in Advanced Solid State and Fibre Optic Sensors** S.M. Vaezi-Nejad (Editor) |
| Volume 12 | **Strained Silicon Heterostructures: Materials and devices** C.K. Maiti, N.B. Chakrabarti and S.K. Ray |
| Volume 13 | **RFIC and MMIC Design and Technology** I.D. Robertson and S. Lucyzyn (Editors) |
| Volume 14 | **Design of High Frequency Integrated Analogue Filters** Y. Sun (Editor) |
| Volume 15 | **Foundations of Digital Signal Processing: Theory, algorithms and hardware design** P. Gaydecki |
| Volume 16 | **Wireless Communications Circuits and Systems** Y. Sun (Editor) |
| Volume 17 | **The Switching Function: Analysis of power electronic circuits** C. Marouchos |
| Volume 18 | **System on Chip: Next generation electronics** B. Al-Hashimi (Editor) |
| Volume 19 | **Test and Diagnosis of Analogue, Mixed-Signal and RF Integrated Circuits: The system on chip approach** Y. Sun (Editor) |
| Volume 20 | **Low Power and Low Voltage Circuit Design with the FGMOS Transistor** E. Rodriguez-Villegas |
| Volume 21 | **Technology Computer Aided Design for Si, SiGe and GaAs Integrated Circuits** C.K. Maiti and G.A. Armstrong |
| Volume 22 | **Nanotechnologies** M. Wautelet *et al.* |
| Volume 23 | **Understandable Electric Circuits** M. Wang |
| Volume 24 | **Fundamentals of Electromagnetic Levitation: Engineering sustainability through efficiency** A.J. Sangster |
| Volume 25 | **Optical MEMS for Chemical Analysis and Biomedicine** H. Jiang (Editor) |
| Volume 26 | **High Speed Data Converters** A.M.A. Ali |
| Volume 27 | **Nano-Scaled Semiconductor Devices** E.A. Gutiérrez-D (Editor) |
| Volume 28 | **Security and Privacy for Big Data, Cloud Computing and Applications** L. Wang, W. Ren, K.R. Choo and F. Xhafa (Editors) |
| Volume 29 | **Nano-CMOS and Post-CMOS Electronics: Devices and modelling** S.P. Mohanty and A. Srivastava |
| Volume 30 | **Nano-CMOS and Post-CMOS Electronics: Circuits and design** S.P. Mohanty and A. Srivastava |
| Volume 32 | **Oscillator Circuits: Frontiers in design, analysis and applications** Y. Nishio (Editor) |
| Volume 33 | **High Frequency MOSFET Gate Drivers** Z. Zhang and Y. Liu |
| Volume 34 | **RF and Microwave Module Level Design and Integration** M. Almalkawi |
| Volume 35 | **Design of Terahertz CMOS Integrated Circuits for High-Speed Wireless Communication** M. Fujishima and S. Amakawa |
| Volume 38 | **System Design with Memristor Technologies** L. Guckert and E.E. Swartzlander Jr. |
| Volume 39 | **Functionality-Enhanced Devices: An alternative to Moore's law** P.-E. Gaillardon (Editor) |

| | |
|---|---|
| Volume 40 | **Digitally Enhanced Mixed Signal Systems** C. Jabbour, P. Desgreys and D. Dallett (Editors) |
| Volume 43 | **Negative Group Delay Devices: From concepts to applications** B. Ravelo (Editor) |
| Volume 45 | **Characterisation and Control of Defects in Semiconductors** F. Tuomisto (Editor) |
| Volume 47 | **Understandable Electric Circuits: Key concepts, 2nd Edition** M. Wang |
| Volume 48 | **Gyrators, Simulated Inductors and Related Immittances: Realizations and applications** R. Senani, D.R. Bhaskar, V.K. Singh, A.K. Singh |
| Volume 49 | **Advanced Technologies for Next Generation Integrated Circuits** A. Srivastava and S. Mohanty (Editors) |
| Volume 51 | **Modelling Methodologies in Analogue Integrated Circuit Design** G. Dundar and M.B. Yelten (Editors) |
| Volume 53 | **VLSI Architectures for Future Video Coding** M. Martina (Editor) |
| Volume 54 | **Advances in High-Power Fiber and Diode Laser Engineering** I. Divliansky (Editor) |
| Volume 55 | **Hardware Architectures for Deep Learning** M. Daneshtalab and M. Modarressi |
| Volume 58 | **Magnetorheological Materials and Their Applications** S. Choi and W. Li (Editors) |
| Volume 59 | **Analysis and Design of CMOS Clocking Circuits for Low Phase Noise** W. Bae and D.K. Jeong |
| Volume 60 | **IP Core Protection and Hardware-Assisted Security for Consumer Electronics** A. Sengupta and S. Mohanty |
| Volume 64 | **Phase-Locked Frequency Generation and Clocking: Architectures and circuits for modem wireless and wireline systems** W. Rhee (Editor) |
| Volume 65 | **MEMS Resonator Filters** R.M. Patrikar (Editor) |
| Volume 67 | **Frontiers in Securing IP Cores: Forensic detective control and obfuscation techniques** A. Sengupta |
| Volume 68 | **High Quality Liquid Crystal Displays and Smart Devices: Vol. 1 and Vol. 2** S. Ishihara, S. Kobayashi and Y. Ukai (Editors) |
| Volume 69 | **Fibre Bragg Gratings in Harsh and Space Environments: Principles and applications** B. Aïssa, E.I. Haddad, R.V. Kruzelecky, W.R. Jamroz |
| Volume 70 | **Self-Healing Materials: From fundamental concepts to advanced space and electronics applications, 2nd Edition** B. Aïssa, E.I. Haddad, R.V. Kruzelecky, W.R. Jamroz |
| Volume 71 | **Radio Frequency and Microwave Power Amplifiers: Vol. 1 and Vol. 2** A. Grebennikov (Editor) |
| Volume 72 | **Tensorial Analysis of Networks (TAN) Modelling for PCB Signal Integrity and EMC Analysis** B. Ravelo and Z. Xu (Editors) |
| Volume 73 | **VLSI and Post-CMOS Electronics, Volume 1: Design, modelling and simulation and VLSI and Post-CMOS Electronics, Volume 2: Devices, circuits and interconnects** R. Dhiman and R. Chandel (Editors) |

# Cross-Layer Reliability of Computing Systems

Edited by
Giorgio Di Natale, Dimitris Gizopoulos, Stefano Di Carlo, Alberto Bosio and Ramon Canal

The Institution of Engineering and Technology

Published by The Institution of Engineering and Technology, London, United Kingdom

The Institution of Engineering and Technology is registered as a Charity in England & Wales (no. 211014) and Scotland (no. SC038698).



First published 2020



The Institution of Engineering and Technology
Michael Faraday House
Six Hills Way, Stevenage
Herts, SG1 2AY, United Kingdom

www.theiet.org



**British Library Cataloguing in Publication Data**
A catalogue record for this product is available from the British Library

**ISBN 978-1-78561-797-3 (hardback)**
**ISBN 978-1-78561-798-0 (PDF)**

Typeset in India by MPS Limited

# Contents

*Part I*

# Design techniques to improve the resilience of computing systems

# Chapter 1
# Technological layer

*Antonio Rubio*[1] *and Ramon Canal*[2]

## 1.1 Introduction

This chapter describes the fundamental characteristics of Complementary Metal–Oxide–Semiconductor (CMOS) technology, and how it can be assessed for system reliability studies. After some definitions, the dominating manufacturing technologies are described together with its advantages and disadvantages. Then, the core memory circuits present in today's computing systems are presented. Finally, the chapter provides an evaluation of these memory circuits when considering reliability across technology nodes.

### *1.1.1 Faults, errors and failures*

Faults, errors and failures are terms that are often confused but have different meanings [1]. A fault is a defect that may trigger an error, stay dormant or simply disappear. Faults in hardware structures could arise from defects, imperfections, or interactions with the external environment. Examples of faults include manufacturing defects in silicon chip or bit flips caused by cosmic ray strikes.

Faults are usually classified into three categories: permanent, intermittent and transient. Permanent faults remain for indefinite periods till corrective action is taken. Oxide Wearout leading to a transistor malfunction is an example. Intermittent faults appear, disappear and then reappear and are often early indicators of permanent faults. Finally, transient faults are those that appear and disappear in a very short period of time (typically one cycle). Bit flips or gate malfunctions due to a neutron strike are examples of transient faults. A fault in a particular system layer may not show up at the user level. This may be because the fault is being masked in an intermediate layer, a defective transistor may affect performance but not the correct operation, or because any of the layers may be designed to tolerate some faults.

Errors are the manifestation of faults and can be classified in the same way as faults. Faults could cause an error, but not all faults show up as errors. The final term, failure, is defined as a system malfunction that causes the system not to meet

[1]Department of Electrical Engineering, Universitat Politècnica de Catalunya, Barcelona, Spain
[2]Department of Computer Architecture, Universitat Politècnica de Catalunya, Barcelona, Spain

*Figure 1.1 Summary of fault, error and failure terms*

its correctness, performance or other guarantees. Figure 1.1 summarizes these terms in the way of when they can arise.

## 1.2 Technology overview

This section describes the main technologies available nowadays as well as the near future candidates. As modern available technologies such as Bulk CMOS, FDSOI (Planar Fully Depleted Silicon On Insulator) and FinFET technologies are introduced in this section while new emerging technologies Carbon Nanotube (CNT) transistors, nanowire transistors and Memristive behavior devices are briefly described. Technologies can be structured depending on the physical magnitude used to describe information states: electric charge is the key one for nowadays modern available technologies (Bulk, FDSOI, FinFET, CNT, nanowires) while other domains are considered for other emerging ones ((a) resistance: RRAM (Resistive Random Access Memory); (b) material phase: PCM (Phase Change Memory); (c) magnetic domain: MRAM (Magnetic RAM); (d) spin of electrons: STT (Spin-Transfer Torque)).

### *1.2.1 Technologies based on electric charge*

Modern available technologies have been and are the core of electronic industry evolution. Today, they are all based on electric charge. The basic device used in these technologies is the transistor, basically a conductive dipole (with nodes typically named Drain and Source), the conductivity of which can be controlled by a specific third electrode (Gate). In Bipolar technology (BJT devices [2]) is the electric current entering in the control electrode that modules transistor conductivity while in MOS [3] technology is the voltage in the Gate electrode that controls the conductivity and consequently the current through Drain and Source. The word "transistor" comes from a transference resistor (resistor with controlled resistance). Modern digital computers (including processors, memories and accelerators) are fully based on MOS transistors technology, the one we will mainly consider in this book. MOS transistors are integrable and scalable what explains the evolution of Integrated Circuits (ICs) in the last decades. MOS technology contemplates two types of devices: PMOS (MOS with semiconductor channel type P) and NMOS (MOS with semiconductor channel type N). Both transistors have the same functional principles and behavior, but they work with dual voltage polarities. The next section focuses only on NMOS (positive voltages) in order not to replicate the same discussion.

#### 1.2.1.1 The MOS transistor

Figure 1.2 shows the vertical cut of a silicon NMOS transistor implementation in a typical planar integrated technology. Main terminals Drain (D) and Source (S) are

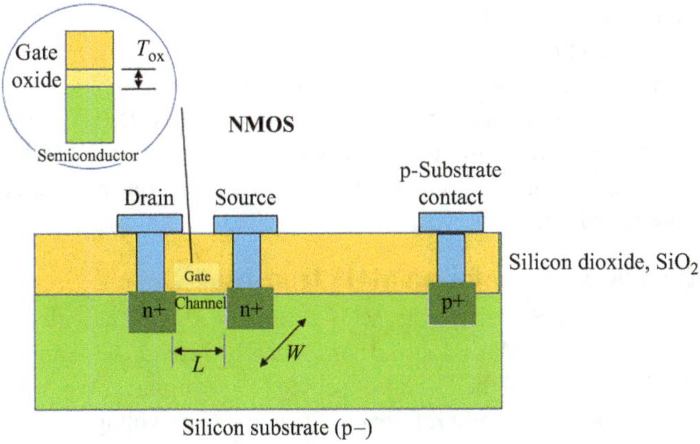


*Figure 1.2 Vertical cut of an NMOS transistor*

connected through metallic vias to two n+ regions embedded in the semiconductor crystal. The Gate (G) electrode, the controlling terminal, is located at a very short distance to the oxide–semiconductor interface (gate oxide thickness $T_{ox}$, with a size of few nanometers). When the gate is metallic, the structure phases are Metal (gate)–Oxide–Semiconductor (substrate) what give the name to the technology, MOS. The originally p-type (for NMOS, n-type for PMOS) region down the gate–oxide–semiconductor (substrate) region is named after the channel of the transistor (the conductive film). The structural parameters of a MOS transistor are given by its dimensions, basically the length ($L$) of the channel region (the one in the substrate under the gate, see Figure 1.2), the width ($W$) of the channel region (dimension perpendicular to the page) and the oxide thickness ($T_{ox}$) in between the gate and the channel. The principle of work of the MOS transistor is based on the fact that when the gate is not biased the electrical equivalence of the drain–source branch is just two serial diodes, one in the reverse position with respect to the other, not allowing significant current conduction. When a positive voltage is applied at the Gate terminal (exceeding a certain threshold voltage $V_{th}$, a key electrical parameter of the transistor), an inverted layer of carriers (n-equivalent instead of its originally p) is created in the interface oxide–semiconductor of the channel region, allowing the electrical conduction between D and S terminals through a thick film located in the interface oxide–semiconductor. The higher the Gate voltage, the higher the conductivity of the channel. Consequently, the Gate terminal controls the resistance of the D–S branch (from practically absolutely no conduction for $V_{GS} < V_{th}$, OFF, to a given intense conduction level for $V_{th} \geq V_{th}$, ON). This simple device is the base of modern ICs allowing the implementation of a wide variety of circuits, amplifiers, logic gates, memories, etc. Generally, the substrate crystal is biased to Ground voltage in the case of NMOS ($V_{DD}$ in the case of PMOS) except in particular cases where a substrate biasing effect is desired (reverse and forward body biasing techniques, RBB and FBB respectively [4]).

The progress of the technology is based on the continuous size reduction or scaling of the transistor (all dimensions, $L$, $W$ and $T_{ox}$) through a progressive photolithographic-based manufacturing precision causing an increase in the number of transistors in a given silicon crystal area or IC (Moore's Law [5]). With technologies of more than 10 μm of resolution in 1971 – the beginning of the MOS technology era; today, advanced technology nodes have reached resolutions of 7 nm, and 5 nm is expected in the near future.

#### 1.2.1.2 Electrical model for an MOS transistor

An analytical electrical model for the MOS transistor allows the calculation of the current through the terminal D and S (and consequently through the channel) where $I_{DS}$ is a function of the Gate and the Drain-to-Source ($V_{DS}$) voltages. We will assume the grounded source node as voltage reference. Thus, the Gate voltage is noted $V_{GS}$. So, the analytical model solves the function $I_{DS} = f(V_{DS}, V_{GS})$. Modern models include the main factors that affect advanced technologies as the channel-length modulation and the carrier velocity saturation. The model we will use in this book is given by [6]

$$I_{DS} = \mu_n C_{ox} \left(\frac{W}{L}\right) \left((V_{GS} - V_{th}) V_{DSAT} - \left(\frac{V_{DSAT} - V_{DSAT}^2}{2}\right)\right)$$

where the parameters can be defined and classified as the following:

*Electrical variables*
$I_{DS}$, current through the channel (Drain–Source current); $V_{GS}$, Gate–Source voltage; $V_{DS}$, Drain–Source voltage.
*Design variables* (can be defined by designer)
$W$, transistor channel width; $L$, transistor channel length; $V_{DSAT}$, Drain-to-Source voltage that saturates the carrier velocity in channel.
*Technologic variables (can be defined by manufacturer)*
$C_{ox}$, gate-to-substrate capacitance per unit of area; $V_{th}$, threshold voltage; $\mu_n$, carrier mobility.

#### 1.2.1.3 Planar bulk technology

Planar bulk technology has been the main choice in ICs manufacturing since the beginning of the MOS technology until the node of 22 nm (2011) where the 3D MOS devices or FinFET were introduced by Intel [7]. In planar bulk technology (Figure 1.3), all transistors PMOS and NMOS forming the IC are embedded in a single silicon crystal in a similar way as the description of the planar MOS in the previous section. In order to get an N-type substrate to include PMOS transistors, N-well regions are created to accommodate PMOS transistors (introducing additional n-type doping dominant on the original p-type substrate). The biasing of the substrate and N-wells is performed through specific contacts. Circuits formed by using both PMOS and NMOS transistor types are named CMOS circuits.

#### 1.2.1.4 Planar FDSOI technology

FDSOI [8] is an alternative planar process technology that relies on two primary innovations. First, an ultrathin layer of insulator, called the buried oxide, is positioned

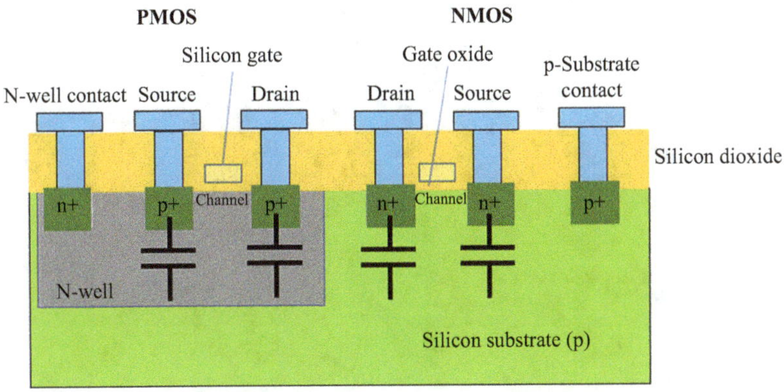


*Figure 1.3 Planar bulk technology*

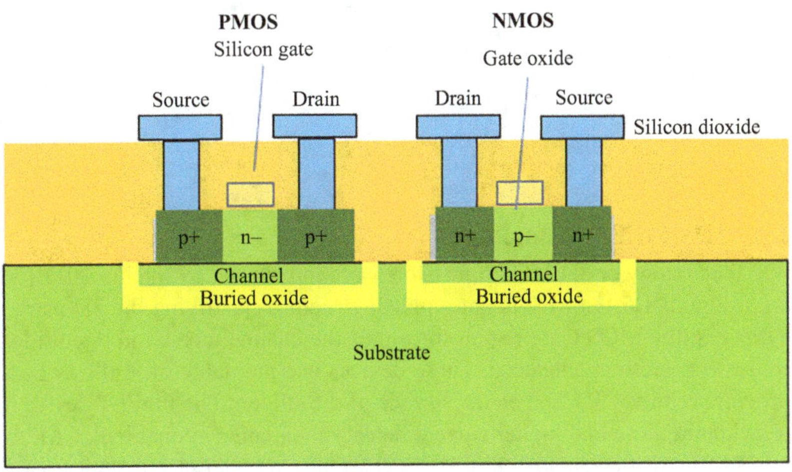


*Figure 1.4 FDSOI technology*

on top of the silicon substrate (see Figure 1.4). Then, a very thin and controlled silicon film implements the transistor regions and channel. Thanks to its thinness, there is no need to dope the channel, thus making the transistor Fully Depleted.

By construction, FDSOI enables transistor electrostatic characteristics much better versus conventional bulk technology what means a dramatic reduction in leakage currents. The buried oxide layer lowers the parasitic capacitance between the source and the drain (parasitic capacitances due to p–n junctions), as can be seen comparing bulk technology cut in Figure 1.3, with the corresponding for FDSOI in Figure 1.4. This makes FDSOI faster than bulk CMOS. One of the main characteristics of FDSOI is that it allows the use of RBB and FBB techniques. This generates adaptive circuits that can dynamically select between high-speed/high-power or low-speed/low-power configurations.

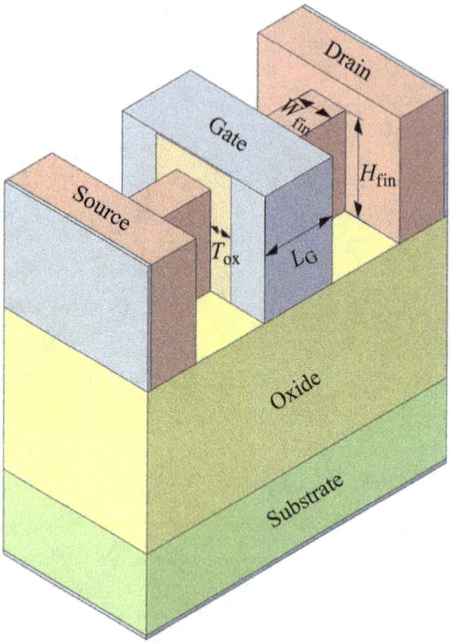


*Figure 1.5 FinFET structure*

#### 1.2.1.5 3D FinFET technology

FinFET Technology or 3D MOS transistors were introduced by Intel in 2011 [7]. As shown in Figure 1.5, FinFET is not a planar structure (justifying the 3D name). In these devices, the MOS Gate region surrounds the channel crystal in three sides, so the equivalent width is enhanced. These devices use the name of FinFETs because the structure reminds 3D fins on the surface of the silicon. The FinFET device has a faster switching time and higher current density than conventional bulk CMOS. As the gate surrounds the channel, FinFET devices have a better electrostatic control of the channel and reduced leakage currents. Nowadays, consequently, the more direct alternatives to conventional bulk MOS are the FinFET and the FDSOI technologies.

#### 1.2.1.6 Carbon nanotubes technology

Carbon Nanotube Field Effect Transistors (CNFETs, Figure 1.6 [9]) are promising candidates as a potential extension to silicon-based transistors (bulk, FDSOI and FinFET). With extraordinary electrical properties, such as quasi-ballistic transport or higher carrier mobility, CNFETs exhibit characteristics that could make this technology a rival of state-of-the-art Si-based MOSFETs. In a CNFET, the role of the channel is played by one or more CNTs for reliability reasons. A CNT is a graphene sheet rolled up to form a hollow cylinder. It can exhibit a metallic or a semiconducting behavior, and it can present different diameters (at nanometer range) depending on its chirality (angle of the atom arrangement along the tube). MOSFET-like CNFETs are composed of three regions. The region below the gate is intrinsic in nature and the two ungated regions are doped with either p-type or n-type. The ON-current is limited by the amount of charges that can be induced in the channel by the gate and not

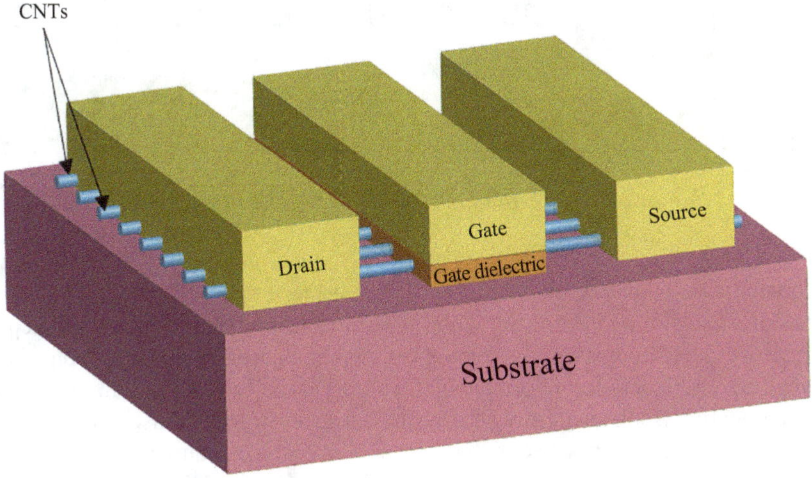


*Figure 1.6 CNT FET structure*

by the doping in the source. They operate in a pure p- or n-type enhancement-mode or in a depletion mode, based on the principle of barrier height modulation when applying a gate potential. This type of CNFET is promising because (1) they show unipolar characteristics; (2) the absence of Schottky Barrier reduces the OFF leakage current; (3) they are scalable; and (4) in ON-state, the source-to-channel junction has a significantly higher ON current.

#### 1.2.1.7 Nanowire (Gate All Around) transistors

A potential future technology is based on nanowire transistors, far still to be considered for the ICs manufacturing. In this case, the channel is implemented on a nanowire, a nanometric filament (with a diameter of the order of one nanometer) that exhibits quantum properties. This conductor, usually made with silicon, is surrounded with an insulator layer (that actuates as the gate oxide of the transistor) first and a silicon-based gate finally. The advantage of these devices is that they do not use semiconductor junctions and the dopant concentrations are uniform, not requiring typical profile gradients as in the previous technologies. They exhibit a near-ideal subthreshold slope, dramatic leakage current reduction and much higher robustness in terms of degradations and temperature sensitivity [10] when compared to previous technologies.

#### 1.2.1.8 Beyond CMOS: technologies based on other magnitudes (Memristive behavior devices (PCM, RRAM, MRAM))

The memristor (short for memory resistor), a potential candidate to implement new paradigmatic circuits of computing units and mainly Nonvolatile Memory (NVM) cells, is a device predicted from theoretic arguments near 45 years ago [11] and demonstrated in a physical device in 2008 by HP [12]. The memristor is a very basic device formed by a sandwich of metal–insulator–metal with only two electrodes, connected to the two metals, consequently is simple, scalable and in many cases compatible with silicon technology manufacturing. Basically, the behavior of the

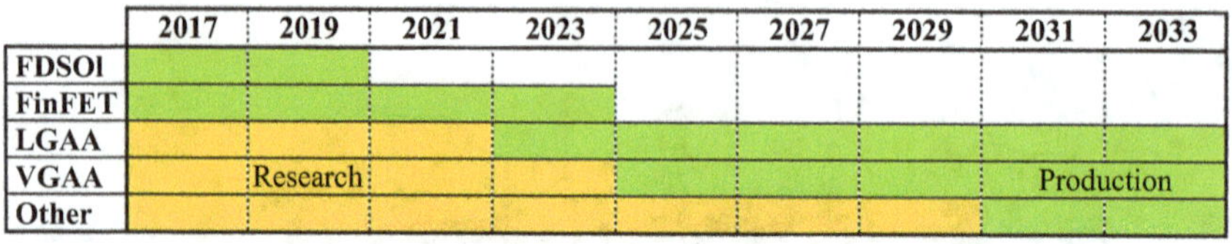


*Figure 1.7 Excerpt of the IRDS 2018 technology outlook*

memristor is based on the modification of its resistance (memristance) by the application of signals in its electrodes, with the unprecedented characteristic of keeping its resistive device permanently when the signal excitation disappears (nonvolatile device). Several works have been published showing the capability of implementing basic computing Boolean units, where the logical states (true and false) are equivalent to ON and OFF resistances. Yet, memristors exhibit a strong industrial potential (with already nowadays commercial market) in the field of massive NVMs [13]. The memristive behavior is based on different physical principles (and domains) depending on the isolator used. An extensive bibliography can be found for filamentary-based (resistive RAM devices [14]), phase-change (PCM [15]), magnetic domain (MRAM [16]) and spin torque orientation (STT [17]).

### *1.2.2 Roadmap for adoption*

The IRDS (International Roadmap for Devices and Systems) is an IEEE-based effort to continue the dissemination of technology outlooks for devices that the ITRS (International Technology Roadmap for Semiconductors) was performing until 2015. The 2018 edition of the IRDS [18] includes the prediction made by the committee of experts. Figure 1.7 shows an excerpt of the information contained in the 2018 edition of IRDS. It is clear that FDSOI and FinFET (or a combination of both) will be the prevailing technology for the most advanced nodes in the coming years. Then, they will be replaced by Gate All Around (GAA) devices (either lateral GAA–LGAA or vertical GAA–VGAA). After that, it is not clear what is to come, the previous section, points to some disruptive technologies (beyond CMOS), but also the IRDS identifies 3D stacking (at the transistor level) as a possibility. IRDS names this alternative 3DVLSI and it describes it as stacked LGAA devices. Overall, at this point in time, it seems clear that GAA devices will replace eventually FinFETs. Beyond that, it is still under research what will be the next.

### *1.2.3 Sources of unreliability in technology*

ICs, as electronic elements, are subject to electrical interactions that may reduce the reliability of the final IC. While design rules, among other, reduce the possibility of the manifestation of such events, the manufacturing process, degradation as well as variations in the environmental conditions can increase the number of faults sporadically or permanently during the lifetime of an IC.

Table 1.1 summarizes the main types of failure mechanisms giving a brief description of each one.

*Table 1.1 Summary of failure mechanisms*

| Sources | Description | Group |
|---|---|---|
| Parameter deviations (PD) | Process variations cause the deviations of designed dimensions (e.g. length, width and thickness) which can affect timing, energy consumption and area occupancy | Manufacturing |
| Random Dopant Fluctuations (RDFs) | Process variation caused by the random fluctuation in the number of dopants in the channel gate and their placement, which results in threshold voltage ($V_{th}$) variations producing permanent failures | Manufacturing |
| Line Edge Roughness (LER) | Process variation caused by the change in the shape of the gate along the channel width direction, which results in $V_{th}$ variations | Manufacturing |
| Random Telegraph Noise (RTN) | Random fluctuation in the device drain current due to the trapping and detrapping of channel carriers in the dielectric traps at the oxide interface, which causes intermittent variations in the $V_{th}$ | Manufacturing |
| Metal Stress Voiding (MSV) | Voids in metal lines due to different thermal expansion rates of metal lines and the passivation material they bond, causing permanent failures | Degradation |
| Electromigration/ Stress Migration (EM) | Voids in metal lines or interconnects caused by the electron flow and resulting in permanent failures | Degradation |
| Hot Carrier Injection (HCI) | Arises from impact ionization when electrons in the channel strike the silicon atoms around the drain-substrate interface. HCI results in a reduction of the maximum operating frequency of the chip | Degradation |
| Gate Oxide Wearout (GOW) | Sudden discontinuous increase in conductance causing a reduction in the current of the transistor, which may initially lead to intermittent faults but may eventually cause a permanent fault | Degradation |
| Negative/Positive Bias Temperature Instability (NBTI/PBTI) | Reduction in mobility of holes and shift in $V_{th}$ when the device is under stress, like high temperatures, slowing down the transistor. NBTI affects PMOS transistors while PBTI affects NMOS transistors | Degradation |
| Time-Dependent Dielectric Breakdown (TDDB) | Dielectric breakdown of the gate dielectric film or the interlayer dielectric in metal lines. Dielectric stress for a long time increases leakage currents and eventually leads to breakdown or short-circuits between metal lines | Degradation |
| Radiation-Induced Failures (RIF) | Can be produced by alpha particles from packaging and neutrons from the atmosphere producing transient failures | Radiation |
| Self-Heating (SH) | Especially important in SOI where transistors have a buried oxide layer to achieve the desired electrical isolation. This layer is a barrier to heat flow from the channel consequently anomalous behavior is observed | Permanent |

*Table 1.2 Qualitative impact of unreliability sources in each technology*

| Source/ device type | Bulk planar | Bulk FinFET | SOI planar | SOI FinFET | Nanowires |
|---|---|---|---|---|---|
| RDF | High | Low | Low | Low | Low |
| NBTI/PBTI | High | High | High | High | High |
| LER | High | High | High | High | High |
| Particle strikes | High | High | High | High | High |
| $dI/dt$ | High | High | High | High | High |
| $dV/dt$ | High | High | High | High | High |
| Self-heating | Low | Low | High | High | Low |
| Hot electron effect (hot carriers) | Low | High | Low | High | Low |

Most of the faults described in this summary can be taken care before a chip is shipped as most of them are related to process variations. Therefore, the most challenging failures to track are the ones related to aging and radiation.

At the same time, the geometry of the transistors as well as the manufacturing steps (and materials) differs for the different device types available. Table 1.2 shows a qualitative analysis of the impact of the different unreliability sources on the different device types.

Overall, the better the channel control, the better the device. Consequently, SOI and FinFETs are dominating the market nowadays, being nanowires the most sought after device to replace FinFETs (as Figure 1.7 corroborates).

#### 1.2.3.1 Environmental considerations

Environmental factors can impact the characteristics or behavior of a source of failure. Table 1.3 shows different environmental factors and describes how these factors impact on the different types of errors. While several depend on the specific deployment location, they must be part of the design considerations.

## 1.3 CPU building blocks

CPUs are usually a mix of combinatorial circuits and memories. Analyzing the reliability of each block is highly dependent on the circuit layout apart from the underlying technology. Consequently, this section describes the most common building blocks, then, Section 1.4 evaluates the initial raw fault rates for each of these blocks.

### *1.3.1 Combinatorial circuits*

Being a combination of logic gates, combinatorial circuits implement a Boolean function for each of its outputs (given the set of inputs). This category includes all

*Table 1.3 Environmental factors and their effects on different types of errors*

| Factors | Transient errors | Intermittent errors | Permanent errors |
|---|---|---|---|
| Temperature | Increased leakage | Device degradation (e.g. NBTI effects) and thermal stress | Device degradation (e.g. electromigration) and thermal stress (e.g. WearOut) |
| Humidity/Dust Acid/Salt | Not affected | Not affected | Corrosion/shorting on contacts |
| Vibration/Pressure Gravity | Not affected | May cause intermittent failures depending on the strength | Mechanical stress and contact/solder breaks |
| EMC/EMI Radiation/Altitude | Increased interferences | May cause intermittent failures for unshielded components | Oxide failure or metal melt and device degradation effects |

functional units (e.g. adders, shifters, multipliers, etc.), decoders and control logic. Given that these blocks do not store any state, they are not prone to store intermittent faults (as the output is recomputed every cycle). On top of that, they usually occupy a smaller footprint than memories in an IC design. Consequently, it is of higher interest to designers to evaluate the effect of faulty memories (which occupy a larger area and store faulty vales). Still, there are some works that focus on combinatorial circuits, as we will see in Chapter 2.

### *1.3.2 Memories*

Memory is the crucial for computation, as we need a place to store data. Read and write memories can be classified into two categories: Static RAM (SRAM) and Dynamic RAM (DRAM). SRAM can hold data as long as the power supply is uninterrupted. The core building block is a loop of two inverters. Most modern SRAM cells are made of six CMOS transistors (although 8, 10 and 12 versions exist). SRAM is the fastest type of memory available.

In contrast, DRAM does not hold data indefinitely. The core building block is a capacitor that (logically) holds the value but is subject to leakage currents (so the value stored leaks away). Consequently, the capacitor needs to be refreshed after a specific period to keep the charge in the capacitor. This introduces some unavailability time for the cells while they are being refreshed.

While DRAM has an obvious size advantage over SRAM, its speed cannot even get close to those offered by static memory cells. In a CPU, SRAMs are used in the register file and caches (and other latency critical storage structures), and DRAM is mainly used in main memory. Table 1.4 describes and compares the most widely used SRAM and DRAM designs.

*Table 1.4 Comparison of various memory technologies for on-die caches*

| | (A) SRAM | eDRAM | |
|---|---|---|---|
| | | **(B) 1T–1C** | **(C) Gain cell** |
| **Cell schematic** | 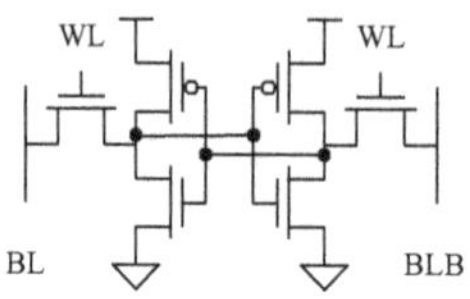  | 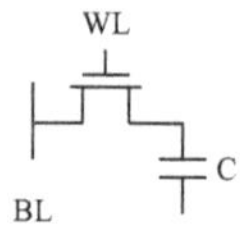  | 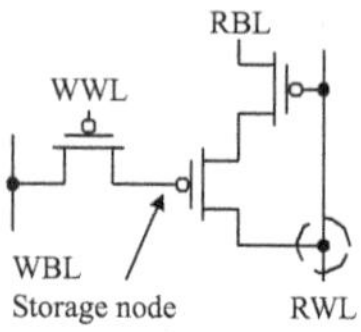  |
| **Data storage** | Latch | Capacitor | **MOS gate capacitance** |
| **Read time** | Short | Short | **Short** |
| **Write time** | Short | Short | **Short** |
| **Read energy** | Low | Low | **Low** |
| **Write energy** | Low | Low | **Low** |
| **Leakage** | High | Low | **Low** |
| **Features** | **(+) Fast**<br>**(−) Large area**<br>**(−) Leakage** | **(+) Low leakage**<br>**(+) Small area**<br>**(−) Extra process**<br>**(−) Destructive read**<br>**(−) Refresh** | **(+) Low leakage**<br>**(+) Decoupled read/write**<br>**(−) Refresh** |

#### 1.3.2.1 Static Random Access Memory

The first traces to SRAM date back to 1964, when 64-bit MOS SRAM was developed at Fairchild Semiconductor. However, the breakthrough came when Intel developed its first 256-bit SRAM, the 1,101 chip in 1969 and formally launched it in 1971.

The SRAM cell consists of a bistable flip-flop connected to the internal circuitry by two access transistors (Table 1.4A, the ones connected to the WordLine (WL)). When the cell is not addressed (WL=0), the two-access transistors are closed and the data is kept to a stable state, latched within the flip-flop. The cell needs the power supply to keep the information. The data in an SRAM cell is volatile (i.e. the data is lost when the power is removed). However, the data does not "leak away" like in a DRAM, so the SRAM does not require a refresh cycle.

The *de facto* SRAM 6T design is shown in Table 1.4A. It is built as any combinational logic in the circuit; thus it does not require any more process steps. It is fast but costly in terms of area and leakage power. As technology shrinks, several other designs have been proposed that increase the cell robustness (i.e. susceptibility to noise, couplings, etc.) and/or they reduce leakage power. The most known designs are the 8T cell and the 10T cell. Obviously, adding more transistors has a cost—at least—in area and power. Nevertheless, 8T and 10T cells increase memory robustness significantly. Eventually, they are a good design choice in low-voltage operation devices and harsh environment systems.

SRAM memories are fundamental components in any computing system nowadays. Almost all on-chip memories in all processing chips are SRAMs (e.g. caches, register files, buffers, tables, etc.).

#### 1.3.2.2 Embedded Dynamic Random Access Memory

In 1996, Mitsubishi took a standard 16-Mbit DRAM and wedged a RISC CPU into the middle. The M32R/D costs more than a separate processor and DRAM, and it did not catch on. Almost 20 years later, Embedded DRAMs (eDRAMs) allow system designers to use high bandwidth, high performance and high-density memory near the processing core (for high-performance chips) or within the System-On-Chip. Logic-based eDRAM technologies are now present in some microelectronics devices as well as in some high-performance devices from IBM and Intel cores. eDRAM is therefore a powerful tool if it can be made cost-effective. Logic or stand-alone DRAM technologies have been used to realize eDRAMs. Logic-based technologies offer the advantage of high performance, and compatibility with existing cores, which is essential. The main challenge of logic-based eDRAM is to find the right compromise between added process cost and memory density.

For trade-offs between process cost and DRAM density, the stacked capacitor lies between two other architecture choices: planar cells and deep trench cells. The planar cell allows a very easy integration with only one added mask, but cell sizes remain two to three times larger than stacked or trench cells, restricting the use to small DRAM capacity. On the other side, the trench cell is very competitive in terms of size, allowing high memory density, but with added process complexity, close to a stand-alone DRAM process.

### *1.3.3 Main memory and storage*

DRAM is currently the technology of choice for main memory. Similarly, NVM is the choice for storage.

The NVM discussion in this section is limited to devices that can be written and read many times; hence, Read-Only Memory and One-Time-Programmable memory are not included although many such memories are important both for standalone and embedded applications (as they usually hold the boot and setup data). The current mainstream NVM is Flash memory. NAND and NOR flash memories are used for quite different applications data storage for NAND and code storage for NOR flash. Other non-charge-storage types of NVM are also considered, including Ferroelectric RAM (FeRAM), MRAM and Phase-Change RAM (PCRAM). These emerging memories promise to continue NVM scaling beyond Flash memories. However, because NAND Flash and to some extent NOR Flash are still dominating the applications, emerging memories have been used in specialty applications and have not yet fulfilled their original promise to become dominating mainstream high-density NVM.

#### 1.3.3.1 Dynamic Random Access Memory

Robert Dennard invented DRAM at the IBM Thomas J. Watson Research Center in 1966/1967. The next year, the first known DRAM chip ever developed was a 256-bit device created by Lee Boysel at Fairchild Semiconductor. Later, Boysel founded Four

*Table 1.5 DRAM technology outlook (excerpt from IRDS [18])*

| Year of production | 2017 | 2019 | 2021 | 2024 | 2027 | 2030 | 2033 |
|---|---|---|---|---|---|---|---|
| DRAM 1/2 pitch (nm) | 18 | 17.5 | 17 | 14 | 11 | 8.4 | 7.7 |
| DRAM cell size factor: $aF^2$ | 6 | 6 | 4 | 4 | 4 | 4 | 4 |
| DRAM bit/chip target | 8G | 8G | 16G | 16G | 32G | 32G | 32G |

Phase Systems in 1969 and developed 1,024-bit and 2,048-bit DRAMs. Intel released the 1,103, the industry's first mass-produced DRAM device, in 1970.

Aside from the energy consumption and the high-performance (and power) dependence on ambient temperature, there are still plenty of technical challenges as well as issues of increased process steps to sustain the cost scaling. Fundamentally, there exist several significant process flow issues from a production standpoint, such as process steps of capacitor formation, or high aspect ratio contact etches requiring photoresists with hard mask pattern transferring layer that can stand up for a prolonged etch time. Furthermore, continuous improvements in lithography/hard mask and etch will be needed. In addition, lower wordline/bitline resistance is necessary for getting the same or better performance.

For DRAM, the main goal is to continue to scale the footprint of the 1T-1C cell, to the practical limit of $4F^2$ [18]. The main challenges in manufacturing are the creation of vertical transistor structures, the introduction of dielectrics to improve the capacitance density and meanwhile keep the leakage low.

In general, technical requirements for DRAMs become more difficult with scaling (see Table 1.5 with the IRDS projections). In the past couple of years, DRAM has been subject to many new improvements in the manufacturing technology and materials. Due to new technologies, DRAM will continue to scale doubling the size (bits/chip) every 4 or more years.

#### 1.3.3.2 Flash memory

Toshiba's Fujio Masuoka invented Flash memory in the early 1980s [19,20]. Masuoka discussed and detailed flash (NOR and NAND) for the first time, as an evolution of the floating gate devices. Flash memories are based on simple One-Transistor (1T) cells, where a transistor serves both as the access (or cell selection) device and the storage node.

Floating gate Flash devices achieve non-volatility by storing and sensing the charge stored "in" (on the surface of) a floating gate. The NAND array consists of bitline strings of now 64 devices or more with a string selection device at each end. This architecture requires no direct bitline contact to the cell thus allows the smallest cell size. During programing or reading, the unselected cells in the selected bitline string must be turned on and serve as "pass" devices, thus the data stored in each device cannot be accessed randomly. Data input/output are structured in "page" mode where a page (on the Wordline) is of several kB (8–16 kB today) in size. Both programing and erasing are by Fowler–Nordheim tunneling of electrons into and out of the floating gate

*Table 1.6 FLASH technology outlook (excerpt from IRDS [18])*

| Year of production | 2017 | 2019 | 2021 | 2024 | 2027 | 2030 | 2033 |
|---|---|---|---|---|---|---|---|
| FLASH 1/2 pitch (nm) | 15 | 15 | 15 | 15 | 15 | 15 | 15 |
| FLASH highest density | 512G | 1T | 1T | 1.5T | 3T | 4T | >4T |
| FLASH 3D maximum memory layers | 64 | 96 | 128 | 192 | 384 | 512 | >512 |

through the tunneling oxide. The low Fowler–Nordheim tunneling current allows the simultaneous programing of many bits (page) thus gives high programing throughput, suitable for handling a large amount of data. Since devices in the same bitline string serve as pass transistors, their leakage current does not seriously affect programing or reading operation (up to a limit), and without the need for hot electrons, junctions can be shallow. Thus the scaling of NAND flash (Table 1.6) is not limited by device punch through and junction breakdown as in NOR flash.

### *1.3.4 Emerging memories*

Since the ultimate scaling limitation for charge storage devices is too few electrons, devices that provide memory states without electric charges are promising to scale further. Several non-charge-storage memories have been extensively studied and some commercialized, and each has its own merits and unique challenges. Some of these are uniquely suited for special applications and may follow a scaling path independent of Flash. Some may eventually replace flash memories. Logic states that do not depend on charge storage eventually also run into fundamental physics limits. For example, small storage volume may be vulnerable to random thermal noise, such as the case of super-paramagnetism limitation for MRAMs.

One disadvantage of this category of devices is that the storage element itself cannot also serve as the memory selection (access) device (transistor) because they are mostly two-terminal devices. Therefore, these devices use 1T-1C (FeRAM), 1T-1R (MRAM, PCRAM and ReRAM) or 1D-1R (PCRAM and ReRAM) structures. Similar to DRAM, the cell consists of an access transistor, plus, the extra capacitor, resistor or magnetic material. In any of these designs, it is challenging to achieve a small ($4F^2$) cell size without innovative access devices. In addition, because of the more complex cell structure that must include a separate access device, it is more difficult to design 3D arrays that can be fabricated using just a few additional masks like those proposed for 3D NAND FLASH.

## 1.4 Characterization

Given the long list of fault sources and the different characteristics, researchers have grouped these sources (Table 1.1) into three blocks: manufacturing, degradation and radiation. There are not many works that perform a comparison of several technology

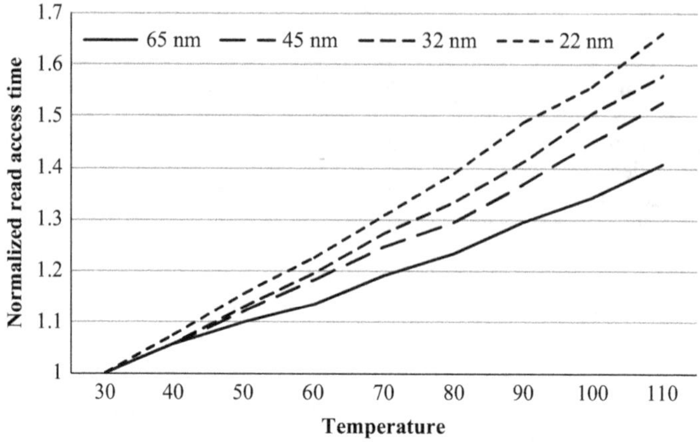


*Figure 1.8 Read Access Time for a 32 kB direct-mapped cache for different technology nodes*

nodes under the same conditions and tools. In this book, we report three of these works [21]. This helps us illustrate the trends and show the effects of the sources of unreliability.

## *1.4.1 Manufacturing*

Most of the works that analyze the impact of unreliability sources are based on the multilevel quadtree model [22]. The methodology is widely used but key parameters (i.e. the extent or severity of variations) are left to the user to insert. As an example, Ganapathy *et al.* [21] explore the effects of manufacturing variations as well as temperature, in a conventional memory array (composed of 6T cells) for bulk planar CMOS technologies (up to the 22 nm) using the Predictive Technology Models [23]. When analyzing solely the effects of technology and temperature (Figure 1.8), authors report a maximum of 1.7 × difference in access time across temperatures, technology introduces a 1.2 × variation in the worst case.

When considering manufacturing variations, authors report the effects on the 22 nm technology node. The effect of manufacturing variations clearly degrades the design. Nominal design points (at the different temperatures) are drawn in diamonds (toward the left of the figure). Any cache at the right side of these points is above (in terms of access time) the nominal one. It is clear from the figure that the majority of caches are above the nominal design. For some temperatures, there is no timing for some caches. This means that the cache is unable to work. Thus, manufacturing variations do not only slow the circuits designed, but it can yield a large part of them unusable (i.e. 20% unusable at 110°C) (Figure 1.9).

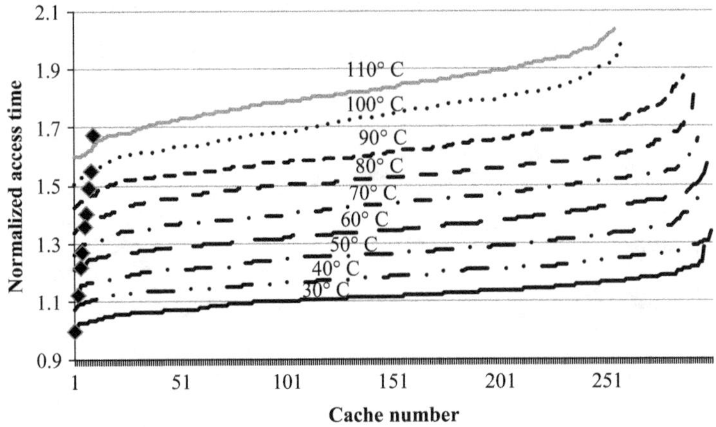


*Figure 1.9 Read Access Time for a 32 kB direct-mapped cache for 22 nm under manufacturing variations*

## 1.4.2 Radiation

There are different sources of radiation. In comparison to outer space systems (i.e. satellites), system under the atmosphere is mostly subject to high-energy neutrons, thermal neutrons and alpha particles. In [24], it is shown that high-energy neutrons contribute the most to faults (77%), while the other two have a smaller contribution (16% and 7%, respectively).

Two industrial works from Intel [25] and Samsung [26] analyze the impact of radiation-induced faults in several fundamental circuits (memories, latches and logic gates) across technology nodes. These works help one to compare the impact of radiation-induced faults across the latest technologies under the same evaluation methodology and framework. Both are based on actual chip measurement. Sadly, these studies use arbitrary units and, beyond showing the trend, specific failure rates are not reported. In contrast, the study of Riera *et al.* [25] is based on electrical (i.e. Spice) simulations of PTM technology models [23]. Apart from sustaining the trends shown by industry, the authors show the methodology and compute the failure rate for SRAM memories, flip-flops and logic gates. Figure 1.10 shows the single-bit soft-error rate (SER) computed as Failures In Time (FIT) for different components and technologies.

As technology scales and transistors scale down, the impact of the radiation is transmitted to multiple transistors for each given strike. Thus, although each component is stronger (i.e. smaller SER), it is now also affected by neighboring impacts. This effect is called Multi-Cell Upset—MCU—if the MCU changes the values of logically together bits then it is called Multi-Bit Upset (MBU). Hubert *et al.* [27] analyzed this effect across technologies. Figure 1.11 shows the rising increase of the appearance of MCU in smaller technology nodes. It also shows that bulk technology is the most vulnerable to MCUs far ahead from FDSOI and FinFET bulk.

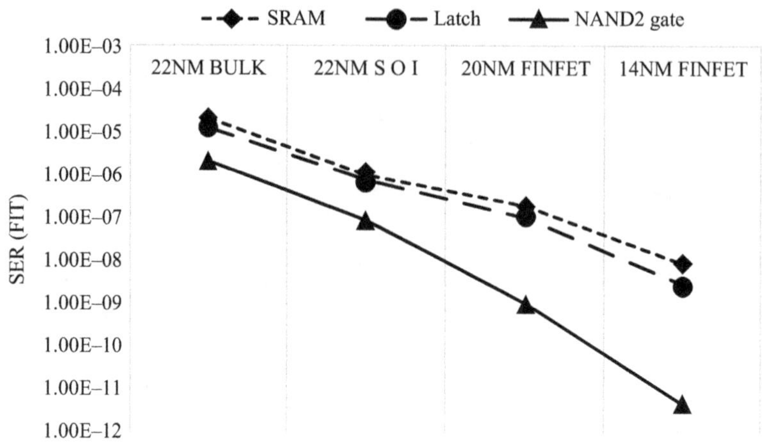


*Figure 1.10 Radiation-induced error rate (soft-error rate computed as FIT rate)*

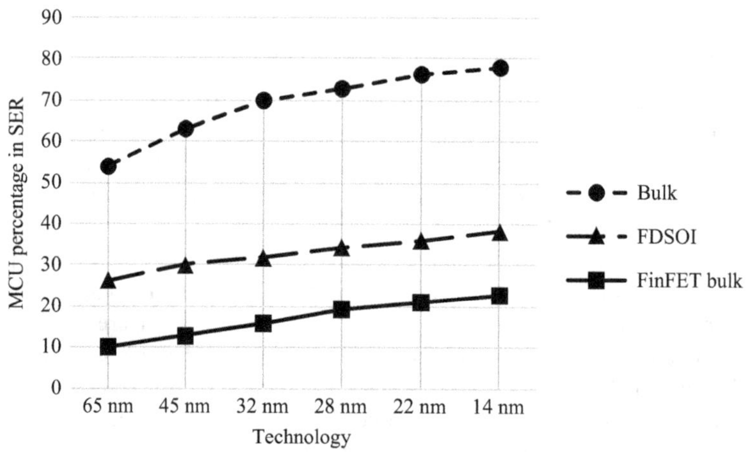


*Figure 1.11 MCU percentage of SER for different technologies*

## 1.5 Conclusions

Reliability is and it will continue to be an important hurdle in lower technology nodes. The limitations of the manufacturing process together with the environmental conditions pose a big challenge to IC designers. Historically, reliability was handled at the circuit level. As technology evolves, the microarchitecture level has joined forces in this effort.

In the next chapters, we will see how an early reliability assessment of the system can be conducted through the different levels up to the system perspective. This can help one to identify the portions of the IC that need more protection as well as to calibrate how much effort and overhead (area, latency, power) is reasonable to guarantee reliable systems.

## References

[1] D. Sorin, Fault Tolerant Computer Architecture, Morgan & Claypool Publishers, USA, 2009.

[2] A. A. Zekry, "The evolution of the microelectronic bipolar junction transistor," in *The First Egyptian Workshop on Advancements of Electronic Devices, 2002. (EWAED)*, Cairo, Egypt, 2002.

[3] S. Chih-Tang, "Evolution of the MOS transistor-from conception to VLSI," *Proceedings of the IEEE,* vol. 76, no. 10, pp. 1280–1326, 1988.

[4] T. Wang, X. Cui, K. Liao, *et al.*, "Employing the mixed FBB/RBB in the design of FinFET logic gates," in *2015 IEEE 11th International Conference on ASIC (ASICON)*, Chengdu, China, 2015.

[5] S. Adee, "The data: 37 years of Moore's Law," *IEEE Spectrum,* vol. 45, no. 5, pp. 56, 2008.

[6] J. M. Rabaey, A. Chandrakasan and B. Nikolic, Digital Integrated Circuits: A Design Perspective. USA: Pearson; 2nd edition, 2003.

[7] D. James, "Intel Ivy Bridge unveiled—The first commercial tri-gate, high-k, metal-gate CPU," in *Proceedings of the IEEE 2012 Custom Integrated Circuits Conference*, San Jose, CA, USA, 2012.

[8] H. Shang, M. H. White and D. A. Adams, "0.25 V FDSOI CMOS technology for ultra-low voltage applications," in *2002 IEEE International SOI Conference*, Williamsburg, VA, USA, 2002.

[9] C. G. Almudéver and A. Rubio, "A comparative variability analysis for CMOS and CNFET 6T SRAM cells," in *2011 IEEE 54th International Midwest Symposium on Circuits and Systems (MWSCAS)*, Seoul, Korea, 2011.

[10] J. A. Smith, K. Ni, R. K. Ghosh, *et al.*, "Investigation of electrically gate-all-around hexagonal nanowire FET (HexFET) architecture for 5 nm node logic and SRAM applications," in *2017 47th European Solid-State Device Research Conference (ESSDERC)*, Leuven, Belgium, 2017.

[11] L. Chua, "Memristor—The missing circuit element," *IEEE Transactions on Circuit Theory,* vol. 18, no. 5, pp. 507–519, 1971.

[12] M. Hopkin, "Found: The missing circuit element," *Nature,* 2008.

[13] D. Frohman-Bentchkowsky, "Non-volatile semiconductor memories," in *1981 International Electron Devices Meeting*, Washington, DC, USA, 1981.

[14] S. Yu, Resistive Random Access Memory (RRAM). USA: Morgan & Claypool Publishers, 2016.

[15] H. Jeong, "High density PCM (phase change memory) technology," in *2016 International SoC Design Conference (ISOCC)*, Jeju, Korea, 2016.

[16] D. C. Worledge, M. Gajek, D. W. Abraham, *et al.*, "Recent advances in spin torque MRAM," in *2012 4th IEEE International Memory Workshop*, Milan, Italy, 2012.

[17] B. Song, T. Na, H. Jeong, J. P. Kim and S.-O. Jung, "Comparative analysis of using planar MOSFET and FinFET as access transistor of STT-RAM Cell in 22-nm technology node," in *2014 International SoC Design Conference (ISOCC)*, Jeju, Korea, 2014.
[18] IEEE IRDS Technical Council, The International Roadmap for Devices and Systems. USA: IEEE, 2018.
[19] M. Asano, H. Iwahashi, T. Komuro and F. Masuoka, "A new flash E2PROM cell using triple polysilicon technology," in *1984 International Electron Devices Meeting*, San Francisco, CA, USA, 1984, pp. 464–467.
[20] F. Masuoka, M. Momodomi, Y. Iwata and R. Shirota, "New ultra high density EPROM and flash EEPROM with NAND structure cell," in *1987 International Electron Devices Meeting*, Washington, DC, USA, 1987, pp. 552–555.
[21] S. Ganapathy, R. Canal, A. González and A. Rubio, "Circuit propagation delay estimation through multivariate regression-based modeling under spatio-temporal variability," in *2010 Design, Automation & Test in Europe Conference & Exhibition*, Dresden, 2010.
[22] A. Agarwal, D. Blaauw and V. Zolotov, "Statistical timing analysis for intra-die process variations with spatial correlations," in *ICCAD-2003. International Conference on Computer Aided Design*, San José, CA, USA, 2003.
[23] Y. Cao and W. Zhao, "Predictive technology model for nano-CMOS design exploration," in *1st International Conference on Nano-Networks and Workshops*, Lausanne, Switzerland, 2006.
[24] N. Seifert, S. Jahinuzzaman, J. Velamala, *et al.*, "Soft error rate improvements in 14-nm technology featuring second-generation 3D tri-gate transistors," *IEEE Transactions on Nuclear Science,* vol. 62, no. 6, pp. 2570–2577, 2015.
[25] M. Riera, R. Canal, J. Abella and A. González, "A detailed methodology to compute soft error rates in advanced technologies," in *2016 Design, Automation & Test in Europe Conference & Exhibition (DATE)*, Dresden, Germany, 2016.
[26] S. Lee, S. Ha, C.-S. Yu, J. Noh, S. Pae and J. Park, "Radiation-induced soft error rate analyses for 14 nm FinFET SRAM devices," in *2015 IEEE International Reliability Physics Symposium*, Monterey, CA, USA, 2015.
[27] G. Hubert, L. Artola and D. Regis, "Impact of scaling on the soft error sensitivity of bulk, FDSOI and FinFET technologies due to atmospheric radiation," *Integration the VLSI Journal,* vol. 50, pp. 39–47, 2015.

*Chapter 2*

# Design techniques to improve the resilience of computing systems: logic layer

*Lorena Anghel[1] and Michael Nicolaidis[2]*

High-reliability and high-dependability applications require integrated solutions against **random hardware faults** and **transient faults**. Random hardware faults or intermittent faults are generated by process or time-dependent variations, i.e., aging, while transients are induced either by radiation, namely, soft errors, or by extreme operating conditions or electronic interference.

Indeed, nanometric static process variations, voltage and temperature dynamic fluctuations due to chip activity, Bias Temperature Instability caused by the stress on the transistors, and single event effects or soft errors are reported to be very important issues in nanometric technology nodes [1,2]. These phenomena induce performance reduction if not taken care properly and may reduce circuit lifetime and Mean Time To Failure. Hence, onchip accurate yield, reliability and performance monitors that check online or periodically violations of guardbands have become necessary. Adaptive compensation schemes are combined with monitors in the attempt to recover from potential error when timing violation occurs. This chapter presents up-to-date state of the art of performance and reliability monitors, insertion methodology and experimental results of different sensors and monitors used for process and environment variations as well as aging compensation.

## 2.1 Introduction

The reduced CMOS transistor sizes approaching the atomic size exacerbate the impact of variability, transient, and intermittent faults. Static variability also called process variation, dynamic variability due to $V_{DD}$ and temperature fluctuations (*PVT*), and temporal variability—due to aging—are already reported for several technologies [1–4]. The impact of the variability is even higher when the circuit is operating near-transistor thresholds supply voltages.

On the other side, soft errors have become an increasing burden for microprocessor and other high-performance designs, such as system-on-a-chip (SoC) and

[1]TIMA Laboratory, Grenoble INP, Grenoble, France
[2]TIMA Laboratory, CNRS, Grenoble, France

manycore platforms, as the number of on-chip transistors continues to grow and their performance to boost. Moreover, their power consumption has to be limited and thus, they tend to operate at lower $V_{DD}$. Intensive research has been dedicated in the previous years to fault tolerant design and mitigation techniques face to soft errors. Gaisler presented in [5] a fault tolerant 32 bits processor, LEON-FT, based on SPARC V8 instruction set architecture. These architectures are protected against Single Event Transient (SET) and Single Event Upset (SEU) by using massive triple majority voters at registers and flip-flops (FFs), or expensive BCH codes for memory or register file arrays, with an overall area overhead close to 100%. Several other architectures such as Intel Itanium [6], IBM S/390 G5 [7], or [8] are reporting similar area and power overhead, which is acceptable for mission and life critical applications. However, for large majority of applications this cost is not acceptable.

To mitigate process, voltage, and temperature (*PVT*) variations and circuit aging and to decrease the sensitivity to radiation-induced single-event effects (SEUs, SETs), various approaches have been proposed. Designers either take advantage of the existing voltage guard-bands, or they increase them at design time so that circuits can still work for 7–10 years [9]. Alternatively, aging and performance monitors such as canary circuits [10–12] duplicating critical paths delays of the functional circuit can be used. They are inserted inside the circuit at design time to detect circuit performance violations in terms of timing errors occurring at runtime. When an error occurs, circuit or system recovery methods can be triggered to stay in safe operating modes. However, it is to be noted that:

- Adding pessimistic timing margins or guard-bands (or equivalent voltage margins) to guarantee all operating points under worse case conditions is generating huge impact on design costs, with an upward trend as technology moves further.
- The usage of sensors and performance violation monitors is a possible solution to detect and possibly correct timing errors, and in addition to the reduction of design margins, they are usually combined with Adaptive Voltage Scaling (AVS) technique or Dynamic Voltage Frequency Scaling (DVFS). As a matter of fact, performance violation monitors trigger AVS or DVFS techniques, and the system adapts dynamically the frequency and the voltage according to the operating conditions and the application needs [3,4]. Adaptive Body Bias (ABB) voltage operation is also seen as a possible solution to compensate for the variations, especially as they do not affect the power consumption of the circuit. Different approaches such as embedded performance monitors combined with adaptive voltages and/or frequency schemes became popular recently [13]. Thus, the performance degradation can be compensated, further modes of low power can be achieved and the circuit's lifetime can be extended. Indeed, sensors and monitors checking deviations of the electrical characteristics of transistors (like ION sensors) as well as replica-based canary circuits are used to detect circuit degradation induced by aging or timing degradation induced by *PVT* variations. However, these approaches cannot address all failure mechanisms like single event effects and Electromagnetic Interference (EMI), which have to be considered separately. Furthermore, these approach monitors are spatially distributed over the die, not

being part of the operating circuit. As performance degradation due to aging being dependent on the design topology and the activity induced by the workload, these monitor structures may age differently from the transistors of the operating circuit.

- The alternative solution is to check the impact of the failure mechanisms during the operation of the functional circuit, concurrently with the application execution (concurrent error detection). In Situ Monitors (ISMs) are used as low-cost error-detecting schemes, able to detect timing fault and transients with different origins and durations. These schemes use double-sampling methodologies introduced and evaluated in early 2000 [14–16]. Indeed, instead of using hardware duplication, double-sampling technique observes a given signal at two different instants, thus allowing detection of temporary faults (timing faults, transients, and upsets) at very low cost. These types of aging monitors can also be used to predict the age of the system and to prevent timing violations. In this case, the age of the system is predicted from parameters that are monitored during the normal operation of the system. These monitor types can detect delay degradations that are much smaller than the considered guard-band.

However, note that each mitigation technique is subject to several of the following drawbacks: large area, power, or performance penalty; false positives; false negatives; and insufficient coverage of the failures encountered in nanometric domain.

The chapter is organized in four main parts. The principle of double-sampling methodology is presented in Section 2.2. Gains, limitations, and challenges of each architecture are widely discussed. State-of-the-art monitors and up-to-date optimized in situ and replica path monitors are presented in Section 2.3. Typical adaptation and measures are presented in Section 2.4.

## 2.2 Performance and reliability monitors

As stated in the introductory part, timing errors induced by variability and aging induced phenomena can be compensated by imposing strong timing margins or the corresponding voltage supply margins. Adding pessimistic safety margins to guarantee all operating points under worse case *PVT* and aging conditions is not acceptable in many designs, since it will consume all the gains of scaling with respect to performance, area, and design cost. Moreover, due to the fact that the final area of the circuit can be unacceptably larger, the design flow closure can be drastically affected. To get around this limitation and further reduce the voltage margins, AVS or Adaptive Frequency Scaling [4] triggered by delay monitors is usually done. Consequently, the circuit's lifetime is extended under wear-out mechanism, and the power management is more accurately handled.

Various classes of monitors have been proposed in the literature. They can be split into two major classes: first of all, monitors for maximum performance violation error or pre-error detection; and second, monitors for measuring the extra delay induced by variability or aging.

### 2.2.1 Double-sampling methodology and the basic architecture

Double-sampling technology has been proposed in [14–16] to counteract drawbacks of traditional concurrent error-detection scheme usually based on circuit duplication and comparison of the duplicated circuit outputs (Double Modular Redundancy—DMR). Indeed, DMR generates very high able area and power penalties, i.e., more than 100%. In addition, it assumes that only one of the two circuit copies is failing at a given time, which is surely not compatible with certain aging mechanisms.

In double-sampling principle, shown in Figure 2.1, the error detection of timing violations is performed by observing the output signals of a circuit endpoint at two different instants (double sampling the output signal).

The main features of this technique are the following:

- A redundant storage element (latch or FF cell, depending on the design style) named here Shadow FF is added to each combinational output FF named here Capture FF.
- The redundant sampling element is driven by a delayed clock signal CLK + $\delta$ the delay is noted as $\delta$, while the regular FF (Capture FF) is driven by the clock signal CLK.
- A digital comparator element is used to check the difference between the Capture FF and the state of the Shadow FF. This comparator checks all pairs of regular FFs and their corresponding redundant sampling elements. In Figure 2.1, only one output combinational signal is shown.
- At the output of the comparison element, an error-detection FF triggered by the rising edge of the clock provides the global error-detection signal.
- The delayed clock signal CLK+$\delta$ is generated by adding delay elements on CLK, such as inverters and buffer cells. These elements have to be added locally in the regular and redundant sampling elements, as adding them on the global clock signal requires implementing two separate clock trees, which is generating even more area and power penalties and will also induce clock skews between the clock signals delivered by these trees. Another option proposed in [14–16] consists in using the rising edge of the clock as latching event for the Capture FF and its falling edge as latching event for the Shadow FF sampling element. In this case,

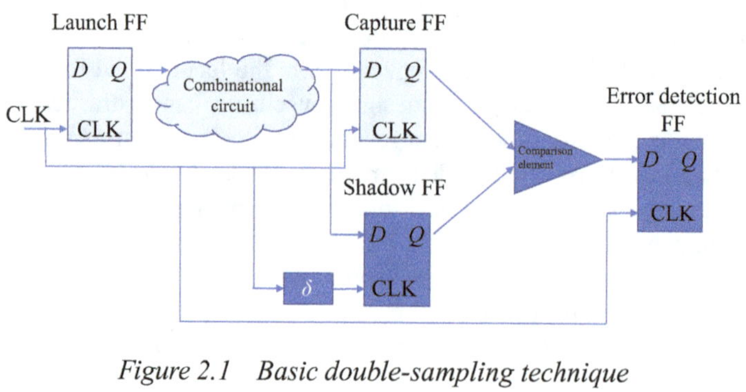


*Figure 2.1 Basic double-sampling technique*

$\delta$ will be equal to the duration of the high level of the clock. The latter option is more advantageous as it eliminates the large number of delay elements required by the former option.

- This scheme does not induce speed penalty and short-time constrains are well mitigated by enforcing constraints on short paths, such as $\delta < D_{min}$ path $- t_{hold}$ (where $t_{hold}$ is the hold time of the Shadow FF and $D_{min}$ path is the minimum delay path duration in any combinational circuit). If this constraint is not enforced, then new input data released by the Launch FF can propagate through a short path and reach the input of the Shadow FF before the end of its hold time. On the other hand, if this constraint is satisfied, then, the input of each redundant Shadow FF sampling element is not affected until the end of the hold time, avoiding hold time violations.
- The outputs of the XOR gates comparing the outputs of Capture FF against the outputs of Shadow FF elements feed an OR tree that produces the global error-detection signal that is eventually captured by an FF.
- As we do not use the individual error signals produced by the XOR gates, we can easily remove the Shadow FF sampling elements; use an XOR gate to compare the input and the output of each regular FF; use an OR tree to compact the outputs of the XOR gates into a global error-detection signal; and use a final error-detection FF to capture the global error-detection signal. More details on these conditions can be found in paper [17].

#### 2.2.1.1 Detection efficiency of double-sampling technique

As the Shadow FF captures data after $\delta$ time with respect to Capture FF, *delay faults of duration lower than* $\delta$ will affect only the value captured by the Capture FF and *will be detected.*

Transient pulses can affect output signal captured by both Capture FF and Shadow FF only if its duration exceeds $\delta$-$t_{setup} - t_{hold}$ (where $t_{hold}$ is the hold time of the Capture FF and $t_{setup}$ is the setup time of the redundant sampling element). Thus, transients with duration smaller than $\delta - t_{setup} - t_{hold}$ are detected.

If a logic error is triggered in the Capture FF, then a potential error non-detection may happen, especially if the error occurs after the instant $tr_{CLK} - t_{setup} -$ DCMP, where $tr_{CLK}$ is the rising edge of the clock, $t_{setup}$ is the setup time of the error-detection FF, and DCMP is the delay of the comparator.

The double-sampling scheme enables only error detection. However, to perform error correction, the double-sampling scheme is usually combined with a local basic retry scheme or global recovery mechanisms, such as checkpoint rollback, which upon error detection, the system repeats the latest operations [14,15]. The retry is very well suited to processor-like design, while checkpoint recovery scheme is more suitable to SoC and many core designs that employ centralized or distributed Reliability Management Units that analyze and decide on consistent checkpoint and roll-back strategies. The double-sampling scheme and the retry mechanism can be declined and employed in various manners to achieve various goals. This will be presented in the next section.

#### 2.2.1.2 Correction of soft errors and timing faults

Soft errors occur randomly; therefore, repeating the latest operation (or operations) most probably leads to a corrected operation. However, if the target faults also include performance faults, different approach must be used. As a matter of fact, timing faults will be reactivated and the error will be triggered again during retry operation.

Thus, contributions [14–16] propose the following correction procedure to perform error correction (see Figure 2.2):

- After each error detection, the latest operation is repeated (operation retry) to correct the error.
- To avoid occurrence of the same error, the clock frequency has to be reduced during retry (e.g., divided by 2 by blocking the high clock level in every two clock cycles).

#### 2.2.1.3 Reliability improvement and adaptive calibration

In the case of *PVT* and aging variation compensations, the double-sampling scheme will cover all potential timing degradations issued by selecting adequate value for $\delta$.

Double-sampling technique can be used also at runtime. In this case, clock frequency of the circuit will be adapted in the case of extra delays that are generated by dynamic variations due to $V_{DD}$ and $T$ fluctuations. This is done by reducing the clock frequency each time when an error is detected by the double-sampling scheme. The self-calibration scheme can also be used after fabrication for adapting the operating frequency of the circuit to the actual circuit delays and compensate for process variability. This is done by using the double-sampling scheme to detect timing errors activated by the fabrication tests and adapting the clock frequency accordingly.

#### 2.2.1.4 Speed increase

Another application of the double-sampling scheme consists in increasing the operating frequency of a circuit by using a clock period smaller than clock delay specification and exploiting the double-sampling scheme for detecting infrequent errors when the particular longest path is activated and correcting them by means of a retry mechanism.

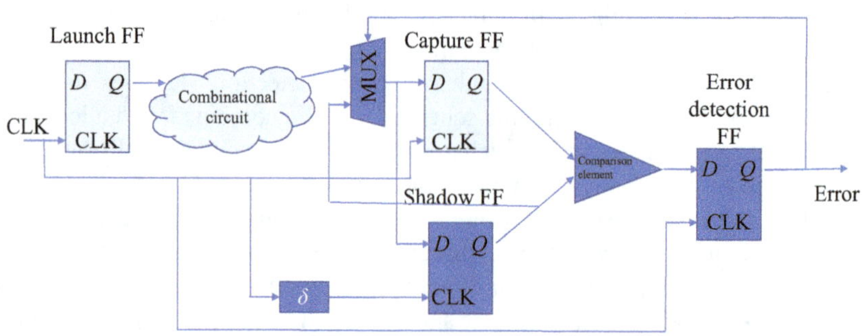


*Figure 2.2 Error correction principle with double-sampling technique*

#### 2.2.1.5 Power reduction

A very important use of the double-sampling scheme was introduced later in contributions [8,9], consisting in reducing power dissipation by aggressively reducing the bias voltage. In fact bias voltage reduction increases the path delays, and the double-sampling scheme together with error correction mechanism will operate the circuit at a clock period shorter than what is actually allowed by the delays of the circuit.

To achieve power reduction, further optimized implementations proposed in [18,19] combine the double-sampling scheme with an error correction scheme, which modifies the double-sampling circuitry with the addition of local error correction. The principle of this mechanism is illustrated in Figure 2.2 by means of a multiplexer, which uses the content of the Shadow FF to replace the content of the Capture FF. The error can be corrected locally by using the contents of the Shadow FF, which is not affected by the timing error. However, this correction takes one extra clock cycle, breaking the temporal coherence with the other pipeline stages. Thus, temporal coherence is enforced [18,19] by using clock gating to stall all pipeline stages in a processor design, for at least one clock cycle. However, as the delay of clock-gating propagation to all pipeline stages may not be compatible with very fast designs, an alternative technique using counterflow pipelining is also proposed in [18,20]. Using this implementation, up to 44% power dissipation reduction was achieved [20] by reducing the supply voltage at a subcritical level implying a mean rate of 1 error for every 10,000 clock cycles.

#### 2.2.1.6 Failure prediction

A second implementation of the double-sampling scheme has been initially proposed in [14] and has been evaluated in [16], and it is shown in Figure 2.3. This implementation got probably more recognition and had been easily embedded within the circuits. This scheme adds the extra delay $\delta$ on the data input of the Shadow FF element rather than on its clock signal.

Thanks to this delay, a second sampling instant is created that precedes by $\delta$ the sampling instant of the regular Capture FF. As this scheme increases by $\delta$, the delay experienced by the Shadow FF, the clock cycle has also to be increased by a timing margin, equal to $\delta$ [14,16]. A very interesting application of this scheme was proposed

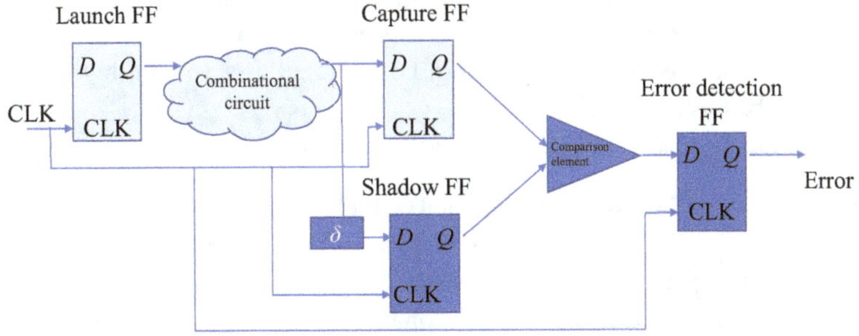


*Figure 2.3 Double-sampling implementation for failure prediction*

later in [21]. It employs the same implementation as Figure 2.3, but the $\delta$ delay is added on the input $D$ of the Capture FF, rather than on the Shadow FF. As explained previously, adding the delay $\delta$ requires increasing the clock period by a $\delta$ time. The circuit is not detecting any anomaly as far as $D_{\text{max}} + t_{\text{setup}} + \delta < T_{\text{CLK}}$, where $D_{\text{max}}$ is the maximum circuit delay, $t_{\text{setup}}$ is the FF setup time, and $T_{\text{CLK}}$ is the clock period. However, if due to aging-induced circuit degradation, the delays of the circuit increase such as it violates $D_{\text{max}} + t_{\text{setup}} + \delta < T_{\text{CLK}}$ condition, the output of the comparison element will go to "1." Thus, this implementation acts performance violation detector (e.g., aging) and can be used to reduce the clock frequency for adapting the circuit to aging-induced degradations.

#### 2.2.1.7 Failure prediction versus error detection

A significant advantage of the failure prediction scheme with respect to the error-detection scheme is that it does not require retry mechanism implementation. Indeed, considering that failure mechanisms are slow and circuit delays degrade gradually, the condition $D_{\text{max}} + t_{\text{setup}} + \delta < T_{\text{CLK}}$ will not be satisfied enough time before $D_{\text{max}} + t_{\text{setup}} < T_{\text{CLK}}$. The circuit will generate alarms that can be called warnings. The alarms indicate that the aging monitor will indicate events before the circuit starts to fail. These events will activate clock frequency reduction or supply voltage increase (DVFS), thus preventing the occurrence of real system failures.

The failure prediction double-sampling scheme will not offer important power savings or speed increase as is the case of error detection by double sampling. Indeed, *PVT* variations and circuit aging induce circuit delay increase that may produce infrequent errors during circuit operation. Even if the frequency of these errors can be quite low, they will still result in some reliability loss that can be inacceptable in some critical level applications. In the failure prediction case, the voltage supply needs to be increased (or the clock frequency reduced) to eliminate these errors. Instead, when the error-detection scheme is used, the retry mechanism is triggered to correct these infrequent errors, without increasing the supply voltage. Therefore, with double-sampling technique used in error-detection mode, we can achieve much power reduction (for instance, [19] demonstrates 52% energy savings at 1 GHz operation for an ARM ISA processor). Instead, for failure prediction implementation, paper [22] presents power saving results in the range of a few percent.

On the other hand, the failure prediction scheme detects the violations of the guard-banding but does not detect the errors produced when the delay of a path exceeds the clock period. Thus, it should be operated at a voltage higher than the Point of First Failure.

Finally, the basic advantages of the double-sampling scheme are as follows:

- The avoidance of massive redundancy, which reduces drastically the area and power penalties with respect to the traditional error-detecting schemes.
- Its capability of detecting the timing faults whose duration will not exceed $\delta$, whatever multiplicity; while the traditional error-detection scheme may be inefficient against multiple timing faults, as these faults invalidate the assumption that only one of the two circuit copies is faulty.

- Though the double-sampling schemes are much more efficient with respect to the traditional error-detection schemes, they have several limitations such as metastability, miss correction, and short path constraints that have to be taken care properly at the design time.
- Using a Shadow-FF sampling element for checking a Capture FF induces undesirable area and power penalties. This is particularly true for the power penalty, because sampling elements are power hungry [23] and power constraints are very tight in modern ICs. Thus, reducing the number of redundant sampling elements as proposed in [16] is highly desirable. To reduce it, we observe the following: The double-sampling architectures provide low-cost solution for *PVT* variations, aging, soft-errors parametric, and logic failures mitigations. As a consequence, it has encountered keen interest in academic research and industrial R&D teams as it can be applied not only to various applications on error detection, failure prediction, self-calibration, reliability, yield improvement, but also power reduction and speed increase; their evaluation and adoption by industrial teams was declined in many solutions. Some of the solutions will be described in the next sections.

## 2.3 Double-sampling-based monitors for detecting performance violations and transient faults

Many implementations have been proposed in the literature for monitors and instruments, all of them exploiting the double-sampling theoretical approach. They can be divided into two classes.

- Monitors such as Ring Oscillator [3], Path Replica [23], Tunable Replica Circuits [24], etc. These monitors are not embedded within the design; they are usually situated in specific parts of the layout, external to the design. Their placement can be random, or closed to specific parts of the circuits that have to be monitored.
- Embedded monitors such as Razor-I, II [19,20], Transition Detector with Time Borrowing (TDTB) monitors [25], Double-Sampling with Time Borrowing (DSTB) monitors, and ISM [26]. These monitors are usually inserted at identified critical endpoints in the design, such as FFs and latches. Their indication is more accurate as they completely follow the activity of the circuit.

In the following sections, the external and embedded monitors are explained with more details.

### *2.3.1 External-design monitors*

**Ring oscillator [3] or Replica Path [26]** aims at replicating the timing behavior of the original circuit longest delay, or certain critical delay regions in a circuit. Self-oscillating paths using various combinations of standard cells are made to mimic critical path frequency as shown in Figure 2.4. The critical path delay and the replica path delay are compared in order to detect static, dynamic variability, and aging.

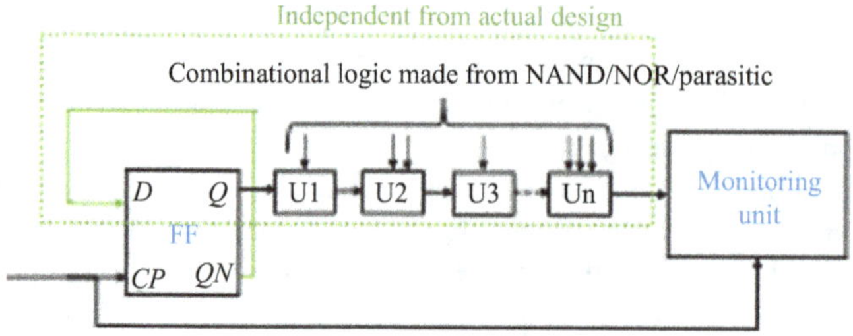


*Figure 2.4 Replica path monitor: the critical path of the design is replicated with U1, U2, U3, etc. standard cells*

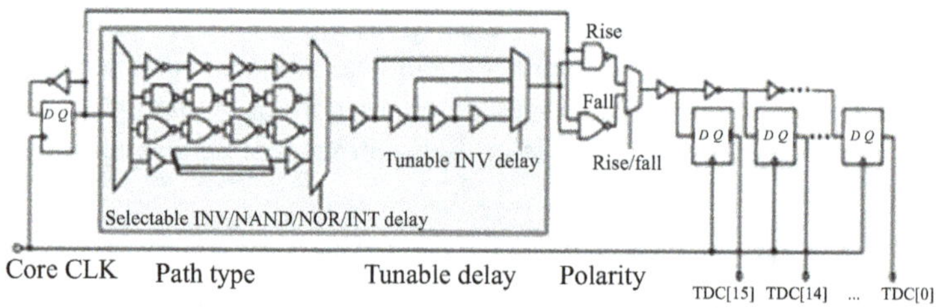


*Figure 2.5 Tunable replica paths monitor [26]*

After fabrication of a given chip, these paths are calibrated to match the maximum operating frequency of the target design. When delays of cells increase due to aging or other variations, the frequency of these self-oscillating paths changes leading to timing violations that can be detected.

Tunable Replica Path Circuit (TRC) [26] is an advanced version of replica path circuits. Different paths are built up with different kind of logic cells. Their outputs are multiplexed and one particular branch is selected to mimic critical path frequency as shown in Figure 2.5 [26]. Degradations can be detected more accurately with respect to generic replica paths, as various cells have different effects under the same stress. Therefore, many mismatches can be detected, also all types of process variability. Tunable Replica Path allows careful calibration of the circuit frequency after fabrication to match reference design's maximum operating frequency. The output of this monitor is usually connected to time-to-digital converter that converts timing margin into digital code. This digital code is then used by the controller included into a self-adapting compensation technique such as Adaptive Voltage and Frequency Scaling (AVFS) or ABB.

Critical Path Sensor (CPS) architecture is another combination of classic replica path monitor and double-sampling technique. The critical path timing of target design is replicated outside the design. Double-sampling monitor added at the end of the path to allow timing violation detections, see Figure 2.6 [20]. As a matter of fact, transient faults, *PVT*, and/or aging degradations, inducing timing errors, are captured by the Shadow FF if the total path delay is exceeding its nominal value. The Capture FF will

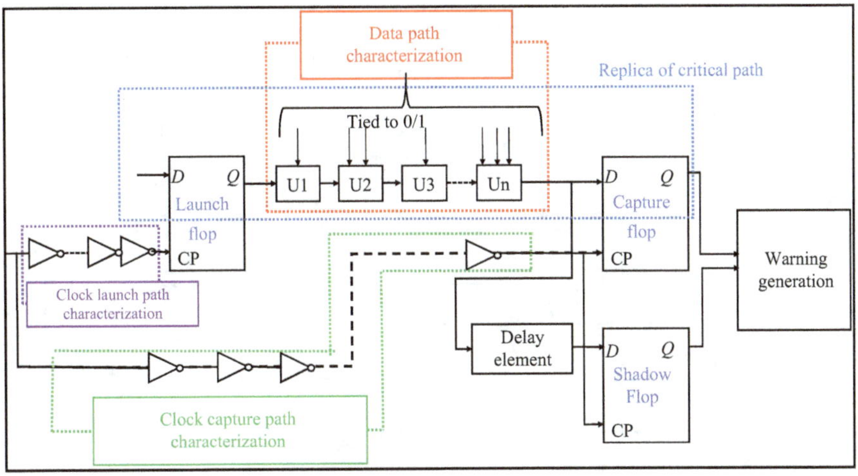


*Figure 2.6 Critical path sensor architecture [20]*

capture the correct data due to additional delay element that is specifically calibrated to cover for the typical delay degradation of a given circuit and application, *PVT*, and aging conditions (see Figure 2.6). The delay degradation is detected by comparing the output of Capture FF with the Shadow FF. Mimicked critical paths show better correlation than generic replica path monitor made of various combination of cells, because they are identical to the critical paths of design, thus having better correlations in terms of delay with the target chip.

One major drawback of all external monitors is related to the fact that their activity is uncorrelated from the system workload, and they do not mimic the real aging experience as the original critical path. Moreover, they do not capture the impact of local variabilities neither. Therefore, internally situated monitors were proposed in the literature to overcome the abovementioned drawback and they are preferred in designs where the sensitivity to workload and local variation is important.

## 2.3.2 Embedded monitors

Embedded monitors coping with previously explained monitoring drawbacks are implemented within the design and are directly linked to the circuit paths delays. They are inserted in specific endpoints of the designs, usually at the end of critical and subcritical paths. Among a large choice of embedded monitors (also called ISM), it is worth mentioning that the well-known error-detection monitors called Razor [18,19]. Razor uses a special Shadow latch or Shadow FF to detect timing failures due to setup violations on the logic stage paths. As shown in Figure 2.7, the timing error is detected by comparing the main FF data and Shadow FF data at the output of the monitored endpoint. When the delay increases due to aging, the Shadow latch and the Slave latch have different outputs as shown in timing diagram. When delay is increased, slave latch fails to capture the correct data and Shadow latch captures the

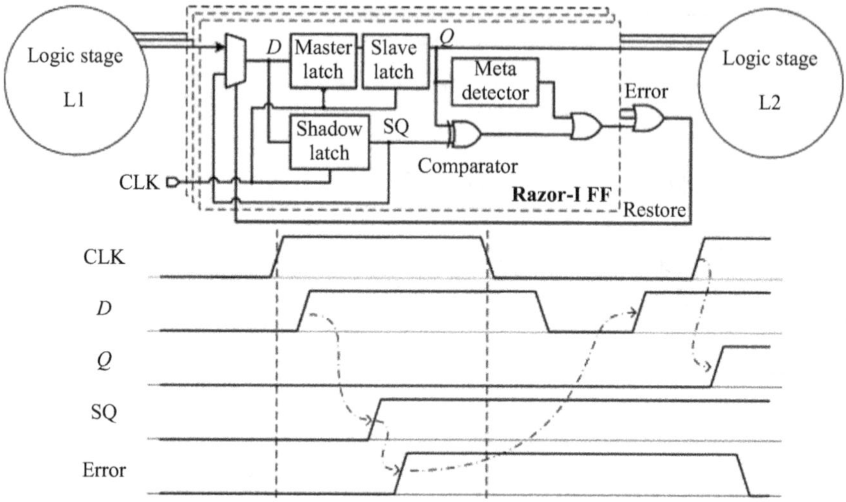


*Figure 2.7 Razor-I structure and timing diagram [18]*

right data. By comparing these two signals, an error signal is activated. A metastability checker is set after the Slave latch that solves any timing issue related to signal timing skews. When error signal is generated, the correct data from Shadow latch is restored to FF. In the Razor-II [19], FF is replaced with level-sensitive latch that reduces area overhead. Also, only error detection is performed in the FF unlike Razor-I where both error detection and correction are performed in the FF. Error correction in Razor-II is done through architectural rollbacks.

Another error-detection monitor was introduced in [25] and is called DSTB. This approach mixes in the same design different redundant storage element latches and/or FFs (see Figure 2.8). When the path delay increases due to aging phenomena and violates the paths slack, the FF captures a wrong value, although the latch will capture the delayed correct output on its active clock level. Thus, the correct output is passed on the subsequent stages of the circuit, and at the same time the error flag is generated and triggers correction strategies, such as voltage adaptation. This monitor implementation generates huge design effort due to the clock signal routing complexity as sometimes timing closure can be an issue. At the same time, the latches and FF strategy mixing can be considered as non-safe for critical application.

TDTB [25] uses time-borrowing concept for error detection. Figure 2.9 shows the architecture. A small pulse is generated by the XOR gate whenever data transitions are occurring on the monitored path connected at *D*. If this pulse occurs when the latch is not transparent, the error signal does not raise. However, when *D*-input data arrives late due to delay degradations on the monitored path, the pulse occurs during the active phase of the clock (the latch is transparent) and the error signal becomes high. TDTB monitor removes metastability from data path to the error generation path, which is a benefit of using TDTB but design complexity of TDTB is high compared to other Electronic Static Discharge (ESD) circuits.

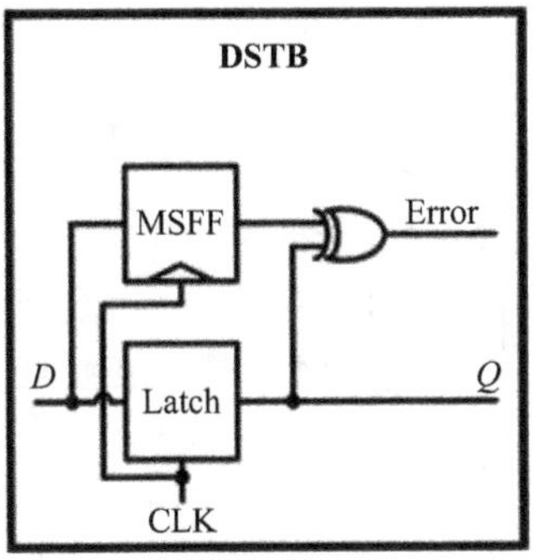


*Figure 2.8 Double sampling with time borrowing monitor [25]*

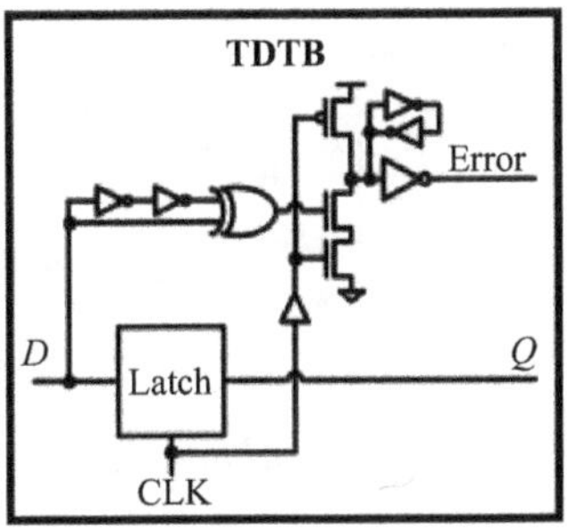


*Figure 2.9 Transition detector with time borrowing monitor [25]*

The issues related to error recovery scheme implementations (i.e., error correction, or rollback) can be avoided by preferring circuit failure prediction monitors [22], i.e., predicting the occurrence of the pre-error before the appearance of any error in system data and state. In this approach, the monitors raise a warning signal when late transition occurs, but the outputs are still error free. In-line with this approach, in [22] stability checker circuit was proposed which detects transitions close to the clock edge thanks to additional delay element in the clock path.

One error-prediction monitor called **Canary FF** [13] is shown in Figure 2.10. It consists in sensing a given propagation path delay violation thanks to a well-sized delay element at the input of the Shadow FF. A pre-error signal is raised when the value at the endpoint Capture FFs differs from the value in the Shadow FF, as the resulting timing slack becomes shorter than the signed-off one. This pre-error signal is the indicator of the Capture FF setup timing violation.

Canary FF design is easy to implement and adding it to a certain design can be done through automation. This scheme is well suited for compensation scheme as indicator of degradation because the added delay element gives enough time to protect design from aging and variation without failures.

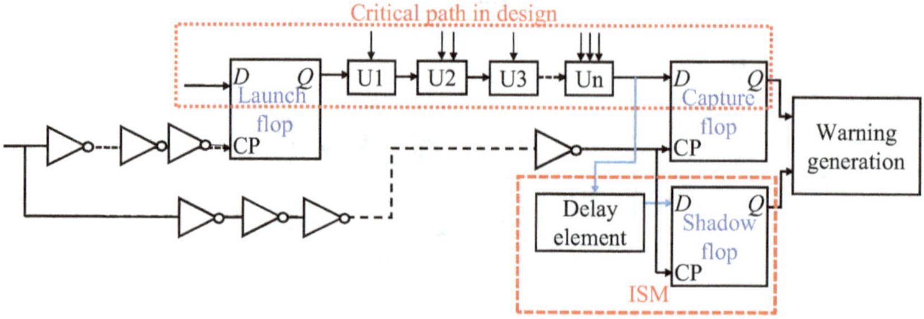


*Figure 2.10 Canary Flip-Flop monitor [13]*

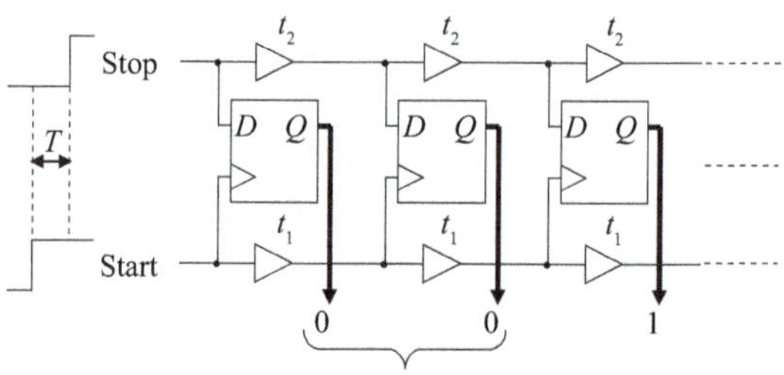


*Figure 2.11 Circuit configuration of Vernier delay line [29]*

### *2.3.3 Other types of monitors*

The second class of monitors use different concept for *PVT* or aging-induced delays detection and measurements [27–32]. Based on different architectures, their purpose is to detect and measure on-chip transient pulses generated by different sources of noise, such as radiation-induced pulses.

It is also possible to use some parts of the hardware that are already used by the circuit for different other purposes. For example, we can use the boundary scan cells for monitoring the path delay of a circuit during normal operation. Papers [30,31] propose to use this for the Small Delay Defect testing for predicting the system's age at runtime. The method proposes to check any transition on output nodes in a specified time window. Similar solutions are presented in paper [32] where Slack-Probe methodology inserts embedded timing slack monitors at a selected set of nets, including intermediate nets along critical paths. The objective is to reduce the number of monitors to be inserted in a given design for better area and power savings.

Paper [27] uses a Vernier delay line for pulse width evaluation followed by a capturing circuit with edge trigger [27]. Figure 2.11 shows the principle of setting up a delay line used to measure timing difference between two transitions. It is composed by two buffer chains and *D* latch gates. Two steps signals (START and STOP) are given to the circuit and the time difference will be measured by the Vernier delay line.

Usually $t_1$ is larger than $t_2$, START signal is feed as clock signals to all $N$ latches, and STOP will be connected to $D$ input of latches. START and STOP signals race and finally STOP signal overtakes START signal. When these signals propagate through a single stage, the time difference between them, which was initially $T$ at the input, is reduced by $t_r = (t_1 - t_2)$. Store 1 for the latches where the time difference becomes 0 or below, and for the other latches, 0. Letting $N$ denote the number of latches storing 1, and $t_{setup}$ being the setup time of latches, the time difference $T$ is estimated by

$$(N - 1) \times t_r \leq T + t_{setup} < N \times t_r$$

This principle can be used to measure also time difference between two paths, one of them being the reference path and the other one the aged path.

### 2.3.4 Discussions

The trade-off between different kinds of monitors is explained in this section. Externally placed sensors in the intended design are suitable for easy implementation and detection of global process centering and average aging, while internally situated monitors are better for fine-grained detection of global as well as local variation and more accurate aging prediction and detection. Also, monitor failure prediction approach is better as they generate alarms or warnings prior to timing failures, giving the system enough time for correction through compensation strategies. Note that external monitors are mainly used without correlations to the real activity of the circuit (e.g., vector less) while still trying to mimic somehow the activity of the circuit intended to be monitored, while ISM alarm activation is related to the real activity of the end-point registers. This last feature can be circumvented by using large number of external monitors located both at high activity and close to timing critical hot spots.

Externally situated monitors are easy to implement, they do not change the reference design netlist so the timing closure is very fast and verification steps take less time. Their usage is very well suited for global variation detection and is not accurate to capture local variations such as within-die, random manufacturing, and circuit aging.

For illustration purposes, external Critical Path Replicas (CPR) and CPSs monitors are implemented in one of chips fabricated in an STMicroelectronics technology. The reference design is considered here as being composed by three ARM A53 microprocessors with four CPRs placed at strategic points within the design, and one CPS. Figure 2.12 shows distribution of the three ARM frequency measurements and their correlations with CPS measurements of frequencies. Figure 2.12 illustrates also the difference between CPR/CPS measures underlining the difficulties to correlate external monitors CPR with the local variability and workload.

As discussed earlier, internally situated monitors are more accurate in terms of global and local variation detection compared to externally situated monitors due to their localization within the design. The drawback for the ISM's insertion is the difficulty to close the timing and fix the critical path rankings. Indeed, some initial selected set of endpoints at early physical synthesis steps or at place and route steps can become subcritical after the detailed routing steps, and in the same time, subcritical

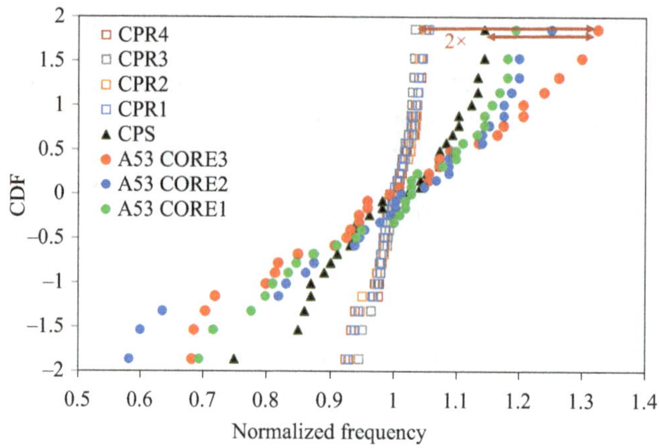


*Figure 2.12 Distribution of normalized maximum operating frequency of ARM A53 microprocessor, Critical Replica Paths (CPRs) and CPS [26]*

endpoints can become critical at the final steps. In that last situation, Engineering Change Order (ECO) loop is carried out to fulfill this new critical path monitoring. The situation is even worse if the monitors are inserted at specific points inside the design, as additional verification steps need to be followed to prove that the monitored design is still equivalent with the initial one.

The activation of ISM inserted on critical paths is another very important issue, because if the path is not activated during the workload execution, the ISM cannot raise an alarm signal. In a complex SoC where multiple functional modes are available, all ISM inserted on the critical paths may not be activated in a particular functional mode. This limitation can be overcome by using a combination of scan design and specific Automatic Test Pattern Generation (ATPG) vectors that are periodically applied to the circuit. Therefore, ISM inserted on critical paths can be activated and potential degradations can be detected.

The situation of ageing monitoring by using internally situated monitors is now presented. Another testchip has been used for this purpose, consisting in a large arithmetic block composed of 2,000 endpoints. This block has been characterized at design time (e.g., fresh) and also in different aged states. Figure 2.13 reports the functional $F_{max}$ and the frequency sweep with the number of rising monitors flags. It is clear that during the aging operation, the $F_{max}$ measured for first ISM flag occurrence decreases with stress time. Moreover, for all critical and subcritical paths, a clear $F_{max}$ reduction with the aging operation is measured. Therefore, for aging monitoring, internally placed monitors are a better option than externally situated monitors.

Effectiveness of monitors also depends on the insertion flow as it helps one to decide the selection of critical paths where monitors need to be inserted. The flow needs to be weakly intrusive with respect to the initial performance.

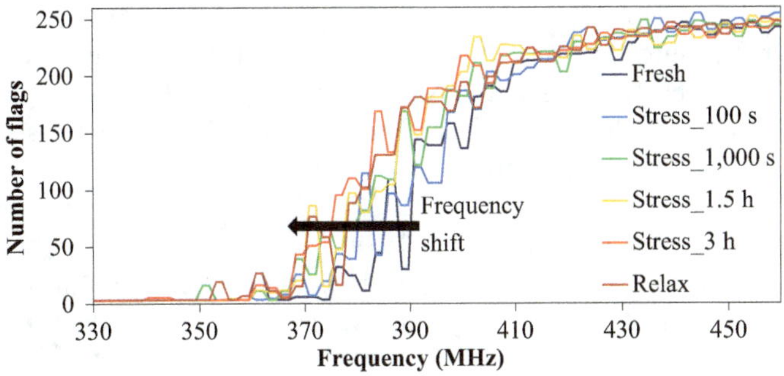


*Figure 2.13 Aging induced frequency shift measured through ISM for increasing dynamic stress versus fresh measurement for digital block [26]*

## 2.4 Conclusions

The usage of sensors and performance violation monitors is a possible solution to detect and possibly correct timing errors, and in addition to the reduction of design margins. The behavior of monitors, their different working principles, and their placement with respect to the circuit have been discussed. Recovery mechanisms are also presented which consist in the combination of monitors with AVS technique or DVFS. As a matter of fact, performance violation monitors trigger AVS or DVFS techniques, and the system adapts dynamically the frequency and the voltage according to the operating conditions and the application needs. Thus, the performance degradation can be compensated, further modes of low power can be achieved and the circuit's lifetime can be extended.

## References

[1] A. Drake, R. Senger, H. Deogun, *et al.*, "A distributed critical-path timing monitor for a 65 nm high-performance microprocessor," in Proc. IEEE of ISSCC Digest of Technical Papers, San Francisco, USA, IEEE, February 2007, pp. 398–399, ISBN 1-4244-0853-9.

[2] P. Gupta, Y. Agarwal, L. Dolecek, *et al.*, "Under designed and opportunistic computing in presence of hardware variability," IEEE Trans. Comput Aided Des. Integr. Circuits Syst., vol. 32, no. 1, pp. 8–23, 2013.

[3] T. Burd, T. Pering, A. Stratakos, and R. Brodersen, "A dynamic voltage scaled microprocessor system," IEEE J. Solid-State Circuits, vol. 35, no. 11, pp. 1571–1580, 2000.

[4] M. Nakai, S. Akui, K. Seno, *et al.*, "Dynamic voltage and frequency management for a low-power embedded microprocessor," IEEE J. Solid-State Circuits, vol. 40, no. 1, pp. 28–35, 2005.

[5] J. Gaisler, "A portable and fault tolerant microprocessor based on the SPARC V8 architecture," in Proceedings of 2002 IEEE/IFIP International Conference of Dependable Systems and Networks (DSN 2002), IEEE, 2002, pp. 409–415.
[6] N. Quach, "High availability and reliability of Itanium processors," IEEE Micro, vol. 20, no. 5, 2000.
[7] M.A. Check, and T.J. Slegel, "Custom S/390 G5 and G6 microprocessors," IBM J. Res. Dev. vol. 43, no. 5/6, 1999.
[8] K.A. Bowman, J. Tschanz, N.S. Kim, *et al.*, "Energy-efficient and metastability immune resilient circuits for dynamic variation tolerance," IEEE J. Solid-State Circuits, vol. 44, no. 1, pp. 49–63, 2009.
[9] A. Tiwari and J. Torrellas, "Facelift: hiding and slowing down aging in multicores," in Proceedings of IEEE/ACM International Symposium on Microarchitecture, Lake Como, Italy, IEEE, 8–12 November 2008, pp. 129–140.
[10] A. Drake, R. Senger, H. Deogun, *et al.*, "A distributed critical-path timing monitor for a 65 nm high-performance microprocessor," in Proc. IEEE of ISSCC Digest of Technical Papers, San Francisco, USA, IEEE, February 2007, pp. 398–399, ISBN 1-4244-0853-9.
[11] M. Nakai, S. Akui, K. Seno, *et al.*, "Dynamic voltage and frequency management for a low-power embedded microprocessor," IEEE J. Solid-State Circuits, E88C, vol. 40, no. 1, pp. 28–35, 2005.
[12] K. Nowka, G.D. Carpenter, E.W. MacDonald, *et al.*, "A 32-bit power PC system-on-a-chip with support for dynamic voltage scaling and dynamic frequency scaling," IEEE J. Solid-State Circuits, vol. 37, no. 11, pp. 1441–1447, 2002.
[13] L. Anghel, A. Benhassain, and A. Sivadasan, "Early system failure prediction by using aging in situ monitors: methodology of implementation and application results," in IEEE 34th VLSI Test Symposium (VTS'16), Las Vegas, NE, USA, DOI: 10.1109/VTS.2016.7477316, 25–27 Apr. 2016.
[14] M. Nicolaidis, "Time redundancy based soft-error tolerant circuits to rescue very deep submicron," in Proc. 17th IEEE VLSI Test Symp., Dana Point, CA, USA, Apr. 1999, pp. 86–94.
[15] M. Nicolaidis, "Circuit logique protégé contre des perturbations transitoires," Patent WO2000054410 A1, 2000.
[16] L. Anghel and M. Nicolaidis, "Cost reduction and evaluation of a temporary faults detecting technique," in Proc. of IEEE DATE Conf., Paris, France, Mar. 2000, pp. 591–598.
[17] M. Nicolaidis, "Double-sampling design paradigm—a compendium of architecture," IEEE Trans. Device Mater. Reliab., vol. 15, no. 1, 2015.
[18] D. Ernst, NS Kim, S. Das, *et al.*, "Razor: a low-power pipeline based on circuit-level timing speculation," in Proceedings of 36th Annual IEEE/ACM International Symposium on Microarchitecture, 5 December, San Diego, USA, IEEE, 2003.
[19] S. Das, C. Tokunaga, S. Pant, *et al.*, "Razor-II: in situ error detection and correction for PVT and SER tolerance," IEEE J. Solid-State Circuits, vol. 44, no. 1, pp. 32–48, 2009.

[20] S. Das, S. Pant, D. Roberts, *et al.*, "A self-tuning DVS processor using delay-error detection and correction," in Digest of Technical Papers of 2005 Symposium on VLSI Circuits, IEEE 2005, pp. 258–261.

[21] M. Agarwal, B.C. Paul, M. Zhang, and S. Mitra, "Circuit failure prediction and its application to transistor aging," in Proc. 5th IEEE VLSI Test Symp., Berkeley, CA, USA, May 6–10, 2007, pp. 277–286.

[22] V. Huard, F. Cacho, F. Giner, *et al.*, "Adaptive wear out management with in-situ management," in 2014 IEEE International Reliability Physics Symposium (IRPS 2014), June 1–5, 2014, Hawai, USA, IEEE, pp. 6B.4.1–6B.4.11.

[23] T. Kuroda, K. Suzuki, S. Mita, *et al.*, "Variable supply-voltage scheme for low-power high-speed CMOS digital design," IEEE J. Solid-State Circuits, vol. 33, no. 3, pp. 454–462, 1998.

[24] M. Cho, S.T. Kim, C. Tokunaga, *et al.* "Postsilicon voltage guard-band reduction in a 22 nm graphics execution core using adaptive voltage scaling and dynamic power gating," IEEE J. Solid-State Circuits, vol. 52, no. 1, pp. 50–63, 2017.

[25] K.A. Bowman, J.W. Tschanz, N.S. Kim, *et al.*, "Energy-efficient and metastability-immune timing error detection and recovery circuits for dynamic variation tolerance," in 2008 IEEE International Conference on Integrated Circuit Design and Technology and Tutorial, 2–4 June 2008, Austin, USA, IEEE.

[26] R. Shah, F. Cacho, R. Lajmi, *et al.*, "Aging investigation of digital circuits using In-Situ monitor," in IEEE International Integrated Reliability Workshop (IIRW 2018), Stanford Sierra, fallen Leaf, USA, 13–18 October 2018.

[27] J. Keane, X. Wang, D. Persaud, and C.H. Kim, "An all-in-one silicon odometer for separately monitoring HCI, BTI, and TDDB," IEEE J. Solid-State Circuits, vol. 45, pp. 817–829, 2010.

[28] T.T.H. Kim, P.F. Lu, K.A. Jenkins, and C.H. Kim, "A ring-oscillator-based reliability monitor for isolated measurement of NBTI and PBTI in High-k/metal gate technology," IEEE Transactions on Very Large Scale Integration (VLSI) Systems, vol. 23, pp. 1360–1364, 2015.

[29] Y. Zhao and H.G. Kerkhoff, "Highly dependable multi-processor SoCs employing lifetime prediction based on health monitors," in 2016 IEEE 25th Asian Test Symposium (ATS), 21–24 November 2016, Hiroshima, Japan.

[30] X. Wang, L. Winemberg, D. Su, *et al.*, "Aging adaption in integrated circuits using a novel built-in sensor," IEEE Trans. Comput. Aided Des. Integr. Circuits Syst., vol. 34, no. 1, pp. 109–121, 2015.

[31] S. Jin, Y. Han, H. Li, and X. Li, "Unified capture scheme for small delay defect detection and aging prediction," IEEE Trans. Very Large Scale Integr. Syst. vol. 21, no. 5, pp. 821–833, 2013. https://doi.org/10.1109/TVLSI.2012.2197766.

[32] L. Lai, V. Chandra, R. Aitken, and P. Gupta, "SlackProbe: a flexible and efficient in situ timing slack monitoring methodology," IEEE Trans. Comput. Aided Des. Integr. Circuits Syst., vol. 33, no. 8, 2014.

# Chapter 3

# Design techniques to improve the resilience of computing systems: architectural layer

*Aviral Shrivastava[1], Kyoungwoo Lee[2], Hwisoo So[2], Jinhyo Jung[2], and Prudhvi Gali[1]*

Unreliable hardware components will affect computing system at several levels—all the way from incorrect transistor outputs, to incorrect values in memory elements, incorrect program variables and control flow, finally causing application failure. Resilience is the ability of the system to tolerate errors when they occur and comprises two main aspects—(i) how to detect the errors and (ii) how to recover from the errors. The lower the level of abstraction at which we can detect and correct the error, the less disruption it causes to all the upper layers of computing abstraction. This chapter gives the overview of all the techniques at processor architecture level to detect and correct the errors.

There are three basic approaches to protect data: (i) Error Detection Code (EDC), (ii) Error Correction Code (ECC), and (iii) replication. EDCs or parity codes are essentially XOR of data and are easy to compute and test, but they can only detect if an odd number of bits in the data (including the parity bit) are faulty and provide no error correction. The second option is to use ECCs. These codes require more hardware and time to compute while they cannot only detect errors but also fix some errors. SEC-DED (Single Error Correction and Double Error Detection) is a commonly used code. The third option is to hold replicas of the data to be protected. While this saves the efforts of generating the EDC or ECC, the redundant data takes extra space, and the comparison of the redundant copies still incurs time. As it becomes important to consider multi-bit errors, researchers attempt to deal with complex cases by combining multiple techniques across system layers.

One natural way to classify soft error protection techniques at the architectural level is based on the architectural component that they intend to protect. Figure 3.1 gives an overview of the architectural components, including cache, register file, and pipeline. Of all the microarchitectural components, caches are probably the most vulnerable since a large fraction of processor transistors are in the cache, and they

[1]School of Computing, Informatics, and Decision Systems Engineering, Arizona State University, Tempe, AZ, USA
[2]Department of Computer Science, Yonsei University, Seoul, South Korea

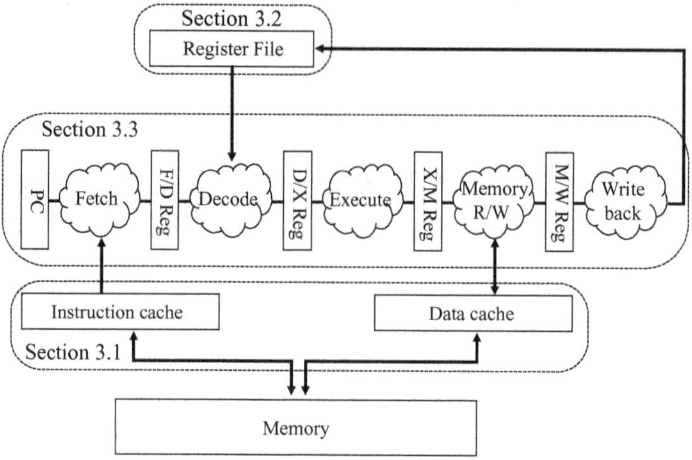


*Figure 3.1 Architectural view of a processor includes the pipeline, the register file, and the memory subsystem (including caches). Sections 3.1–3.3 give an overview of the various architectural techniques to detect and recover from errors in the data in the caches, register file, and the processor pipeline, respectively.*

operate at lower voltage swings than the rest of the components. Section 3.1 discusses the techniques to protect the data in the caches. While the basic mechanism to detect faults is through data replication, ECC, and parity bits (also called Error Detection Codes or EDCs), the techniques here deal with applying them in ways so as to maximize the protection while minimizing the overheads. Section 3.2 explains the different mechanisms to protect the data in the Register Files (RFs). Even though applying ECC in RF is possible, it incurs severe power and performance overheads. Therefore, RF protection approaches focus on innovative ways to apply ECC protection in noncritical paths. Also there are partial protection approaches to protect the most vulnerable values in the RF. Section 3.3 discusses techniques to protect the values in the latches of the processor pipeline, as instructions are executing. While RF and caches are memories and can be protected by ECC, processor pipeline contains computation logic (pipeline stages) as well as memory (latches), and thus, redundant execution is required to protect the computation in pipelines. As a result most techniques are based on executing the instructions on redundant pipeline and then comparing the results at certain synchronization points.

## 3.1 Cache protection techniques

In a processor, caches are the most susceptible to errors. First off, caches occupy the majority of chip real-estate as shown in Figure 3.2 and therefore are the most exposed to error sources such as high-energy particles [2]. Second, they have high transistor density in SRAMs and operate at low supply voltage, making them increasingly less resilient to errors as technology scales [3]. Further, errors in caches can easily and

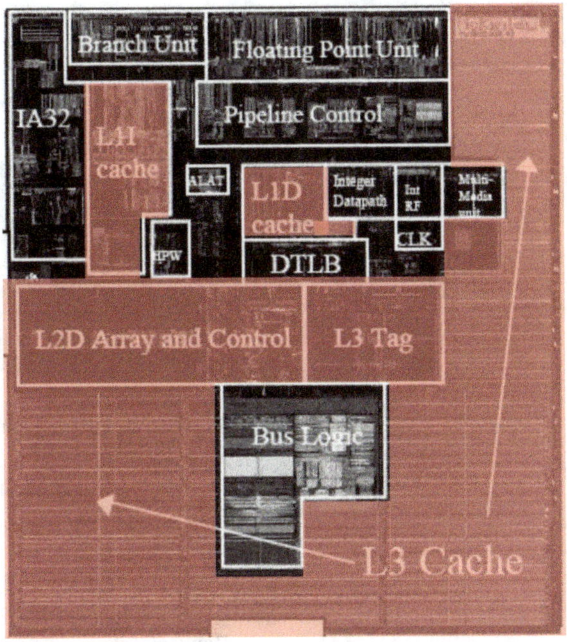


*Figure 3.2 Area of cache components (shaded as red) in Intel Itanium-2 Processor[1]*

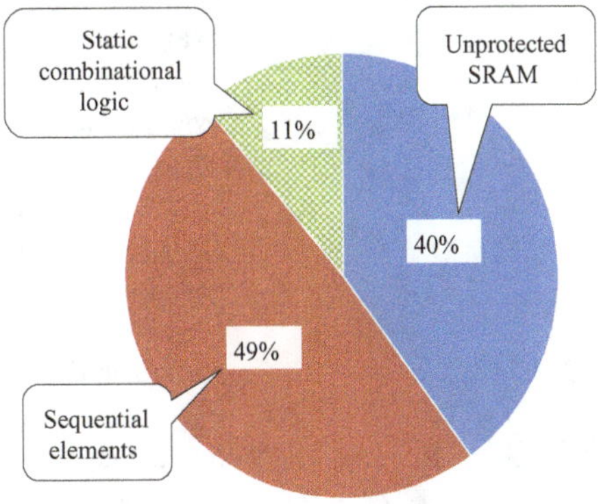


*Figure 3.3 Decomposition of overall soft-error rate for designs such as microprocessors, network processors, and network storage controllers [4]*

quickly propagate to the CPU and main memory, which makes cache protections more important in system designs. Indeed, Mitra *et al.* [4] noted that soft errors in unprotected caches (SRAM) contribute to around 40% of those in processors as shown in Figure 3.3, and Shazli *et al.* [5] have shown that 92% of machine checks are triggered by soft errors in the L1 and L2 caches. Table 3.1 summarizes the attributes

*Table 3.1 Summary of cache protection techniques*

| Techniques | Related to | | | | Summary |
|---|---|---|---|---|---|
| | **Replication** | **Parity** | **ECC** | **Multi-bit Error protection** | |
| Li *et al.* [6] | | O | O | | Applies EDCs to clean lines and ECC to dirty lines, and early write-back |
| Hong and Kim [7] | | O | O | | Flexibly applies EDC to clean data and ECC to dirty data in word level |
| Jeyapaul and Shrivastava [8] | | O | O | | Manages write-backs based on memory profiling to minimize vulnerability |
| Ko *et al.* [9] | | O | O | | Presents guidelines on how to design data caches with parity and ECC |
| Zhang *et al.* [10] | O | O | O | | Replicates data in active use within the L1 data cache |
| Zhang [11] | O | O | O | | Replicates data of L1 data cache in a separate small cache at the same level |
| Lee *et al.* [12] | O | O | O | | Partially applies ECC to small region of data cache containing important data |
| Mukherjee *et al.* [13] | | | O | O | Periodically cleans up single bit errors to avoid temporal multi-bit errors |
| Farbeh and Miremadi [14] | | O | O | | Applies per-set EDC/ECC by exploiting fast access mode |
| Chen *et al.* [15] | | O | O | | Aligns ECC of LLC to the unused of cache compression |
| Hong and Kim [16] | | O | O | | Groups cache sets of LLC and designate a few lines as ECC bit lines |
| Manoochehri *et al.* [17] | | O | | | Implements error correction with parity and two special XOR registers |
| Yoon and Erez [18] | | O | O | | Stores ECC bits of parity protected LLC into the memory as cacheable data |
| Hung *et al.* [19] | | | O | O | Exploits SEC-DED for soft and hard errors based on the number of defective cells |
| Sun *et al.* [20] | O | | O | O | Applies multi-bit ECC to selectively based on the number of defective cells |
| Alameldeen *et al.* [21] | | | O | O | Applies multi-bit ECC to most vulnerable sections of the cache during low voltage |
| Kim *et al.* [22] | | O | O | O | Combines two-dimensional error codes to cover large-scale multi-bit errors |

and features of several existing techniques to enhance the resilience of caches against errors.

Conventional cache protection techniques such as ECC can incur high overheads in terms of area, power, and performance by protecting entire blocks with the expensive ECC only. Several techniques [6–9] have been proposed for efficient cache protections by judiciously applying different levels of protections such as EDC and ECC into blocks with different attributes such as dirtiness.

Li *et al.* [6] proposed an adaptive error coding scheme. Unlike conventional caches that use the same error detection/correction scheme throughout all cache blocks, their technique applies error detection mechanisms to clean cache blocks and error correction mechanisms to dirty blocks. The key observation is that faults in clean cache block can be recovered by reloading the data from secondary caches or memory, and therefore detection-only schemes such as parity are sufficient for protecting clean cache blocks. On the other hand, dirty cache blocks hold values that are different from the ones in secondary caches. Therefore, they cannot reload the data from the secondary caches or anywhere in the memory hierarchy and should be protected by error correction schemes such as ECC. Based on this observation, Li *et al.* proposed a cache protection scheme to protect clean and dirty cache blocks using different complexities of error protection mechanisms as described in Figure 3.4. This adaptive error protection scheme is more energy-efficient than using error correction schemes throughout and is more reliable than using parity protection only. They also proposed the early write-back policy, which writes back dirty blocks to the lower level memory after a fixed time gap from the last write operation. This early write-back policy can effectively improve the error resilience compared to conventional write-back policy, as it reduces the time after the last update and before the write-back (i.e., causing the reduction of dirty blocks), during which the block should be protected with the more expensive error correction schemes. In addition, in the presence of frequent cache block updates, it can outperform the conventional write-through policy by avoiding frequent accesses to the lower level cache, which is significantly expensive.

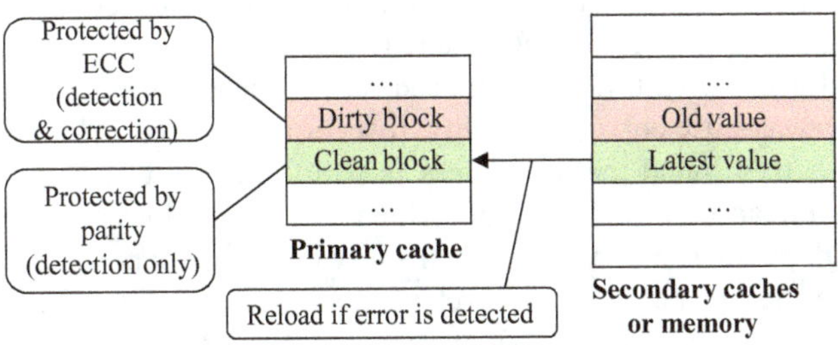


*Figure 3.4 Adaptive protection of Li et al. with parity on clean blocks and ECC on dirty blocks [6]*

Hong and Kim [7] observed other interesting facts about dirty words in the Last-Level Cache (LLC). First off, most data in dirty cache lines are actually clean. In dirty lines of the LLC, about 64% and 86% cache words are clean for integer and floating point benchmarks, respectively. In addition, 88% of dirty cache lines are changed at most twice before the eviction. In other words, if a line in the LLC is modified more than twice, it is reasonable to assume that no additional update will be made on the line before its eviction. Based on their observations and the work of Li *et al.* [6] (error detection mechanisms are sufficient for clean data), Hong and Kim proposed a flexible word-level cache management scheme for the LLC. They applied ECC to the dirty cache words of LLC in the word level, while Li *et al.* did in the line level. Figure 3.5 illustrates the implementation of flexible ECC for the last level cache [7]. For each set of LLC, the number of dirty lines ($n$) and the number of dirty words ($m$) required to be protected with ECC are restricted to optimize the area overhead. As shown in Figure 3.5, the ECC string is added for every set of LLC. The ECC string consists of three chunks: (i) dirty line mapping bit (DLMB) chunk, which is used to indicate the dirty lines of the set. Each entry of DLMB chunk indicates one dirty line in the set. (ii) ECC mapping bit (EMB) chunk, which is used to indicate the dirty words in the dirty line. The number of the bits for each entry in EMB is decided by the number of words in each cache line, and each bit in the entry is raised as 1 when the corresponding word is modified. To maintain EMB chunk, every L1 cache line should maintain dirty bits to identify which bits in the line are modified, and dirty bits are utilized to update EMB entry. The $i$th entry of EMB indicates the dirty words in the line that is indicated by $i$th entry of DLMB. (iii) ECC check bit chunk contains the corresponding ECC bits for dirty cache words. For a line read operation to the LLC, flexible ECC can detect the faults by parity and recover the corrupted line either reloading the replica from memory for clean line or exploiting ECC from ECC string for dirty line. For a line write operation to the LLC, flexible ECC should update the ECC string of the set. If the total number of dirty lines or the total number of dirty words exceeds a predefined threshold, a victim line is decided and is written back to the memory. Choosing the victim line carelessly may result in huge performance degradation. Since a dirty line that is updated twice will not be rewritten in most cases, a line that has been updated more than once is considered as stable and is more likely to be selected as a victim. A stable bit is added for every cache line to record the updates, as shown in Figure 3.5. If a dirty cache line is updated, its stable bit is raised since the line has been updated more than once.

Jeyapaul and Shrivastava [8] proposed Smart Cache Cleaning (SCC) in order to achieve power-efficient vulnerability reduction of caches. In their previous research [23], they found that dirty lines in parity protected cache are vulnerable when they are read by a processor or written back to the lower level of memory. Write-through policy can improve the reliability of the parity protected cache since it can ensure the update of lower level memory. However, it suffers the performance and energy overheads from high memory traffic from frequent write-backs to the lower level memory subsystem. Early write-back cache architecture [6], which writes back all dirty blocks with periodic intervals or events, can reduce the frequency of write-back operations compared to write-through policy by sacrificing

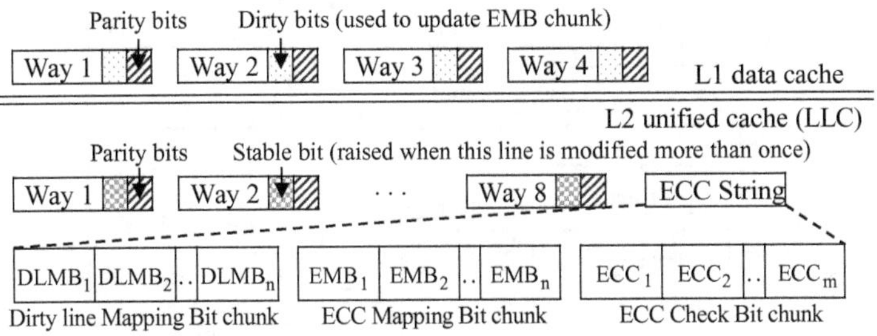


*Figure 3.5 Implementation of flexible ECC management for each set of LLC [7]*

cache vulnerability. Early write-back cache can explore the trade-off between vulnerability and performance/energy overheads by adjusting the write-back intervals. However, both write-through and early write-back are hardware-based techniques and thus they cannot be sensitive to the data access patterns of the application. Predefined periodical write-back schemes such as early write-back can be inefficient in two cases. If the write-back interval is too long, a dirty line may induce vulnerable periods while the cache line is unused after the last update. On the other hand, if a write-back occurs just before an additional update, only few vulnerable intervals after the write-back and before the new update are removed, while the memory traffic increases. These inefficiencies can be improved if write-back is scheduled with the prior knowledge of data access patterns such as the timing of last update or iterative updates in a loop. Accordingly, Jeyapaul and Shrivastava proposed SCC, which efficiently customizes write-back process based on data access patterns of applications. The main idea of SCC is to perform cache cleaning or write-backs only if cleaning is necessary. For example, it is inefficient to perform write-back while a dirty line is being reused repeatedly by the program. On the other hand, write-back is necessary right after the last update to prohibit additional vulnerability from unused data. SCC exploits the memory profile information to evaluate the vulnerability profit per cache write-back for each store operation. For each of such operations, its intermediate vulnerable time, i.e., time after the write operation and before the eviction or before last read operation, is calculated. If the intermediate vulnerable time exceeds a certain threshold, SCC decides that cleaning is required. Based on the decisions of cleaning on the store operations with the same instruction address (scc_insn_addr), SCC generates bitwise pattern of cleaning (scc_pattern), where one means cleaning is essential and zero means cleaning is not required. Figure 3.6 shows the SCC architecture, where additional blocks required by the SCC technique are shaded. In this architecture, a SCC register pair, the scc_insn_addr and scc_pattern, should be updated by profiling analysis. For every execution of targeted store operation (denoted as csw), SCC architecture iterates the pattern in scc_pattern and decides to clean the data by raising the CleanEN flag.

Ko *et al.* [9] investigated polices of parity protections in level 1 data caches and presented guidelines on how to design data caches with parity protections in an

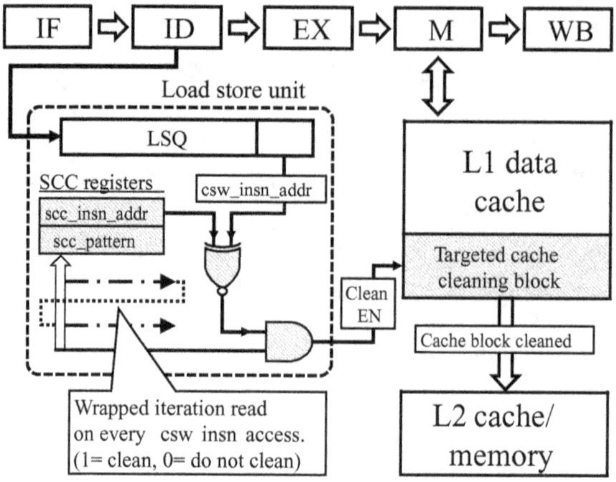


*Figure 3.6 Smart Cache Cleaning (SCC) analyzes instruction patterns to find the most power-efficient write-back operations [8]*

efficient manner. In their work, they come up with two important questions: (i) when parity should be checked? and (ii) what should be the granularity of parity bits? Since full-scale fault injection experiments on the cache are infeasible, they first defined the Cache Vulnerability Factor (CVF) as a metric to estimate the reliability of a cache configuration. In smaller scale fault injection experiments, the estimated reliability calculated by CVF is matched perfectly with the failure rate in fault injection campaigns. Using this metric with in-depth study, they found out counter-intuitive answers to the questions asked. First off, they found out that checking parity at only reads is the best option in terms of both energy efficiency and reliability. Processors such as ARM Cortex-A8 implement parity checking at both read and write operations, based on the naive belief that more checks lead to more resilience. However, this implementation is in fact suboptimal. As shown in Figure 3.7, checking parity at write operations makes the period between the last access to the next write vulnerable. Therefore, checking at writes not only increases energy and performance overheads but also damages the reliability of the cache. Further, they found out that the granularity of dirty bits needs to be equal to that of parity bits in order to achieve the maximal reliability. The widely used ARM Cortex R4 processor is implemented with the parity bits at the byte level and dirty bits at the block level. Since these processors implement parity bits at a finer level, i.e., a byte level rather than a block level, one would assume that they are more reliable than caches implemented with both parity and dirty bits at the block level. However, when using such schemes, a write to any word in the block makes the entire block dirty due to the dirty bit per block. In this case, the finer grained parity bits are not helpful at all, since it still cannot identify where in the block the error occurred. In conclusion, Ko *et al.* suggest checking parity only at reads and implementing the parity bits and dirty bits at the word level. They have also studied the level 1 data cache protections with ECC in an effective manner.

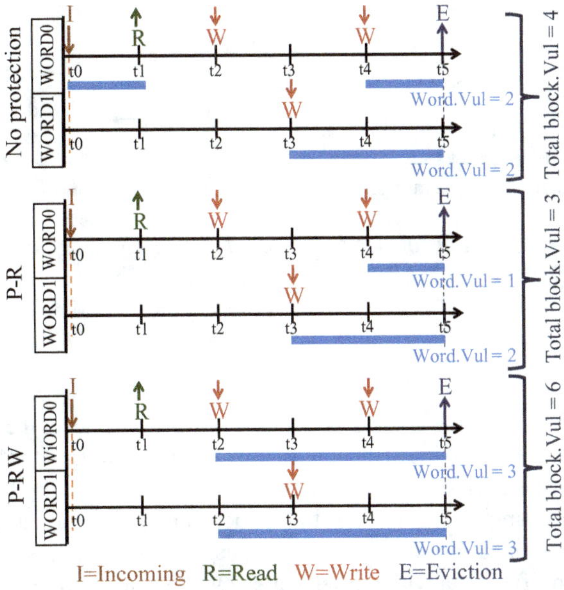


*Figure 3.7 Assume that a cache block consists of two words such as WORD0 and WORD1 and Read from WORD0 occurs at t1 followed by Writes to WORD0 at t2, WORD1 at t3, and WORD0 at t4. Note that P-R and P-RW imply parity checking at reads and parity checking at both reads and writes, respectively. If parity checking happens at reads only, the vulnerability decreases since WORD0 is no longer vulnerable from t0 to t1. On the other hand, if parity is checked at writes also, the vulnerability increases. This is because an error detected at write at t3 renders both words vulnerable, since it is unclear which word contains the error [9]*

Cache replication is another choice for designers to enhance the cache resilience by recycling unused cache blocks [10] and introducing extra space [11]. The main idea behind these replication schemes is to exploit the asymmetric usage of caches, i.e., active and non-active blocks, and the inherent locality of caches where small portion of spaces can accommodate the infrequent error occurrence in expensive caches.

Zhang *et al.* [10] presented ICR (In-Cache Replication), a cache replication technique with ECC and parity. The key idea of this technique is to exploit existing cache space to hold replicas of cache blocks. However, evicting blocks that may be needed in near future to achieve space for replication can degrade the performance. To evict blocks that are unlikely to be used, they adopted a dead block prediction strategy [24] with a 2-bit counter for each cache block. Figure 3.8 plots the diagram of gray-coded counter for dead block predictor. This 2-bit counter is incremented at a timer tick. When this counter saturates, the corresponding cache block is considered as

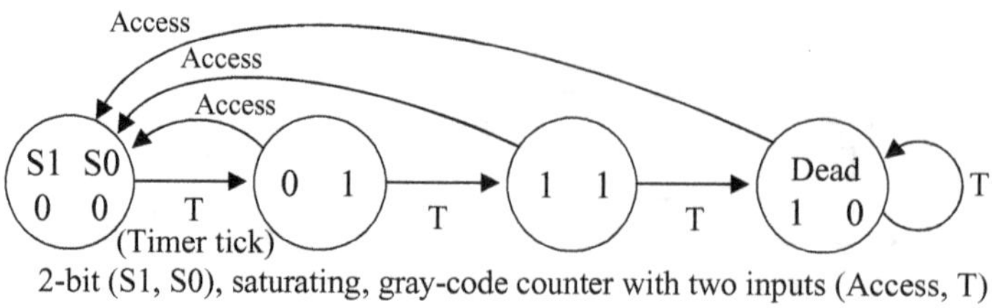


*Figure 3.8 A 2-bit counter for dead block prediction in ICR [8], based on the technique in [24]*

dead. On the other hand, the counter is reset, i.e., active, whenever the corresponding cache block is accessed. To implement ICR, they attempted to find the best answers to various questions. The first one is to find when to replicate the data. They considered two options: replication only at writes to L1 cache and replication at both reads and writes. Of these, replication at writes only has been chosen. The reason is that it could protect more modified data with the same area overhead as it protects only modified data. In addition, although this method needs to visit the L2 cache in cases of errors in unmodified data, the performance overhead is considered minimal since errors do not occur frequently. Another question is where to place the replicates. The replicates are placed in the set $(m + N/2) \bmod N$, where $m$ is the set holding the original data and $N$ is the total number of sets in the cache. Within the set, the data is placed using the LRU (Least Recently Used) policy, and only if the set has an empty block or a dead block. This is because failing to replicate data does not harm reliability significantly, and other methods require more complex cache architectures. It should also be noted that each block needs to be marked as primary or replicated since the tag of a replicate block may match another block in the set. More questions follow such as how to protect and check the replicated and non-replicated data. They choose to protect the replicated data with parity and consider the choice between parity and ECC for non-replicated data. When checking the data, sequential and parallel checks are considered. The sequential checking method checks the replicate only if an error is detected in the original data, and the parallel checking method checks both the original and replicated data in parallel. The authors exploit codes to represent the different schemes answering these questions. Depending on the mechanism used to protect non-replicated data, they use P for parity and ECC for ECC. Depending on the checking mechanism, they use PS for sequential checking and PP for parallel checking. Finally, depending on when to replicate data, they use (LS) for replicating at both reads and writes, and (S) for replicating at writes only. Among many configurations, they introduced two particularly useful schemes. The ICR-P-PS (S) scheme has similar performance overhead but much higher reliability than simple parity protection. The ICR-ECC-PS (S) scheme has much lower performance overhead than simple ECC protection without compromising too much reliability. An example of ICR-ECC-PS (S) cache is shown in Figure 3.9.

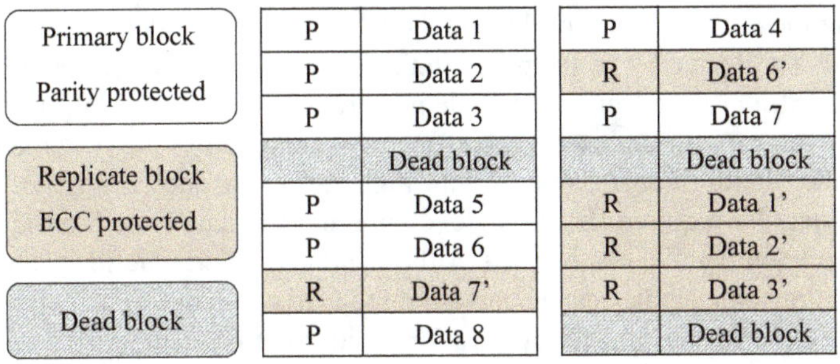


*Figure 3.9 A two-way set-associative ICR-ECC-PS (S) cache [8]. Note that data 5 in set 5 cannot be replicated, although there exist dead blocks in sets 4 and 8, since there is no empty block or dead block in set 1 ((5 + 8/2) mod 8). (Data 8 will be replicated when it is written.)*

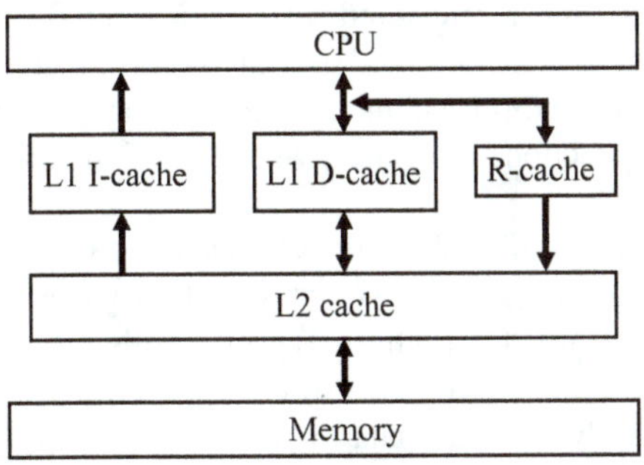


*Figure 3.10 RC (Replication Cache) keeps the replica of every write data to the level 1 cache in a small fully associative cache at the same level [11]*

Though ICR [10] showed acceptable reliability with acceptable performance overhead, it has limited coverage and noticeable performance degradation. This made ICR schemes unacceptable for applications requiring high reliability or strict time constraints. Therefore, extending on the ideas of ICR, Zhang proposed another replication technique [11]. Since ICR could not replicate all writes to the level 1 cache, this technique adds a small fully associative cache at the same level as the level 1 data cache, as shown in Figure 3.10. The additional cache, called the Replication Cache (R-Cache or RC), is designed to hold the replica of every write to the level 1 data cache and to write back dirty data into the level 2 cache. The RC also tries to answer the questions ICR did, and, due to its unique architecture, ends up with similar but different answers. Regarding when to replicate the data, RC reaches the same conclusion as ICR. It is much more important to replicate dirty data since clean data can easily be recovered and therefore RC replicates data only at writes. The replicated data is placed in the RC, replacing blocks using the LRU policy. Regarding the method of the protection, RC considers two possible options: protecting the RC with parity, or comparing the data in the L1 cache and the RC cache in parallel. Since the L1 data cache is also protected by parity, parity in the RC is sufficient to detect all single-bit errors. Indeed, an RC with eight blocks demonstrated 97.3% of read hits in level 1 data cache on average with higher energy-efficiency compared to the ICR method [10]. On the other hand, in environments with high soft error rates, resilience to multi-bit errors may be required. In this case, comparing the data stored in the L1 cache and RC cache in parallel can be used. Although this method requires extra cycles in load operations, the latency can be hidden if the load can proceed speculatively. This method is especially useful for applications with high data integrity requirements. Zhang also proposed storing two replicas in the RC cache for each write operation. Since this method has three copies of each data, it can correct multi-bit errors using the voting algorithm. This version can be used for soft error prone applications with cost constraints.

ECCs based on hamming codes [25] can not only detect single and double bit errors but also correct single bit errors although they in general incur high overheads in terms of area, power consumption, and performance [26,27]. In order to minimize these costs, researchers have investigated several interesting design choices [9,12,13]. Some have even studied hybrid techniques, combining previously proposed techniques for efficient protections of caches, which are significantly sensitive to area, power, and performance.

Lee *et al.* [12] tried to minimize the overheads of ECC through partial protection. The main observation they made is that not all errors cause failures since some data are more failure critical than others. Based on this observation, they partition the data into two categories: failure critical data and failure noncritical data. However, a fine-grained analysis on all data to partition them in two categories can be extremely complex. Lee *et al.* therefore focus on multimedia applications. They make the assumption that all multimedia data, such as image pixels, are failure noncritical, and all other data, such as loop variables, are failure critical. This separation principle is easy to implement and effective, classifying up to 63% of data as failure noncritical. The data is then stored in different caches according to their classification, much like Horizontally Partitioned Caches (HPC) [28]. HPC is a technique that has more than one cache at the same level of memory hierarchy. Each memory address is mapped to only one of such caches. Conventionally, this method was used to improve the performance of the system, since partitioning data to different caches could improve cache behavior. Extending on this concept, Lee *et al.* proposed the Partially Protected Cache (PPC). The PPC consists of two caches in the same level of memory hierarchy: the soft error prone cache and the soft error protected cache, as shown in Figure 3.11. Failure noncritical data are mapped and stored in the soft error prone cache. Since soft errors in multimedia data only result in a minor loss of QoS, the soft error prone cache is left unprotected. On the other hand, failure critical data are stored in the soft error protect cache and are protected using ECCs. To minimize the performance overhead of PPC, the size of the soft error protected cache is limited such that its total penalty due to the ECC is less than that of the soft error prone cache. As data mapped to soft error protected cache show very good cache behavior, the performance penalties of cache misses due to the smaller cache size are minimal. The PPC was able to

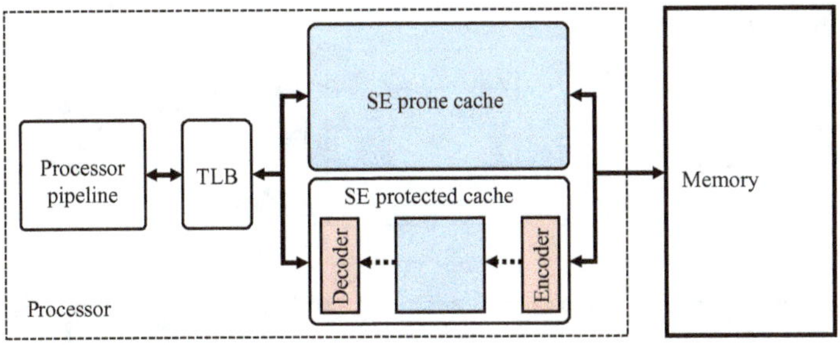


*Figure 3.11 Partially Protected Caches consist of a soft error (SE) prone cache without protection and a soft error (SE) protected cache with ECC and achieves power-and-performance-efficient protection by mapping failure critical data into soft error protected minor cache [12]*

achieve failure rates close to those of fully ECC-protected caches with performance and energy consumption similar to those of unprotected caches.

Protection against multiple-bit errors in caches is very expensive as it requires additional parity codes and areas. Therefore, widely used techniques such as SEC-DED do not provide multi-bit error correction. As caches become larger, however, increasingly many cases of multi-bit errors must be considered. A temporal double-bit error is one of such cases, which are caused by two separate single-bit errors in the same ECC word. One solution to this problem is cache scrubbing [29] that periodically cleans up single-bit errors to prevent temporal double-bit errors. Cache scrubbing involves multiple steps. First, the target bits are read from the cache. Then, the ECC of the bits is computed and compared with the stored ECC in order to correct any existing single-bit errors. Finally, the bits and the computed ECC are written back. If the scrubbing happens frequently enough, virtually all temporal double-bit errors would be eliminated, as seen in Figure 3.12. However, cache scrubbing incurs extra overheads for the cache. These induced overheads from software implementations would be unacceptable. The overheads can be significantly reduced by introducing hardware implementations, but they may still be intolerable. Therefore, Mukherjee *et al.* [13] proposed a technique to calculate the likeliness of temporal double-bit errors. Conventional methods to estimate the mean time to failure (MTTF) of temporal double-bit errors require complicated probabilistic calculations. However, they found a much simpler equation: the MTTF of temporal double-bit errors = $M \times$MTTF of a single-bit error where $M$ represents the mean number of single-bit errors to become a double-bit error and can be easily calculated with a simple computer program. Based on this estimate, they found the reasonable interval for cache scrubbing, if needed at all, for a given cache size.

The number of redundant bits required for parity code is independent of the data length, and the number of redundant bits required for Hamming code based ECC is only logarithmically proportional to the data length. Therefore, a higher granularity of EDC/ECC reduces the number of additional bits required for redundancy.

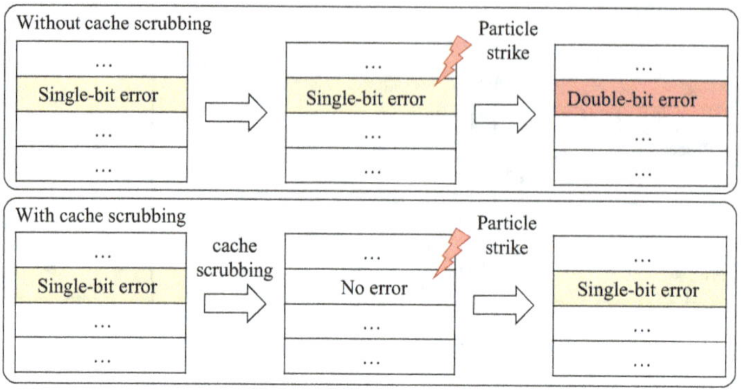


*Figure 3.12 If the cache is scrubbed frequently, the cache is safe from multi-bit errors even when two strikes affect the same cache line [13,29]*

Using this fact, the L2 and L3 cache usually exploit per-line EDC/ECC to reduce the area overhead. However, per-line EDC and ECC schemes for L1 cache require entire line accesses for every read operation. Due to performance and dynamic energy overhead of per-line EDC/ECC, L1-cache applies per-word EDC/ECC. Farbeh and Miremadi [14] noticed that in a set-associative L1 caches, fast access mode enables parallel accesses to all cache ways to minimize the latency of the cache [30]. Based on the fast access mode, Farbeh and Miremadi proposed per-set protected cache, or PSP-Cache to achieve area efficiency of higher granularity while minimizing performance and energy consumption overhead. The PSP-Cache applies a single EDC/ECC to the data words of all cache ways at once, instead of applying separate EDC/ECC to each cache way. Figure 3.13 illustrates the conventional per-word EDC/ECC (Figure 3.13(a)) and PSP-Cache (Figure 3.13(b)) architecture. Unlike the per-word architecture that stores redundant EDC/ECC codes per word, PSP-Cache stores one EDC/ECC code for all words in a set and therefore can reduce the number of redundant bits compared to the conventional per-word architecture. When accessing a PSP-Cache, all words in the cache set are accessed simultaneously in fast access mode, and EDC/ECC checking or generation is proceeded before the way selection.

A couple of techniques exploit some region of data bits in the cache lines as ECC bits, instead of using additional bits for ECC implementation. Chen *et al.* [15] focused on cache compression techniques for the LLC. They proposed the Free ECC that utilizes unused fragments of compressed LLC for implementing ECC, based on the observation that many fragments in compressed cache lines are unused. They also realized that if a cache block cannot be compressed, the corresponding compression encoding code field can be used to keep the EDC. They first revised the $B\Delta I$ [31] compression algorithm to ensure that any compressed cache block has enough unused fragments to implement EDC. Then, they designed Free ECC based on the size of unused fragments for each cache block. The main idea of Free ECC is to store ECCs into unused fragments of the same block of the compressed data if there exist sufficient unused fragments. If not, Free ECC first stores EDC into the unused fragments and

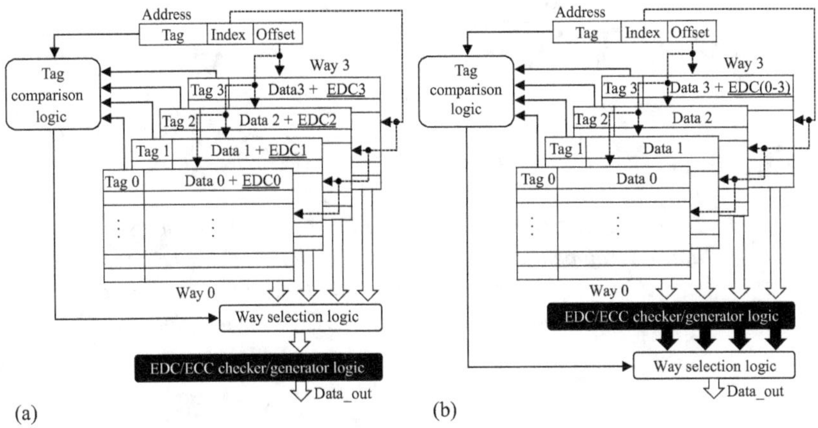


*Figure 3.13 PSP-Cache applies per-set EDC/ECC by exploiting fast access mode of L1 cache that enables one to access all cache ways simultaneously [14]: (a) conventional cache architecture and (b) PSP-Cache architecture*

then stores ECCs to the last cache block in the set. This ECC is used only if an error is detected by the EDC. Similarly, if a block cannot be compressed, Free ECC stores EDC into the compression encoding code field and stores ECC into the last block in the set. Figure 3.14 illustrates an example of Free ECC architecture. Free ECC appends additional bits to each tag field. A single-bit *S* field is used to tell if a block in the set is not compressed. This *S* bit is set if the corresponding set contains one uncompressed block, as seen in way 1 set 1 of Figure 3.14. Since compression encoding code is not required for the uncompressed block, the EDC bits are stored in the corresponding compression encoding code field. The *E* field indicates the location of the ECC segment in the last cache set, if the ECC is not stored within the block. The special case of the *E* field holding 0000 in binary means all compressed blocks have their ECC in the same block, as in the case of way 0 set 1. The 1-bit *D* field indicates which of the two block's ECC is in another cache block. This bit is required if unused fragments of one of the compressed blocks are insufficient to store ECC. For example, way 2 set 1 contains two compressed blocks, and therefore *S* field is set to 0. However, due to the lack of unused fragments, data 3 is protected only with EDC instead of ECC. The location of ECC of data 3 is indicated by *E* field (15), and the *D* field is set to 1 to indicate that the second block overflown.

While Free ECC [15] exploits the unused segments of compressed caches, Hong and Kim [16]'s Smart ECC Allocation (SEA) cache locates all ECC check bits in the cache data space without cache compression scheme. Similar to Free ECC, the SEA cache targets the LLC, in which a small change of cache size does not result in considerable performance overhead. However, naively storing every ECC bit into data cache can sacrifice a huge amount of LLC data space, which can degrade the

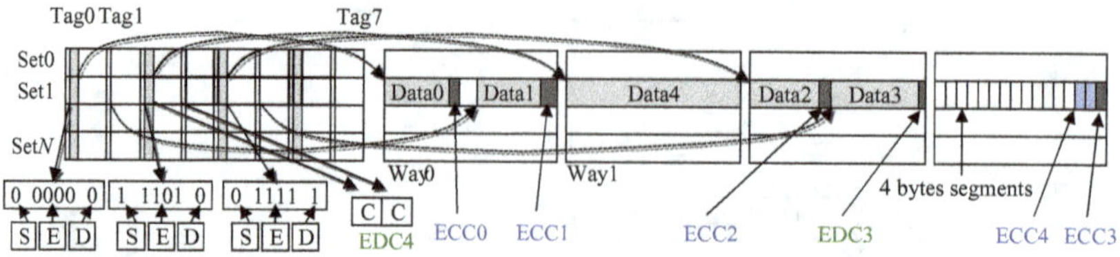


*Figure 3.14 Free ECC stores either ECC or EDC codes of compressed blocks into unused fragments. For the uncompressed blocks, EDC codes are stored in compression encoding code field. ECC codes are stored in the last cache block in the set for EDC-protected blocks [15].*

performance of LLC. Based on the observation of Li *et al.* [6] (clean cache can be protected only with EDC since it can reload the replica in lower level cache or memory), SEA cache adopts parity in the cache line level. The SEA cache groups several cache sets and uses a few lines in each group as ECC lines for the other dirty lines in their group. Figure 3.15 illustrates the implementation of SEA cache. Every data line is protected by EDC and one ECC line status bit (EL bit) is added to each line. This EL bit is used to distinguish data lines from ECC lines that are used to store ECC bits of other data lines. The SEA cache then groups several cache sets. In Figure 3.15, SEA groups cache lines in a modular manner to avoid issues from spatial/temporal localities. After the grouping of cache sets, the ECC lines are dynamically allocated. Each ECC line contains seven ECC batches, and each ECC batch consists of (i) one valid bit that indicates the validity of the batch, (ii) DLMB that indicates the corresponding dirty data line in the group protected by the current batch, and (iii) a set of ECC bits for 64-bit ECC to protect one data cache line (64-bytes). The SEA cache should carefully adjust the maximum number of ECC lines per ECC group. If only one ECC line is allowed for each group, there may not be sufficient area to store all ECC bits, and several dirty lines should be written back causing additional DRAM traffic. On the other hand, too many ECC lines per ECC group can occupy too much cache space, consequently increasing miss rates of the cache. The SEA cache deals with this problem by adjusting the maximum number of ECC lines at runtime, based on the numbers of cache misses and write-backs. If the number of cache misses continuously exceeds that of write-backs, the SEA cache decreases the maximum number of ECC lines, and vice versa.

Efforts have been put to reduce the expensive costs of ECCs in caches by applying cheaper EDCs with special registers [17] or storing ECCs of the LLC to the main memory [18]. These alternative methods expand the interesting design choices at the system level for efficient error protections on caches.

Manoochehri *et al.* [17] proposed an interesting error correction approach with parity schemes. They proposed correctable parity protected cache (CPPC), which adds error correction capability to the dirty words of parity protected cache with the addition of two special purpose registers. The main idea of CPPC is to keep the XOR

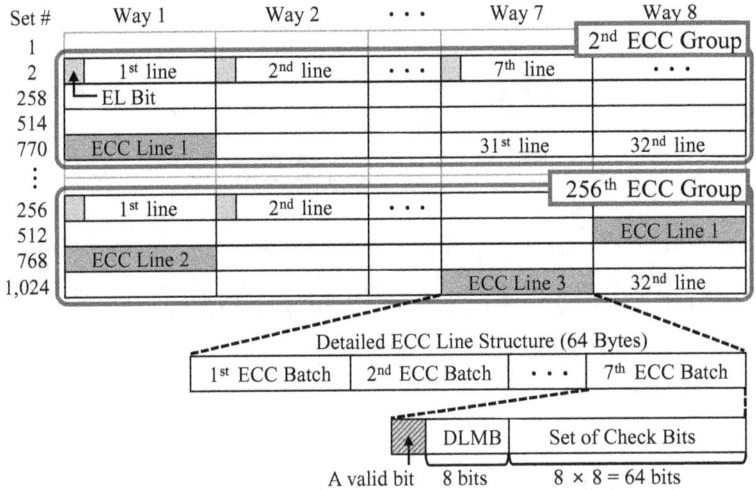


*Figure 3.15 SEA (Smart ECC Allocation) cache presents grouped ECC protections per cache set [16]*

of all dirty words in the cache. This XOR value can be used to recover from errors in dirty words by calculating the cumulative XOR value of the word and all other correct dirty words in the cache. Clean words can be recovered by reloading its replica from the next level cache or memory [6]. Figure 3.16 illustrates the L1 cache with CPPC architecture to maintain the XOR of all dirty words. On every store operation, the new data is XORed into register R1. Whenever dirty data is removed from the L1 cache, it is XORed into register R2. This includes dirty words that are written back or overwritten by new store operations. In other words, R1 holds the cumulative XOR of all writes to the L1 cache, R2 holds the cumulative XOR of all removed dirty data, and therefore the XOR value of R1 and R2 represents the cumulative XOR of all dirty words currently in the L1 cache. When a fault in a dirty word is detected, the CPPC processes XOR operations for R1, R2, and all other correct dirty words to obtain the correct value of the corrupted word. Since one XOR operation is faster than a cache access, these additional XOR operations do not incur much performance degradation. The CPPC can even recover from multi-bit errors in a dirty word, if the error can be detected by parity. However, an even number of faults in a dirty word cannot be detected by parity. The CPPC can choose to adopt parity interleaving, which XORs non-adjacent bits. While interleaved parity can effectively protect the cache against horizontal multi-bit faults, it cannot recover from some vertical multi-bit faults. For example, if two bits with the same offset in two vertically adjacent words are corrupted, the resulting XOR value of the two corrupted words is the same as that of those two words without faults. Protection against vertical multi-bit faults can be implemented with one of two alternatives. The first solution is byte-shifting, which rotates the input of XOR operation for R1 and R2. The data values of vertically

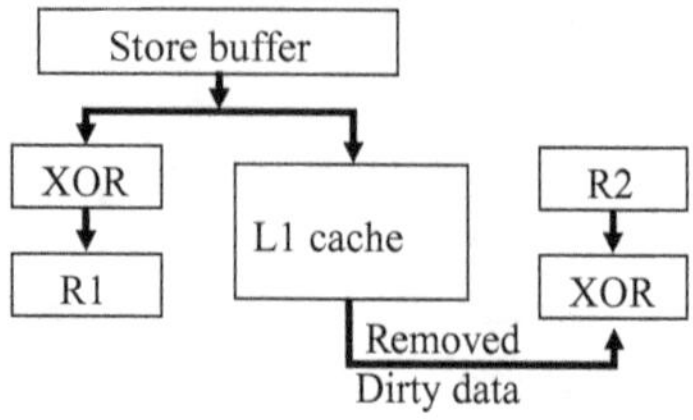


*Figure 3.16 CPPC, or Correctable Parity Protected Cache, introduces two registers, R1 and R2, and supports error correction with parity. The registers store the cumulative XOR value of data written into the cache and the cumulative XOR value of dirty data removed from the cache, respectively [17].*

adjacent words should be rotated differently before the XOR operation. The second solution is to increase the number of XOR register pairs (R1, R2) and to make sure that adjacent words be XORed with different XOR register pairs.

In addition to the observation of Li *et al.* [6], i.e., clean cache can be protected only with EDC since it can reload the replica in lower level cache or memory, Yoon and Erez [18] asserted that error correction is needed infrequently, while error detection is needed for every cache access. Based on these observations, Yoon and Erez proposed Memory Mapped ECC for the LLC. The main purpose of Memory Mapped ECC is to minimize the high energy and area overhead required to maintain ECC bits in the SRAM. To achieve this goal, Memory Mapped ECC protects the LLC with EDCs such as parity and stores the ECC bits of dirty cache lines into the low-cost off-chip DRAM. Memory Mapped ECC leverages on the fact that ECC is only required for error correction, and that error correction is an infrequent event. Therefore, the slowdown of error correction due to the low latency of memory does not impact the performance significantly. Figure 3.17 illustrates the Memory Mapped ECC architecture. The LLC is protected by tier-1 error code (T1EC), which is a low-cost EDC scheme such as parity. On the other hand, tier-2 error code (T2EC) is a higher cost ECC scheme, and the T2EC bits of each dirty lines are stored into the DRAM memory namespace. An on-chip register (T2EC_base) points to the base address of the T2EC. Memory Mapped ECC defines the T2EC memory namespace as cacheable, which is advantageous in two perspectives. First, this allows Memory Mapped ECC to exploit locality in the T2EC addresses, reducing the required DRAM bandwidth. Second, caching the T2EC data allows the granularity of write operations to the DRAM to match that of DRAM bursts, instead of writing few bytes at a time and wasting DRAM bandwidth. This also lets the LLC act as a write-combining buffer for T2EC values. For every read access to the LLC, Memory Mapped ECC detects potential errors in T1EC. If an error is detected on a clean cache line, Memory Mapped ECC reloads a replica from memory. If a fault is detected in a dirty cache line, Memory Mapped ECC finds the corresponding T2EC bits using the T2EC_base and corrects the fault. For every write access to the LLC, Memory Mapped ECC generates T1EC and writes the data into

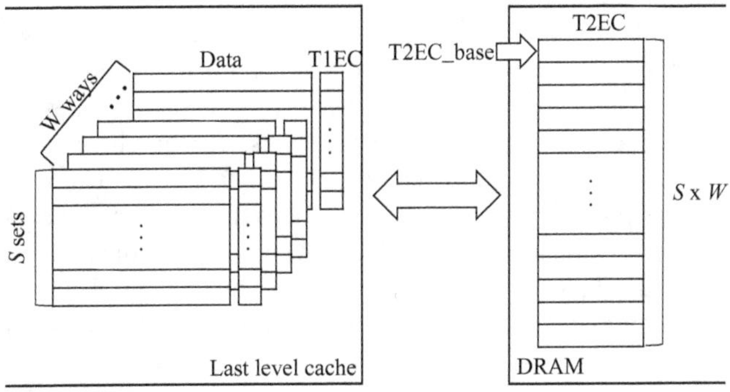


*Figure 3.17 Memory Mapped ECC protects LLC by low-cost EDC such as parity, and stores ECC bits of dirty cache lines into DRAM [18]*

the address. On the other hand, T2EC should be generated only for dirty lines, which are written back from higher level caches. In other words, write operations that fetch data in the memory into the LLC do not make a line dirty, and T2EC for these write operations need not be generated. After Memory Mapped ECC generates the T1EC for a dirty line, it calculates the T2EC and its address and stores the T2EC into either memory or the cached region corresponding to the address.

As error rates increased, many techniques have been proposed to consider multi-bit errors and make caches highly robust against them in a cost-efficient manner [13, 17,19–22].

Hung *et al.* [19] noticed that redundancy techniques exploiting spare elements to replace defective elements were already in wide use [32,33]. Several techniques [34, 35] combined these redundancy techniques with ECC, since only one defective cell in a cache block can be recovered by SEC-DED for every access. These techniques only replace cache blocks in the rare case in which the number of defective cells exceeds the correction capability of the ECC and therefore requires only a small number of redundancy elements. On the other hand, Hung *et al.* observed that SEC-DED does not stop at correcting a cache block with one defective cell but goes on to detect an additional single-bit soft error on cache blocks with one defective cell. By combining this observation with the fact that clean data can be recovered without ECC [6], Hung *et al.* proposed a technique for the combination of SEC-DED and redundancy techniques, which can tolerate hardware defects with high reliability against soft errors. They proposed the assurance update, which is a selective write-through update to ensure that only clean data can be assigned to cache blocks with one defective cell. Hung *et al.* classify cache blocks into good blocks (blocks with no defective cell), tolerable blocks (blocks with one defective cell), and bad blocks (blocks with more than one defective cell). For every dirty write operation to a t-cell (a cell in a tolerable block), the updated data should be propagated into the lower level cache or memory. This assurance update can degrade the performance since it requires additional write-backs. A cache row with at least one t-cell will be replaced with spare elements, based on a Built-In Self-Repair mechanism [33]. Two optimizations are proposed to reduce the performance overhead from the additional write-backs from the assurance updates. The first is to perform assurance updates and maintain dirtiness in the block level, while conventional caches maintain dirtiness at the line level. This reduces unnecessary write-backs of clean data blocks. The second optimization is the data swapping technique, which reduces the number of write-through operations from assurance updates. Figure 3.18 shows an example of data swapping. Without data swapping, the dirty data 3 will be stored in a t-block (tolerable block), and accordingly it requires an assurance update (Figure 3.18(a)). On the other hand, the clean data 4 will be stored in a g-block (good block). This can be considered as overprotection since clean data in a clean block can be sufficiently protected with EDC. Using a swapping function as shown in Figure 3.18(b), the dirty data 3 can be stored into a g-block and the clean data 4 to a t-block. This rids the necessity of assurance updates of data 3 and maintains high reliability since data 4 can be corrected even in t-blocks.

Sun *et al.* [20] claimed that with appropriate architecture design, applying multi-bit correctable ECC to L2 cache to tolerate a large number of random hardware defects in L2 cache while maintaining the reliability against soft errors is possible. While using multi-bit ECC for guarding memory against multiple random hardware defects was already studied [36,37], the use of multi-bit ECC was avoided in cache memory due to its longer latency compared to the SEC-DED. In addition, multi-bit ECC requires significantly more redundant bits than SEC-DED does, which is a huge drawback in high-cost cache memories. Rather than applying multi-bit ECC uniformly, Sun *et al.* proposed a selective multi-bit ECC for the L2 cache, protecting the L2 cache with SEC-DED uniformly and only applying multi-bit ECC when it is necessary. Since not all cache lines are protected by the expensive multi-bit ECC, this can effectively reduce the performance, area, and energy consumption overheads

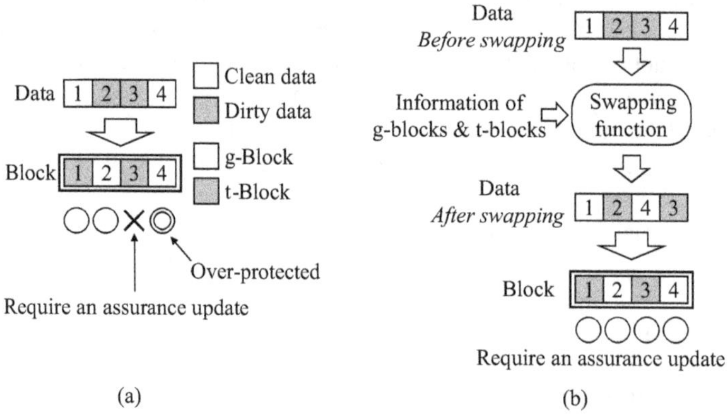


*Figure 3.18 Data swapping can reduce the number of assurance update by swapping the clean data into g-block and dirty data into t-block [19]: (a) without data swapping and (b) with data swapping*

compared to uniformly applied multi-bit ECC. The selective protection applies multi-bit ECC to the subblocks of the cache with more than one defect, where a subblock is a unit of protection in a cache. They defined a g-subblock (good) as a subblock with no hardware faults, a s-subblock (single) with as one with a single defective cell, and an m-subblock (multiple) as a subblock with more than one defective cell. Figure 3.19 illustrates the L2 cache architecture with selective multi-bit ECC proposed by Sun *et al.* A fully associative M-ECC cache stores the subblock addresses of m-subblocks and their corresponding redundant bits for multi-bit ECC. While this M-ECC is sufficient to achieve the main functionality of selective ECC, the performance degradation in terms of instructions per cycle can still be significant. An M-ECC implemented L2 cache accesses the fully associative M-ECC cache for every cache access and in addition needs to decode the multi-bit ECC bits for every access to an m-subblock, which can incur high energy and performance overheads. Sun *et al.* therefore proposed two optimizations for this M-ECC cache, based on the temporal locality of the cache. To reduce the long latency of these accesses, a small pre-decoding buffer with an LRU policy is added to the multi-bit ECC core, keeping the copies of the most recently accessed m-subblocks. If an access to the cache hits in the pre-decoding buffer, the data inside of this buffer is used instead of explicitly accessing the M-ECC cache and decoding corresponding multi-bit ECC, reducing the number of M-ECC accesses. The M-ECC cache may also be explicitly accessed in g-subblock and s-subblock accesses, although these subblocks are not protected with multi-bit ECC. Since most subblocks are not m-subblocks, the overhead from these unnecessary accesses can add up to be significant. For this issue, Sun *et al.* implemented a Fast Look-Up (FLU) buffer as CAM (Content Addressable Memory), which helps one to reduce these accesses to the M-ECC cache. The FLU buffer keeps the addresses of the recently accessed g-subblocks and s-subblocks, so that if a cache access hits on the FLU buffer, the system can skip the access to the M-ECC cache as the target block is not protected with multi-bit ECC. An L2 cache with pure selective multi-bit ECC can detect but cannot correct soft errors on (i) s-subblocks with a single defect, since s-subblocks are protected with only SEC-DED and (ii) m-subblock with multiple defects, since, although the m-subblock is protected with multi-bit ECC, the capability of each protection scheme is fully exploited to compensate hardware defects. Based on the observation that clean data can be corrected from detected errors by reloading replicas, Sun *et al.* add a small fully associate dirty RC as shown on the left side of Figure 3.19. The dirty RC keeps the most recent write-back cache blocks from the L1 cache to L2 cache, and detected but not correctable soft errors can be recovered from by reloading replicas from this cache.

As modern processors reduce the supply voltage in order to lower power consumption, the probability of multi-bit errors in a cache line becomes hard to ignore. Alameldeen *et al.* [21] noted that even in low supply voltages, the chance of double-bit errors is significantly lower than that of single-bit errors. They therefore proposed the Variable-Strength ECCs (VS-ECC), which focuses on scaling the level of protection of each line to its vulnerability. The VS-ECC allocates SEC-DED ECC to all cache lines, eliminating all single-bit errors, while reserving a few more bits per cache set to apply strong multi-bit ECCs to up to four cache lines that exhibit multi-bit induced

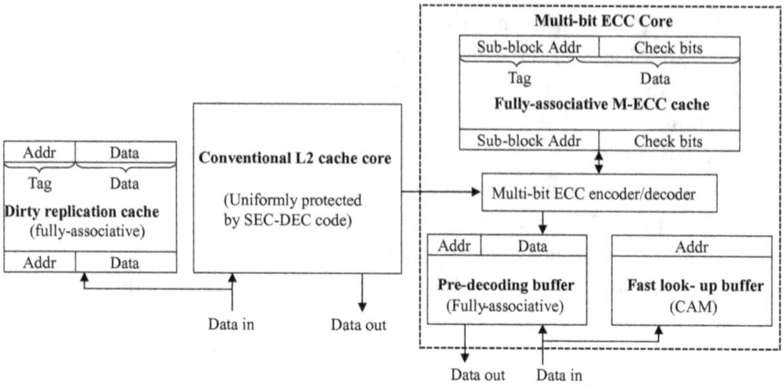


*Figure 3.19 L2 cache architecture with selective multi-bit ECC protection for the subblocks with multiple defects [20]*

failures. Since many cache designs already include SEC-DED, the area overhead of VS-ECC only includes the few bits per cache set and the logic required for the strong multi-bit ECC. The main challenge of VS-ECC is to determine which lines are most vulnerable to multi-bit errors. When the program goes into low-voltage mode for the first time, the VS-ECC goes through a runtime characterization phase. In this phase, four of the cache lines are kept active and protected by the multi-bit ECC. If the line at any point detects a multi-bit error, it is marked as a multi-bit failure line. In parallel, the other inactive lines go through a traditional memory test, written with predefined patterns and read back to check for errors. Lines that experience multi-bit errors during the memory test are also marked. If five or more lines are marked, the protection from VS-ECC is insufficient at the voltage, and the cache is switched to a higher voltage. After the test is complete, the active and inactive regions of the cache are swapped, and the previously active region is tested. With the results from the testing, the VS-ECC can be implemented in one of three designs, as shown in Figure 3.20. This version is implemented in a 16-way set-associative cache. The first design is VS-ECC-Fixed. In this design, each cache set is augmented with four extended ECC fields, used to correct multi-bit failures in any 4 of the 16 ways. The tags are also modified to contain one extra bit called the Extended ECC bit (E-bit), which is used to distinguish lines susceptible to multi-bit failures. The second design, VS-ECC-Disable, is similar to the first design but adds an additional bit to the tag. This bit is called the Disable bit (D-bit) and is used to indicate that the cache line is invalidated. In this version of the cache, all lines are protected with SEC-DED at the minimum level, lines with one or two persistent failures are protected with multi-bit ECC, and lines with more than two failures are invalidated and no longer used in low-voltage mode. The third design is called the VS-ECC-Variable cache. This design adds a two-bit status and a four-bit pointer to the cache tag. The status indicates the level of protection of the line, from 1: SEC-DED to 4: 4EC5ED. For lines with protection level 2 and above, the SEC-DED bits associated with the cache

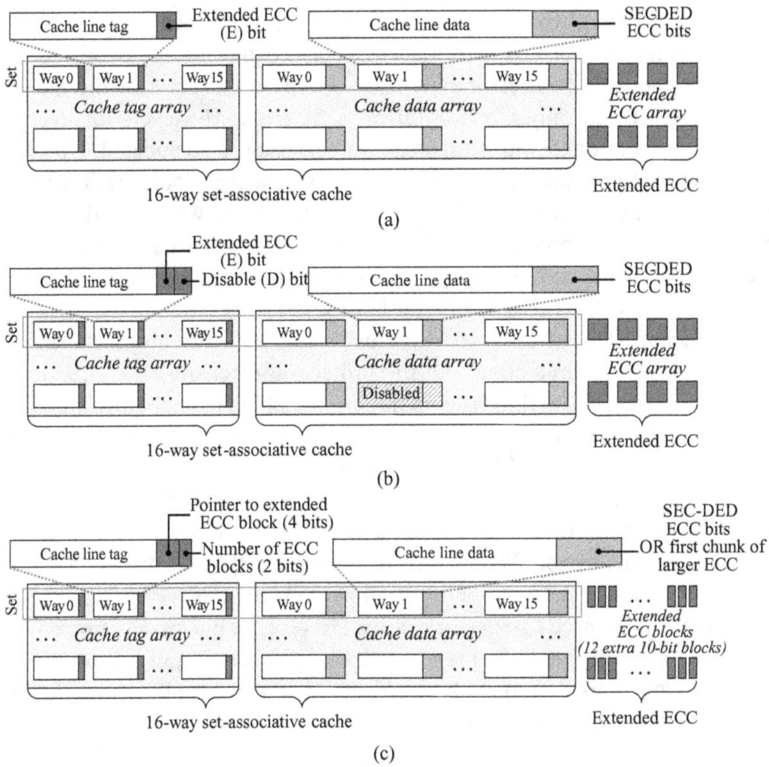


*Figure 3.20 VS-ECC Cache organization with three alternative designs [21]: (a) VS-ECC with a fixed number of regular and extended ECC ways, (b) VS-ECC with line disable capability + a fixed number of regular and extended ECC ways, and (c) VS-ECC with a variable number of correction bits (1–4) for all ways in a cache set*

are used as the first 11 bits of a stronger multi-bit ECC. The rest of the ECC is found using the 4-bit pointer to the extended ECC block.

Kim *et al.* [22] took a new perspective in protecting caches against multi-bit errors. Instead of the expensive multi-bit ECCs, they proposed two-dimensional error coding techniques, either EDC or ECC, to protect the cache architecture against large-scale multi-bit errors. Their technique leverages on the fact that cache cells spend most of time in the absence of errors. They therefore decoupled the detection and correction scheme by combining light-weight horizontal per-word error coding with vertical column-wise error coding. In their two-dimensional error coding, only light-weight horizontal error code is used to detect a fault or even correct a correctable fault if ECC is applied as the horizontal error code. The vertical error code is used only when the horizontal error code detects faults. Combining different types of error codes, such as interleaved error codes and/or stronger multi-bit ECC codes for horizontal and vertical error codes, with additional schemes such as physical bit interleaving can provide trade-offs between error coverage and performance/energy overhead. During a read access, the horizontal error code is used to detect an error, while the vertical error code is accessed only when an error is detected. However, for every write access, both the horizontal and vertical error codes should be updated. Updating the horizontal error code is simple since it only requires the new data. On the other hand, to update the vertical error code, both the new data to be written and the old data that is integrated in the previous vertical error code are required. Consequently, the cache controller should convert every write operation as read-before-write operation. Figure 3.21(a) shows the example of read-before-write operation in the 2D-protected cache architecture. This specific example uses SEC-DED for horizontal protection and parity for vertical protection, but the protection mechanisms can be modified as discussed above. Before the write operation, the old data and vertical parity code are read. Then, the new data can be written to the data row, updating the horizontal ECC code. At the same time, the vertical parity code is recalculated based on the old data, old vertical parity code, and new data. If an error is detected by, but not recoverable from using the horizontal error code, the recovery process is proceeded with the vertical error code. For the previous example that uses parity as a vertical protection, the original data can be restored by XORing the vertical parity row and all other rows that are used to calculate parity. However, if bits in the same position of multiple rows are corrupted, the previous recovery scheme cannot correctly restore the corrupted row. This should be distinguished by the horizontal error code. Figure 3.21(b) shows the corresponding recovery algorithm. If the horizontal error code is based on ECC, there is another chance to correct column-wise multi-bit errors even with the parity-based vertical protection, as shown in the gray area of Figure 3.21(b).

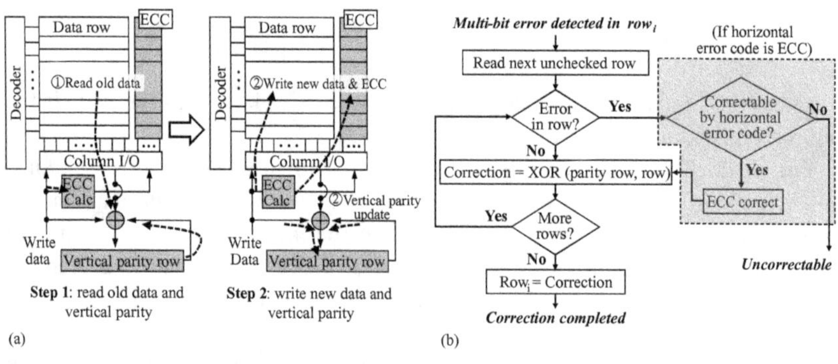


*Figure 3.21 (a) Read-before-write policy and corresponding vertical parity update in the 2D-protected cache and (b) error recovery algorithm in 2D-protected cache against multi-bit errors [22]*

## 3.2 Register file protection techniques

RFs can be responsible for the majority of faults affecting the architectural state of the processor [38]. Moreover, since RFs are accessed very frequently, corrupted data in RFs can quickly spread to other parts of the system, adding to the importance of protecting RFs. While memory structures are routinely protected using parity or ECCs [4], protecting RFs poses unique challenges because an RF is typically in the timing-critical path of a processor and is one of the hottest blocks of a chip. Because the reliability of electronic components decreases exponentially with the increasing operation temperature [39], power-efficient protection of RF is especially important.

Techniques to protect RFs can be classified into hardware approaches, software approaches, and hardware-software hybrids. Commercial processors usually take the hardware approach, protecting their RFs with parity bits. For example, SPARC ERC32 [40] is a SPARC processor that implements parity codes in registers. After the heavy ion testing, the architects of ERC32 processor claimed that parity-based error-detection mechanisms succeeded in detecting more than 97.5% of all injected errors (most of them in registers), and therefore it significantly reduces the MTBF (Mean Time Between Failures) for undetected SEU (Single Event Upset) errors. Other processors such as the Itanium processor [41] also implement the parity protection in the RF. While parity protections can detect errors, it is unable to recover from them.

IBM S/390 G5 Microprocessor designs their RFs with ECCs [42] in them and claims that the processor has essentially 100% detection and recovery from any transient error. However, this advantage comes at the expense of the increased power consumption and latency. Even a simple ECC can take up to three times the delay of a simple ALU operation [43]. Further, even though the latency of ECC operations can be hidden by parallelizing the ECC with other operations, the area and energy consumption overheads cannot be hidden and ignored. The energy consumption of an ECC-based scheme in particular can be as high as an order of magnitude larger than the energy consumed during a register access [27]. Thus, the search for power-efficient protection of RF continued. Researchers [38,44–46] protect RFs either fully or partially using various mechanisms such as ECC, parity, and duplication, in a way that is completely oblivious to the program running on the processor. Software compatibility (i.e., software need not be modified at all in order to protect an RF from soft errors) is a big advantage of the hardware approaches. On the other hand, the additional cost due to the extra circuitry for the hardware approaches, which is also permanent, can be high. Table 3.2 presents the categorization of several register protection techniques, which will be discussed in the following paragraphs.

Montesinos *et al.* [44] made two key observations. First off, the data stored in a physical register is not always useful. Figure 3.22 shows an example of the useful period of a register. In the lifetime of each register version allocated in a physical register, the data of the particular register is invalid from allocation to write. On the other hand, this data will not be used from the last read to de-allocation. Indeed, the data is only useful after the write and before the last read. If a soft error occurs in a physical register when the data is not useful, it will have no impact on the processor's architectural state. Consequently, a register only needs to be protected while it contains

*Table 3.2 Summary of register file protection techniques*

| S. no. | Protection scheme | Techniques |
|---|---|---|
| 1 | Protection by parity | [40,41] |
| 2 | Protection by ECC | [42] |
| 3 | Hardware-managed partial protection | [38,44,47] |
| 4 | Software-managed partial protection | [48,49] |

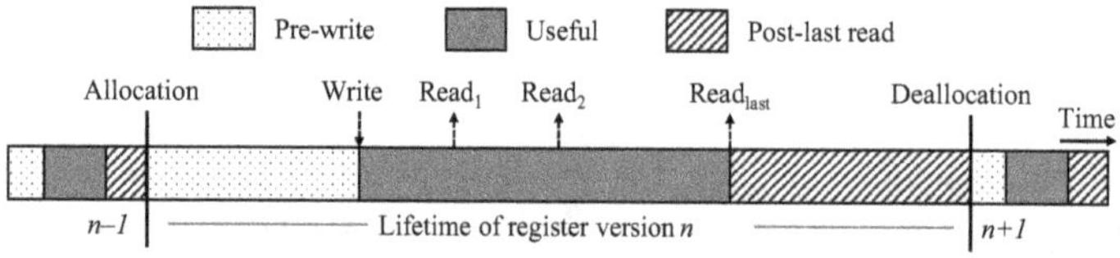


*Figure 3.22 A register version is useful after the write operation and before the last read operation, while pre-write and post-last read intervals are not useful and therefore do not need to be protected [44]*

the useful data. Montesinos *et al.* observed that registers in SPECint applications are useful for only 22% of their lifetime, and those in SEPCfp applications for only 15% of their lifetime. Second, not all the registers are equally vulnerable to soft errors. Indeed, they observed that the longest living 10% of the allocated register versions account for more than 46% of the useful lifetime of all register versions. In other words, most of the other registers are short-lived and their contribution to the total vulnerable time is extremely small. Therefore, giving higher priority to protection of the long-lived registers is definitely a cost-effective approach.

Based on these two key observations, Montesinos *et al.* proposed ParShield [44]. ParShield architecture consists of the Shield architecture for selective protection and parity protection for all RFs. Shield adds three hardware components to a traditional RF for an out-of-order processor: (i) a table that stores the ECCs of some registers, (ii) a set of ECC generators, and (iii) a set of ECC checkers, shown by the dotted area in Figure 3.23. The ECC table is organized as a CAM. It protects the most vulnerable register versions in the RF. When a physical register is about to be written on, a decision is made whether to protect the register or not based on the predicted lifespan of the register. If the decision is made to protect the register, an entry for the register version is allocated in the ECC table, and the ECC generator calculates the ECC of the register data in parallel with the register write operation. When a physical register is read, if there is its entry in the ECC table, the ECC checker is used to verify the data's integrity. If there is no error, the processor proceeds as normal. If an error is detected, the processor stalls and (i) fixes the register data using ECC, (ii) flushes the Reorder Buffer (ROB) from the oldest instruction that reads the register version, and (iii) flushes the whole ECC table and resumes. Finally, ParShield architecture

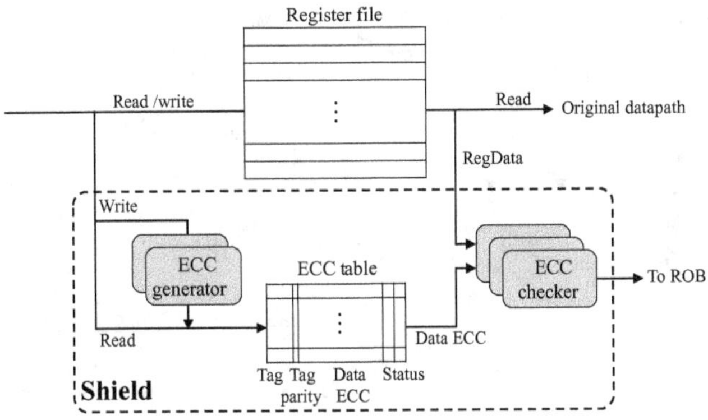


*Figure 3.23 Shield architecture predicts the lifespan of each register version and selectively protects the most vulnerable registers with ECC [44]*

is completed by adding a parity bit to every RF with Shield architecture, and ECC circuits of Shield architecture are re-used to generate and check the parity bit.

Memik *et al.* [47] proposed a technique to utilize unused physical registers as duplicate storage for actively used registers. Consider the case in which a new register (`r1`) is being assigned, and there is an unused free register (`r11`). In this example, the free register (`r11`) is used as a backup copy for the newly assigned register (`r1`). In other words, two writes happen, one to register `r1`, and the other to register `r11`. By adopting parity with this duplication, the replicated value can be used for recovery when a fault is detected. When register `r1` is redefined, the register `r11` is freed up. For the proposed technique, the register value should be copied for every write operation. Figure 3.24 illustrates the basic hardware modifications in the RF and register renaming hardware required for this replication technique. The Register Renaming Circuit (RRC) and the reservation station are augmented to store the copy register name of the destination register. The RRC must also store the copy state to distinguish between copy registers and original registers. The RF is modified to receive the copy register name, which is indicated as Rdcopy in Figure 3.24. At every write operation, the RF should update the copy register indexed by Rdcopy with the same value as the original destination register. The main observation behind this method is that a significant fraction of the physical registers in Superscalar processors is not used during the execution. To exploit unused registers as duplicates of active ones, Memik *et al.* investigated two different strategies: conservative and aggressive. In the conservative strategy, they limit their protection scheme, using only unused registers as duplicate storage. When a new physical register is allocated and decided to be duplicated, either an unused register or a duplicate of another register is selected as the new duplicate storage. On the other hand, the aggressive strategy takes a wider view, and registers that are not used for a long time, can also be selected as duplicate storage of active registers. While the aggressive strategy slightly sacrifices the performance, it can effectively increase the number of protected register accesses.

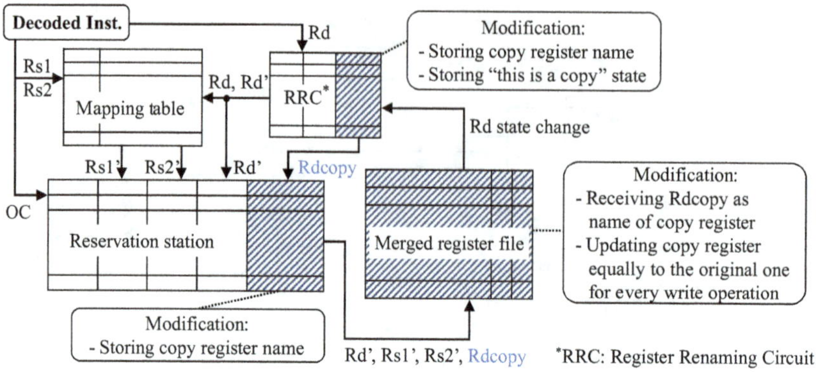


*Figure 3.24 Register Renaming Circuit (RRC), reservation station, and register file are modified to replicate register values into unused registers [47]*

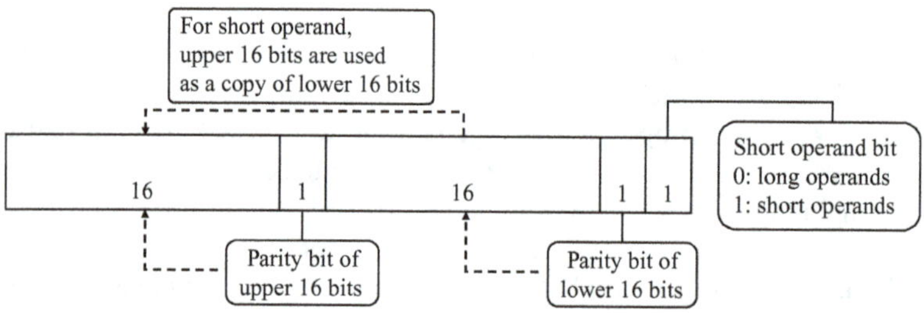


*Figure 3.25 To replicate a value of short operand into upper bits, parity bits are associated with upper and lower bits, and additional one bit is appended to represent the register file with short operand [45]*

Another interesting and orthogonal approach was proposed by Kandala *et al.* [45]. They observed that a large number of register values are narrow, i.e., less than or equal to 16 bits for a 32-bit architecture; therefore, the upper 16 bits of the registers can be used to replicate short operands, enhancing the register integrity. With this observation, they extended the register replication scheme of Memik *et al.* [47] with in-register replication for narrow registers. They add a short operand bit in each register to distinguish short and long operands, and a parity bit for every 16 bits within the RF as shown in Figure 3.25. To read operands from a register, the short operand bit is checked first. For a short operand (16 bits or smaller), the parity bit of the lower 16 bits is checked. If there is no fault, the lower 16 bits will be read with the sign extension. Otherwise, the parity bit of the upper 16 bits is checked. If the upper 16 bits are correct, the contents of the upper bits are copied to the lower bits. In the rare case in which both the upper and lower bits are corrupted, the data cannot be recovered. For a long operand, the value is replicated in unused registers, based on the technique previously proposed by Memik *et al.* [47]. Since short operands can be replicated without any additional unused registers, they can effectively reduce the number of registers that should be replicated in unused registers.

Blome *et al.* [38] proposed an RF protection technique that uses a small cache of live registers called the Register Value Cache (RVC). The RVC maintains duplicate copies of the most recently accessed values within the RF. The inputs to the RF are split up and fed directly to the RVC, as shown in Figure 3.26. Essentially, the read/write control logics for the RF are duplicated. Error checking is performed at the read operation by detecting a mismatch between the values in the main register and in the RVC. The RVC also maintains a Cyclic Redundancy Check (CRC) of the value, so that in the case of an error, it can determine whether the error was in the original register (then the value in the RVC can be used) or in the RVC (then the value in the original register can be used). In detail, when a mismatch between the register and RVC is detected, the stall signal is raised to stall the pipeline. During the stall of the pipeline, the RVC checks the validity of its data with the corresponding CRC

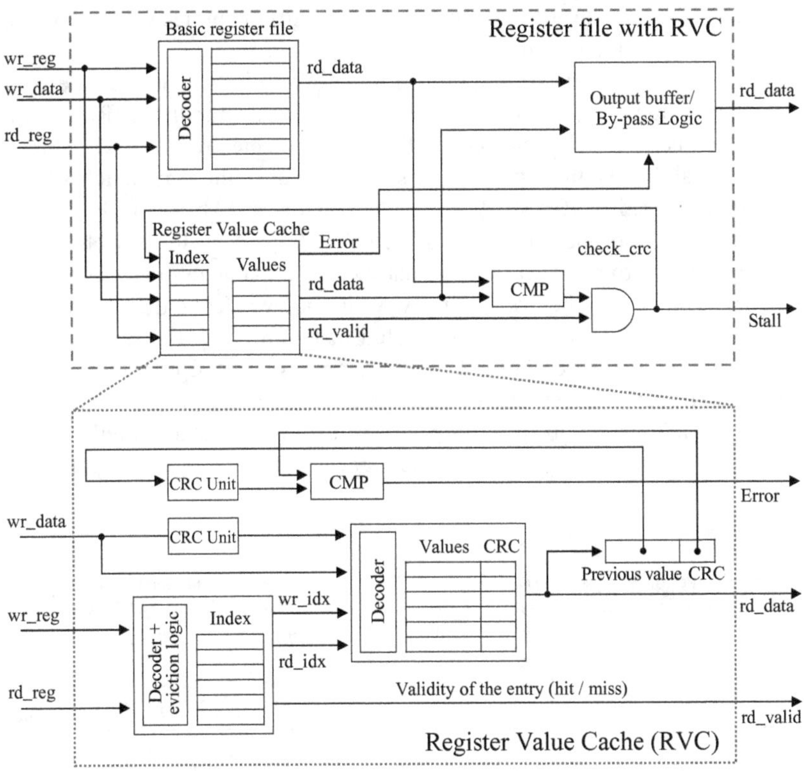


*Figure 3.26 Register Value Cache (RVC) copies the inputs to the register file to duplicate the most recently accessed values and protects them based on the Cyclic Redundancy Check (CRC) [38]*

value. Since the RVC needs to verify its previous data and CRC value, the RVC uses a previous value in the buffer that keeps the data and CRC values of the last cycle. If the CRC check fails, the RVC raises an error signal and the data in the physical register is considered to be correct. Otherwise, the value of the RVC is considered to be correct. Thus, the RVC is capable of detecting and correcting errors that have occurred in both the combinational and sequential logics of the RF.

Some hardware techniques (e.g., Montesinos *et al.* [44]) advocate the concept of partially protected RF that protects only some of the registers to minimize the hardware overhead. For these hardware-based partially protected RFs, the decision of which variables to be mapped to the protected registers must be made at run-time using another piece of hardware, which can dissipate a significant amount of power [49]. To overcome the problem of power overheads, hybrid approaches [48,49] have been proposed. The main idea of these approaches is to expose the hardware protection mechanism to the software so that the decision of which variables to be mapped to protected registers can be made at the compile-time, possibly by the compiler. Therefore, the program binary must be modified. The hybrid approaches can be quite effective, and more power-efficient than their hardware-only counterparts, but they still rely on special hardware components such as partially protected RF, which must be supported by the architecture.

The main challenge of hybrid approaches is to determine which data to be assigned and mapped to the protected RFs. To minimize the vulnerability of RFs, the most vulnerable registers should be placed in the ECC-protected registers. Of course, the proper vulnerability estimation of the RFs is essential for these approaches. Yan and Zhang [48] proposed the Register Vulnerability Factor (RVF) based on the susceptible intervals to soft errors and lifetimes of RFs. The RVF presents the probability of soft error propagation from the RF to other hardware components. They observed that soft errors in the RF can be masked if new values are written to the RF before other components read it. Therefore, as illustrated in Figure 3.27, the RF is vulnerable only during the write–read and read–read intervals since soft errors on read–write and write–write intervals will not be propagated to other components. The proposed RVF information can be obtained by profiling with performance simulation. Then, the compiler assigns the registers with the highest RVF values to ECC-protected RF

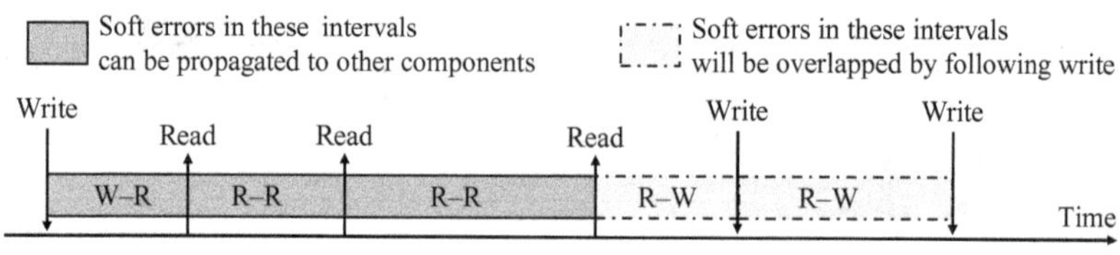


*Figure 3.27 A register file is susceptible against soft errors during the write–read and read–read intervals, while write–write and read–write intervals are unsusceptible [48]*

based on the profiling results. Further, the compiler reschedules the instructions, executing the write operations as late as possible and the read operations as early as possible, to minimize the write–read and read–read intervals, causing the reduction of the RVF.

While Yan and Zhang [48] focused on minimizing the RVF, they did not consider the energy efficiency. Lee and Shrivastava [49] concentrated to optimize both reliability and energy efficiency of software-managed partially protected RFs through static register reallocation. Since optimizing register allocation for performance, vulnerability, and energy consumption at once is extremely complicated, register swapping is applied to the registers after performance-driven register allocations. Lee and Shrivastava take a step-by-step approach to find the best register swapping in terms of the energy consumption and vulnerability for partially protected RF. In particular, they first find and apply the best Function-level Register Swapping (FRS) and then find and apply the Program-level Register Swapping (PRS). Since FRS should respect all calling conventions, caller-saved registers and callee-saved registers can only be swapped within their own groups. Solving the FRS problem for callee-saved registers is much more complicated than the FRS problem for caller-saved ones since the live range of callee-saved registers may span several functions. Figure 3.28 shows an example of one callee-saved register in the case of a function call. Based on the result of the FRS, this register may be used in a callee function, as shown in Figure 3.28(a), or unused, as shown in Figure 3.28(b). Therefore, the vulnerability of the caller function (F1) depends on the callee function (F2). In addition, since the register can either be read or written ($t_5$) after the return of the callee function, the vulnerability of the callee function also depends on the caller function. These inter-dependencies make the FRS problem with callee-saved registers extremely complex, while PRS problem and FRS problem with caller-saved register can be solved by an efficient algorithm. In effort to reduce this complexity, Yan and Zhang made two key observations: (i) callee-saved registers are most likely to be first read after the function is called, and (ii) callee-saved registers are most likely to be written after the function returns. Based on these observations, they proposed a simple heuristic that breaks the inter-dependence between functions in callee-saved registers and solves FRS problem with callee-saved registers as the same as FRS problem with caller-saved registers.

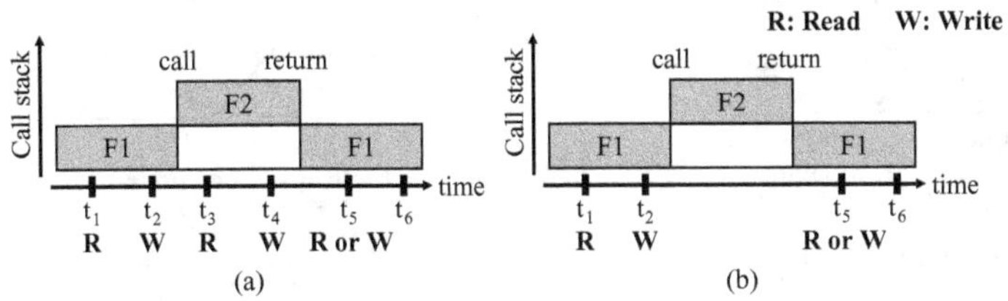


*Figure 3.28 Live range of callee-saved registers with respect to function call: (a) register is used in callee function and (b) register is not used in callee function*

## 3.3 Pipeline and core protection

Pipeline is one of the most important parts in the execution engine that is prone to soft errors causing significant impacts on the system. One way of pipeline protections is to duplicate the pipeline itself. This can be achieved through the spatial redundancy and the temporal redundancy. Temporal redundant techniques exploit the features of Simultaneous and Redundant Threading (SRT) processor [50] and spatial redundant techniques exploit those of Chip-level Redundant Threading (CRT) multiprocessor [51] as shown in Figure 3.29. Both of these processors achieve the fault tolerance by executing and comparing two copies called leading and trailing threads of a given application [52]. Most of the techniques consider soft errors, i.e., transient faults while some of them additionally care for hard errors, i.e., permanent faults. Table 3.3 summarizes the features, required hardware modification, and targeted processor of pipeline protection techniques, which will be discussed in this section.

AR-SMT [53] technique exploits SMT (Simultaneous Multi-Threaded architecture). It executes two streams such as A-stream (Active) and R-stream (Redundant) on an SMT processor simultaneously. The main purpose of exploiting two streams is to compare the results from both the streams in order to detect soft errors. Figure 3.30 illustrates how A-stream and R-stream work on the SMT processor core. Whenever A-stream commits an instruction, it also pushes the result to the FIFO queue called the delay buffer. On the other hand, the committed results of R-stream are compared to those of A-stream in the delay buffer. If they are not identical, an error is detected and it rolls back to the lastly saved checkpoint. Otherwise, committed R-stream states are check-pointed for the later recovery. R-stream can be generated on-the-fly by hardware, but additional supports from Operating System (OS) are required for AR-SMT. OS should make R-stream to maintain the separate memory image from A-stream. For this, OS maintains two separate memory images and handles the address translations separately for both the streams to replicate the operations with some delay in the R-stream. Synchronization of A-stream and R-stream is also required to handle

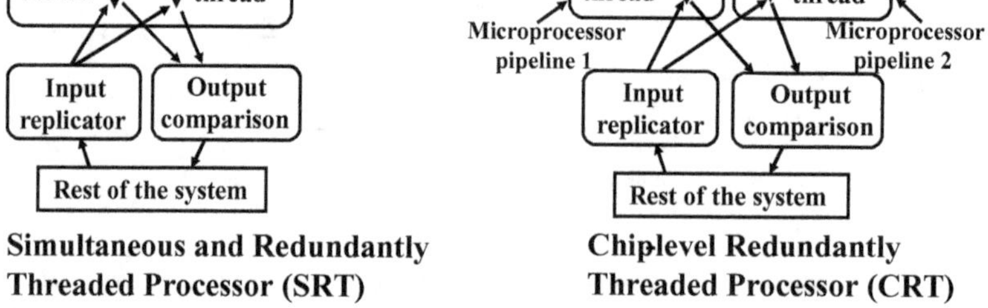


*Figure 3.29 SRT processor has single core that executes both the threads whereas CRT processor has a core for each thread [51]*

*Table 3.3 Summary of pipeline protection techniques*

| Technique | Additional hardware | Processor architecture | Fault types | Recovery |
|---|---|---|---|---|
| AR-SMT [53] | Delay buffer | SMT-based | Transient | Yes |
| SRT [50] | Active Load Address Buffer or Load Value Queue, Branch Outcome Queue Check store buffer | Out-of-order, speculative SMT | Transient | No |
| SRTR [54] | SRT [50] + Register Value Queue, Dependence chain queue | Out-of-order, speculative SMT | Transient | Yes |
| BlackJack [55] | SRT [50] + Dependence Trace Queue | Out-of-order, speculative SMT | Transient, Permanent | No |
| SlipStream [56] | Delay buffer, IR predictor and detector, recovery controller for IR-misprediction | Out-of-order, speculative SMT or CMP | Transient | No |
| CRT [51] | Load Value Queue, Line prediction queue, Store comparator | Chip Multiprocessor (CMP) | Transient, Permanent | No |
| CRTR [52] | CRT [51] + Dependence chain queue | CMP | Transient | Yes |
| FingerPrinting [57] | Result capturing and fingerprinting unit buffer to compare fingerprints | CMP | Transient | Yes |
| Reunion [58] | FingerPrinting [57] + Synchronizing handler | CMP | Transient | Yes |
| DCC [59] | FingerPrinting [57] + Age table | CMP | Transient, Permanent | Yes |
| DIVA [60] | Additional pipeline stages | Heterogeneous custom cores | Transient, Permanent, Design | Yes |
| UnSync [61,62] | Communication Buffer, Error Interrupt Handler | CMP | Transient, Permanent | Yes |

exceptions, traps, and context switches. This can be implemented by stalling A-stream until R-stream completely consumes the delay buffer.

Reinhardt and Mukherjee [50] proposed SRT processor based on SMT processor. While AR-SMT [53] doubles the memory space for two streams, SRT excludes caches and memory from the sphere of replication since caches and memory are usually

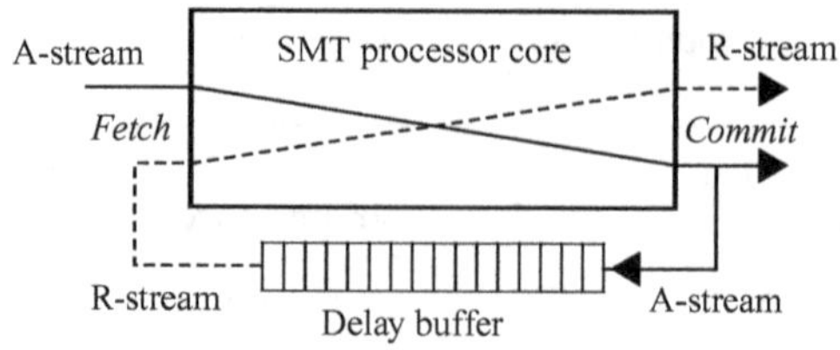


*Figure 3.30 High-level view of AR-SMT technique, which duplicates a stream into two streams and compares their results at the commit stage of R-stream to detect a fault [53]*

protected by parity or ECC. SRT can even exclude RFs from the sphere of replication if they are already protected by parity or ECC; otherwise, SRT also needs to replicate register files. SRT processor dynamically schedules its hardware resources among the redundant copies to improve the performance. Redundant instructions in an SRT processor are executed at different cycles in a different order due to the dynamic instruction scheduling. Therefore, the lock-step techniques are inappropriate since they replicate the input and compare the output by cycle-level synchronization for redundant threads. SRT proceeds the output comparison between redundant threads for the data and addresses of store operations. Further, the address of un-cached load operations also should be verified since they have side effects in I/O devices that are outside of the sphere of replications. On the other hand, addresses of cached load operations are not necessary to verify since they do not modify the architectural state of the machine. Faulty result of cached load operation can be detected via other output comparison. In addition, output comparison on values to RFs should also be verified if they are not replicated. Memory inputs such as load values should be carefully replicated in SRT since data values may be updated by other processors or DMA (Dynamic Memory Access) I/O. Dynamically scheduled redundant instructions can get different values if cached data is updated between redundant load accesses. This divergence can lead to different outputs from redundant threads and false failure even if there is no soft error. Reinhardt and Mukherjee adopted two alternative mechanisms to duplicate the cached load data in SRT, Active Load Address Buffer (ALAB), and Load Value Queue (LVQ) data structures. ALAB ensures the cached load operations from both the threads to receive the same value in the presence of the dynamic instruction scheduling, by delaying cache replacements and invalidations after the load of the leading thread until the corresponding load of the trailing thread. On the other hand, LVQ stores the cached load addresses and values of the leading thread and makes it useful for the trailing thread instead of making the leading thread to wait on the trailing thread's operations. Figure 3.31 illustrates the idea of LVQ. For cached load operations, the trailing thread compares the effective address in the LVQ from the leading thread and its own effective address for fault detection and copies the data from LVQ instead of accessing data cache. While LVQ is much easier to implement than ALAB and can accelerate the fault detection of faulty addresses, LVQ constraints the loads of the trailing thread to be scheduled after the corresponding loads of the leading thread.

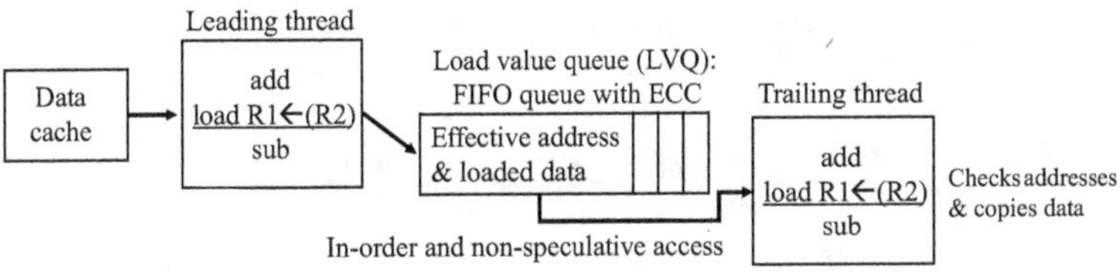


*Figure 3.31 Load Value Queue stores the effective addresses and data values of load operations in the leading thread, which will be used by the trailing thread [50]*

SRT performance can be improved by two mechanisms; the first one with slack fetch inserts the constant slack between two threads so that the trailing thread always can check the outcomes of the branch instructions and cache hits/misses of the leading thread, and the second one maintains the Branch Outcome Queue (BOQ) that stores the results of the branch instructions in the leading thread and makes it available for the trailing thread.

Vijaykumar *et al.* [54] proposed Simultaneously and Redundantly Threaded processors with Recovery (SRTR) by extending SRT [50] with recovery schemes from transient faults. They observed that non-store instructions of the leading thread may be committed before the execution of corresponding instructions in the trailing thread. In SRTR, the leading thread is not allowed to commit an instruction before it is checked by the trailing thread since it is impossible to undo the committed faulty instruction. To reduce the performance overhead induced by stalling the leading thread, SRTR should check the results of redundant instructions as early as possible. For the right recovery, SRTR should check every instruction unlike to SRT, and register values often have been written to the physical registers by the time when the results of redundant instructions are checked. Therefore, bandwidth pressure of RF is increased significantly. To avoid this register bandwidth pressure, SRTR presents Register Value Queue (RVQ) to hold register values for checking after the trailing thread completes the execution, which in turn increases the pressure of RVQ. They proposed the Dependence-Based Checking Elision (DBCE) that checks the last instruction in the dependency chain and skips the other instructions in the dependency chain to reduce the number of checks and to reduce the RVQ pressure. Figure 3.32 illustrates how DBCE works. Since instruction 5 uses the result of instruction 3 and instruction 3 uses that of instruction 1 in Figure 3.32, they are in the same dependence chain. Any soft error during this chain can be detected by verifying the output of instruction 5 since errors on instruction 1 and 3 will also affect the output of the last instruction. However, DBCE should carefully form a chain with masking instructions such as bitwise instructions or compare instructions. Even if the inputs of these masking instructions are corrupted, they may still generate correct outputs. Suppose that one instruction generates the wrong output due to soft errors. If a masking instruction uses corrupted results and produces the correct output, the last instruction in the chain uses the output of masking instruction, and it cannot detect the error of previous corrupted instruction. Thus, the output

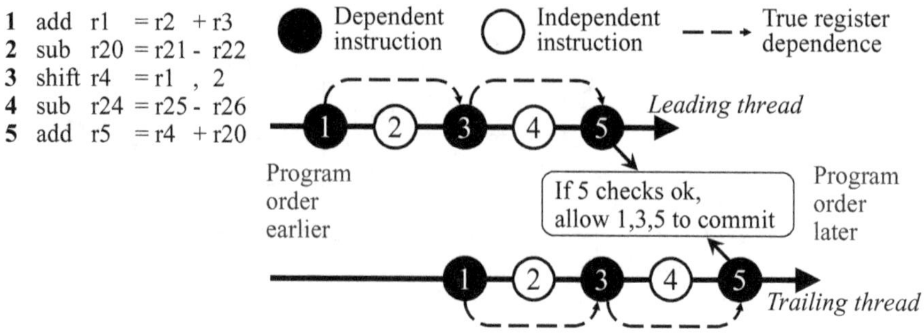


*Figure 3.32 Example of dependent instructions for Dependence-Based Checking Elision (DBCE) in SRTR [54]*

of corrupted instruction will be committed to the RF, and then SRTR cannot recover it correctly. This problem can be solved by disallowing masking instructions in the middle of any chain since the source operands of the first instruction in a chain are verified by the last instructions of other chains.

On the other hand, Schuchman and Vijaykumar [55] noted that the error coverage of SRT [50] against hard errors is limited since leading and trailing threads are almost identical and in many cases they will use the same hardware resources. To achieve spatial diversity against hard errors on a single SMT core, they proposed BlackJack [55] microarchitecture. The naive approach to provide spatial diversity for SRT is to shuffle trailing thread instructions so that they will go to different execution ways. However, it has several challenges; shuffling after rename cannot provide spatial redundancy for the front-end ways from the issue queue, such as fetch, decode, and rename units. This is because instruction locations of both threads within instruction cache block are not changed, and as follows both instructions may go to the same front-end way of issue queue. On the other hand, shuffling the trailing thread instructions before rename can violate the program correctness since the dependence information is missing. Schuchman and Vijaykumar observed that the leading thread can determine the dependencies before the trailing thread executes. They proposed safe-shuffle for BlackJack, which provides dependence information for the trailing thread from the execution of the leading thread so that shuffling the trailing instructions can be done before the fetch stage without violating dependencies. Information about dependencies, rename maps, and pipeline resources are stored in a Dependence Trace Queue (DTQ) as shown in Figure 3.33. Shuffling is performed on the trailing thread by using the information from the DTQ before they are fetched.

Sundaramoorthy *et al.* [56] observed that a full dynamic instruction stream can be shortened without changing the output of the original program by removing ineffectual computations and computation related to highly predictable control flow. Based on the observation, they proposed SlipStream architecture for both SMT and Chip Multiprocessor (CMP). The main idea is to shorten one of redundant streams while previously discussed techniques always execute almost full dynamic instructions for

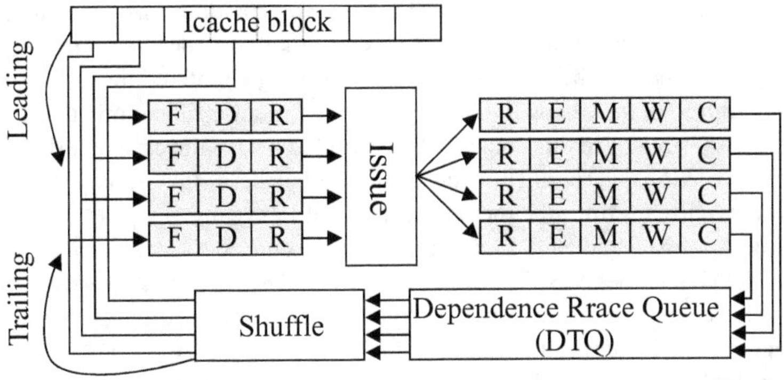


*Figure 3.33 In BlackJack architecture, instructions of trailing thread are shuffled before the fetch stage to achieve spatial redundancy in both front-end and back-end way of issue stage [55]*

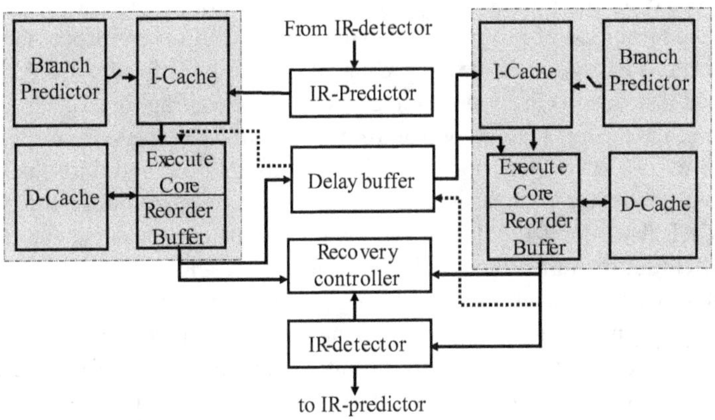


*Figure 3.34 In SlipStream architecture, IR-predictor can skip dynamic instructions of A-stream based on the prediction with information from IR-detector, which monitors R-stream to find the computation in-effective instructions [56]*

both streams. Sundaramoorthy *et al.* named the leading stream as the advanced stream, or A-stream, and the trailing thread as the redundant stream, or R-stream. Specifically, ineffectual computation and computation related to highly predictable control flow of A-stream are skipped, based on the prediction. Figure 3.34 illustrates the SlipStream architecture. Similar to the AR-SMT, SlipStream adopts the delay buffer to compare the control and data flow outcomes from both A-stream and R-stream. SlipStream requires additional hardware resources to shorten A-stream. Instruction Removal predictor or IR-predictor generates the PC of next block of instructions to A-stream, similar to the branch predictor, but can skip any number of dynamic instructions that are predicted to be ineffectual. To provide information for prediction, Instruction Removal detector or IR-detector monitors the R-stream and detects the instructions that can be removed when A-stream encountered them later. Finally, recovery controller maintains the addresses of the memory locations that might be corrupted in A-stream. In this context, the corrupted means the case that IR-predictor removes instruction that should be executed for the correct output. Recovery controller can recover A-stream from the memory context of R-stream.

While AR-SMT and SRT focused on a single core SMT, Mukherjee *et al.* [51] extended RMT schemes of SRT to two-way CMPs. They noted that two-way CMP can adopt the lockstep, where both processors perform the same instruction on a cycle-by-cycle and provide better fault coverage than SRT, specifically in terms of hard errors. However, lockstep uses hardware resources inefficiently since both copies should waste resources together on mis-speculation and cache misses. Mukherjee *et al.* [51] proposed CRT to achieve both hard error coverage of lockstep and efficiency of SRT. CRT is similar to SRT but redundant threads are executed in separate processor cores. For example, if the leading thread of program A and the trailing thread of program B are executed on the processor pipeline 1, then the trailing thread of program A and the leading thread of program B should be executed on the processor pipeline 2, as shown in Figure 3.35. Like SRT, CRT adopts the LVQ to forward the result of cached load instructions from the leading thread to the trailing thread. For the output comparison, CRT adopts store comparator that is similar to the store buffer in SRT. CRT receives the data of store operations in the trailing thread and finds the matching entries from the leading thread, and proceeds output comparison for the matching. Further, CRT also adopts the line prediction queue that has the same concept of the BOQ in SRT, but forwards line prediction results since CRT assumes a processor to access the instruction cache. By forwarding line prediction results from the leading thread to the trailing thread, CRT can completely eliminate mis-fetches in the absence of faults. LVQ, line prediction queue, and store comparator in CRT should consume the inputs from different processors in CRT since leading and trailing threads are distributed in different cores. In fact, they are outside of the sphere of replication in CRT and should be protected by hardware techniques such as ECC and parity codes.

Gomaa *et al.* [52] proposed CRTR (Chip-level Redundantly Threaded processor with Recovery) by extending CRT [51] with recovery scheme. CRTR starts from the

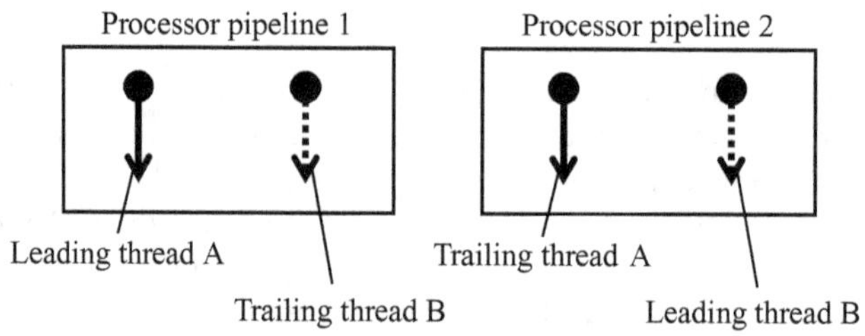


*Figure 3.35 Unlike SRT [50], CRT distributes redundant threads in separate processor cores [51]*

similar idea of SRTR [54], which prevents committing of leading thread instructions before they are checked by the trailing thread. Gomaa *et al.* noticed that this shortens the slack size between the leading thread and the trailing thread, and as follows it is inappropriate in CMP due to inter-processor delay. Rather, they proposed the asymmetric commit, which allows the leading thread to commit register updates before the checking. On the other hand, the trailing thread of CRTR should commit their register updates only after the completion of corresponding checks to keep the register values of the trailing thread for the recovery. Similar to the CRT, the memory update of CRTR should be committed after the checking to ensure that memory always keeps the safe data. Thus, CRTR can recover from transient faults by exploiting correct register values from the trailing thread and memory. To resolve the bandwidth for result checking, CRTR also adopts the ideas of RVQ and DBCE that are discussed in SRTR. However, they found that disallowing masking instructions in the middle of a chain in DBCE is inefficient since many integer and almost all floating-point instructions are masking ones. Figure 3.36(a) illustrates the example of masking instructions. Since i3 is a masking instruction that may produce the correct output with corrupted input operands, it cannot be in the middle of a chain. Therefore, DBCE cannot make a chain as (i1, i3, i4). Gomaa *et al.* observed that masking instruction can be dangerous if it consumes an input operand, and the corrupted operand is also used in any non-masking instruction later. Therefore, if a masking instruction is a last consumer of the operands, it can be in the middle of a chain without increasing the vulnerability. CRTR adopts Death- and Dependence-Based Checking Elision (DDBCE) that extends DBCE by tracking register death to determine whether masking instructions are the last consumer of their input operands. Figure 3.36(b) shows the example of DDBCE. Since i5 overwrites the value of r6 after i3 reads r6, i3 is the last consumer of r6. While i3 cannot be in the middle of the chain in DBCE (Figure 3.36 (a)), DDBCE can form a chain with (i1, i3, i4) so that only check for i4 is needed for these three instructions without harming the reliability.

Smolens *et al.* [57] proposed another interesting optimization for dual modular redundant multithreading techniques. They proposed FingerPrinting to optimize detection schemes for redundant streams. The key idea of FingerPrinting is to summarize the execution history of each stream in a hash-based signature. In a FingerPrinting

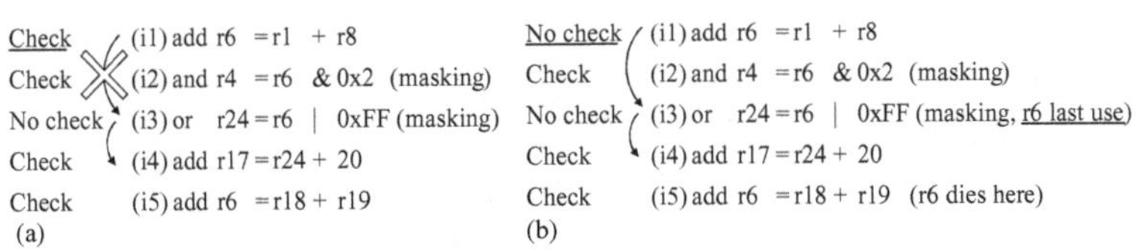


*Figure 3.36 (a) DBCE in SRTR does not allow masking instructions (i2 and i3) to be in the middle of chain. (b) On the other hand, DDBCE in CRTR allow masking instructions to be in the middle of a chain, if they are the last consumer of their operands [52].*

scheme, only summarized signatures are needed to be compared while previous techniques based on SRT [50] and CRT [51] compare all address and data information of store operations. Therefore, FingerPrinting can reduce inter-processor communication bandwidth for output checking significantly by reducing the number of output comparisons. FingerPrinting can be implemented in two alternatives: (i) capturing all committed states from ROB and Load-Store Queue (LSQ), as shown in Figure 3.37(a). Since ROB usually does not record the result of instructions, ROB of this alternative should be modified to record the result of each entry. (ii) Capturing results of all executed instructions right after the execution, as shown in Figure 3.37(b). The FingerPrinting in this alternative may contain potentially speculative states. FingerPrinting can be adopted with checkpointing to provide both fault detection and recovery. Starting from the previous checkpoint, FingerPrinting keeps two fingerprints by accumulating the results of two redundant streams. To keep safe state for checkpoint, fingerprint of two streams should be compared when both streams reached to the next checkpoint. If there is a mismatch between two streams, FingerPrinting roll backs two streams to the safe checkpoint.

Due to shared-memory programming model in CMP, independent threads may observe different values for the same dynamic load. To guarantee that redundant threads always see the identical view of memory, previously discussed techniques such as CRT [51] adopts LVQ; LVQ keeps the result of cached load operation in leading thread, and trailing thread consumes the load results from LVQ. However, Smolens *et al.* [58] pointed out that strict input replication with custom LVQ of CRT-based techniques requires significant changes to highly optimized microarchitectures. They observed that relaxed input replication, which permit redundant threads to issue memory operation independently to existing cache hierarchies, can produce correct output. When there is an input coherence, it can be resolved by fault detection and recovery mechanism that are adopted for soft error detection. Based on this key observation,

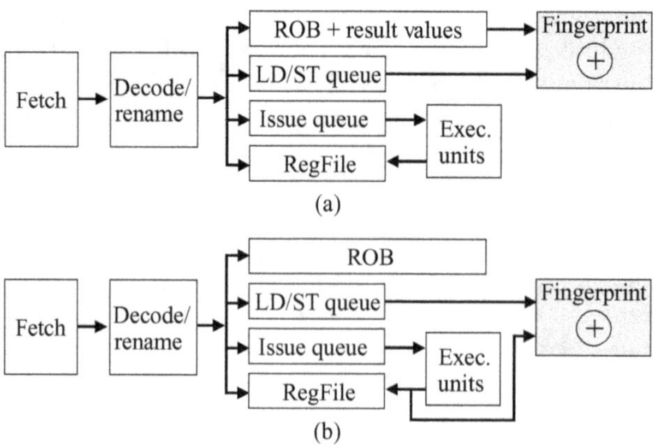


*Figure 3.37 Fingerprint can be implemented by (a) capturing LSQ and committed states from ROB or (b) capturing LSQ and the results from execution units [57]*

Smolens *et al.* proposed Reunion execution model with relaxed input replication and fault detection and recovery scheme for both soft errors and input incoherence. In the Reunion model, two processor cores that execute redundant threads are considered as one logical processor pair. Each logical processor pair consists of two cores such as vocal and mute cores. The vocal core exposes all the updated values to the system with abiding coherency and memory consistency requirements. If and only if there is no soft errors, the state of vocal core is considered as safe-state. On the other hand, mute cores are not exposed to the update of systems and only access to its private cache, as shown in Figure 3.38. Memory access of mute core is considered as phantom request, which is noncoherent memory request. Reunion compares execution results of a logical processor pair before the results are visible to other logical processor pairs, based on FingerPrinting [58]. If the comparison matches, the state of vocal core is considered as a new safe state. Otherwise, Reunion roll-back schemes restore the architectural states of vocal core to the previous safe-state for the recovery. At the same time, the mute core should copy the register states from the vocal core for recovery after the roll-back process. In addition, if there is an input incoherency, the incoherent values in private cache of mute core should be replaced. The Reunion proposes re-execution protocol that includes mute register initialization and synchronizing request to update private cache of mute core.

LaFrieda *et al.* [59] noticed that previous techniques such as CRT [51], CRTR [52], and FingerPrinting [57] statically bind core pairs at design time for redundant threads. They pointed out that both cores of core pair should be disabled in the presence of a permanent fault on one core, as shown in Figure 3.39(a). In addition, in the presence of process variations, both cores of core pair may have to run at the speed of the slower core. To address these limitations of static binding, LaFrieda *et al.*

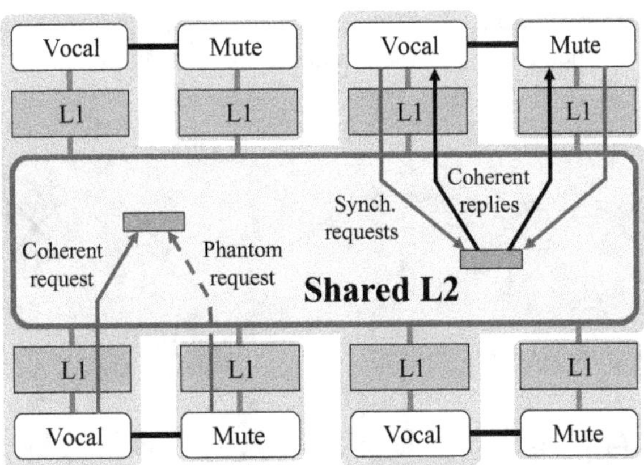


*Figure 3.38 In Reunion architecture, vocal core is exposed to system update by abiding memory consistency, while mute core only accesses its own private cache. In the presence of input incoherence, synchronizing request deals with the update of private cache in mute core [58]*

proposed Dynamic Core Coupling (DCC), which allows any core to form a virtual Dual Modular Redundancy. As illustrated in Figure 3.39 (b), DCC does not need to disable two cores in a pair when there exist permanent faults on one core of the pair. In addition, DCC also can issues redundant threads of the high IPC thread on distant cores to reduce hot spots, as shown in thread A of Figure 3.39(b). Further, DCC can provide recovery of a program from permanent fault on a core by assigning another core dynamically when a permanent fault is detected. For a dynamically coupled core pair, system bus of a shared memory CMP is used for communication between two cores in a pair. DCC adopts several optimizations to reduce the overhead from intercore communications. To reduce the communication bandwidth from the output comparison, DCC adopts FingerPrinting [57] scheme for output comparison of two cores. In addition, they keep long interval for checkpointing, to reduce the frequency of fingerprint comparison. However, due to the long interval of checkpointing, a large number of memory stores should be buffered between two checkpoints. DCC adopts cache buffering technique from Cherry [63]. The main idea is to mark private cache lines as unverified when they are written but not verified and ensure that unverified lines will not leave the private cache. After the verification at the end of checkpoint interval, and write to the verified dirty lines should be written back to the low-level cache. Similar to the concept of vocal and mute core in Reunion [58], only one core named as master writes the value to the shared cache, while another core in a virtual pair named as slave just evicts to write the verified cache line without write-back. Input replication is another challenge of DCC. CRT [51] adopts LVQ and [58] adopts relaxed input replication, but they are not inappropriate in DCC due to interprocessor communication on the system bus and long interval of checkpointing. Cache buffering technique of DCC can ensure the master–slave memory consistency for sequential application. In a parallel execution, after one of a virtual pair loads from an address, other core should not update the values in this address until both threads in a pair commit their load operations. To support master–slave memory consistency, DCC adopts age table. When a store or load instruction is committed, the total number of

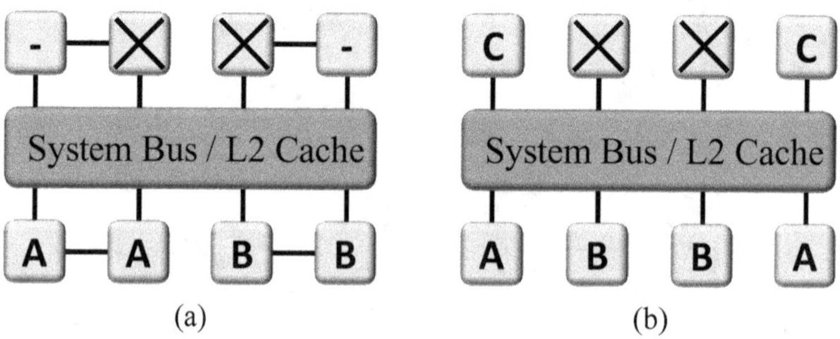


*Figure 3.39 An eight-core CMP with statically coupled for redundant multithreading (a) and dynamically coupled (b) in the presence of two faulty cores [59]*

committed load and store instructions at that time is considered as the age of that instruction and recorded in the age table where the address is used for indexing. DCC can delay the store operations in other cores by exploiting information from age table to prevent memory incoherence.

On the other hand, Austin [60] proposed Dynamic Implementation Verification Architecture (DIVA) architecture with dynamic verification, which can detect not only soft and hard errors but also design errors in the processor. The main idea of DIVA is to split processor design into two parts in different ways. The first one is DIVA core with deeply speculative design, which is designed for high-performance. The second one is functionally and electrically robust DIVA checker. In a DIVA architecture, core processor performs a role similar to the leading thread and checker performs similar to the trailing thread. DIVA core executes the instruction in out-of-order scheduling, similar to the traditional out-of-order cores. However, when an instruction is completed, both inputs and results of the instruction are sent to DIVA checker instead of commit stage, as shown in Figure 3.40. A DIVA checker deals with the verification of the inputs and results of DIVA core. Specifically, the DIVA checker verifies them in a functional checker stage (CHK). CHK is composed with two parallel and independent verification pipelines, as shown in Figure 3.41, the first pipeline CHKcomp verifies that the instruction is executed correctly in the functional units. The execute stage of CHKcomp pipeline can be differently implemented from implementation of DIVA core, in the purpose of costs or reliability. This CHKcomp pipeline has no dependency since precomputed inputs and results are delivered from a core processor. The second pipeline CHKcomm verifies that operand's values are correctly communicated correctly from one instruction to another. CHKcomm simply re-executes all communications in an instruction prior to its retirement, with Read (RD) stage that is a simple load/store instruction set. Similar to CHKcomp, precomputed input and results eliminate most dependencies in CHKcomm. Only a single bypass is required to handle the case that the output of previous instruction is used for the input of the current instruction, since this value is not visible in this stage until it is committed. In addition, the DIVA checker adopts a watchdog timer (WT) to detect that a DIVA core cannot produces any retirement of instruction due to deadlock or livelock. Finally, the DIVA checker commits the instruction from the core processor only if both CHKcomp and CHKcomm pipelines verify that computation from the

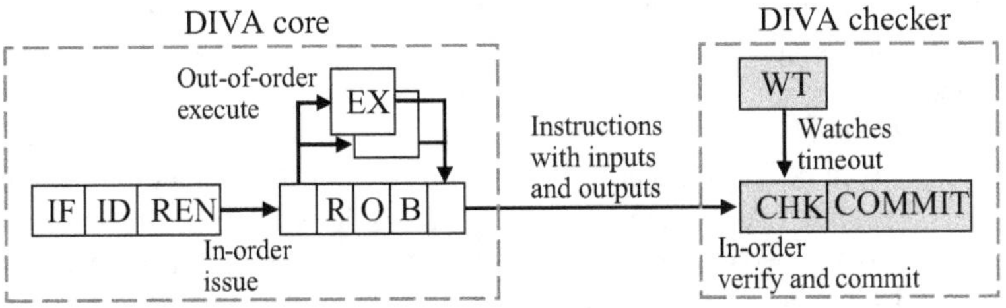


*Figure 3.40 Functionally and electrically robust DIVA checker deals with the verification and commit of deeply speculative DIVA core [60]*

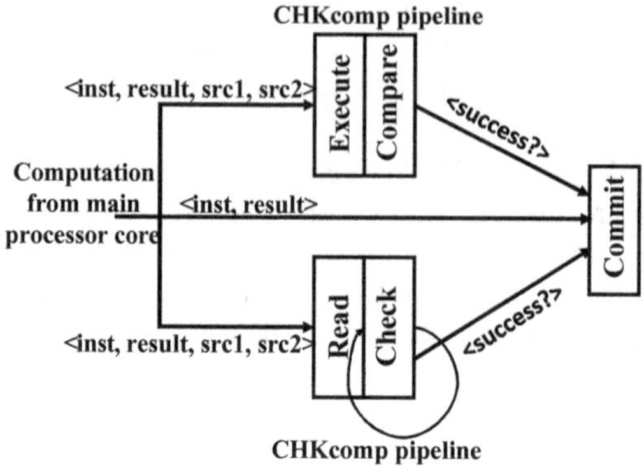


*Figure 3.41 Figure illustrates the DIVA architecture and its interface to the core processor. One of the two pipelines to verify the correctness in computation, the other to verify the operands flow from one instruction to another [60]*

core processor correctly executed the instruction. If any failure in any pipelines of the DIVA checker, DIVA proceeds recovery mechanism based on the results of the checker.

The previously discussed pipeline or core protection techniques assume that caches are protected. While there exist separate cache protection techniques, combining two different protection techniques is neither easy nor simple, and the overheads even add up. To protect all the core components and the private caches in one integrated technique, Jeyapaul *et al.* proposed the UnSync [61,62].

A salient feature of UnSync is that it reduces the frequency of synchronization among the cores. In previously discussed redundancy-based techniques for error resilience, the execution on the cores or within a core is synchronized either by lock-stepping [42] or through memory accesses [50] because (i) error detection is implemented by comparing the execution outputs or by comparing the memory accessed among the redundant threads (on the same or different cores) and (ii) when an error is detected, both the cores can be directed to re-execute a set of instructions from a previously identified error-free position (e.g., checkpoint [57]). However, as suggested by the name of the architecture, UnSync, the need to synchronize executions among the redundant cores is eliminated. This is made possible by (i) hardware-based error detection mechanisms that eliminate the necessity to compare redundant executions, and (ii) the "always forward execution" based recovery mechanism ensures that the cores resume from the last executed position of the correct core and no re-execution is required in either core. When an error is detected on a core, execution on both the cores is stopped and the architectural state from the error-free core is copied onto the erroneous core. While resuming the processor, the execution sequence is altered on only the erroneous core since PC is copied from the error-free core. The error-free core resumes from where it was stopped. Another interesting aspect of this technique is that the amount of instructions retraced (if any) by the erroneous core depends on the difference in the execution speeds between the two cores. In the case when the erroneous core was executing at a slower speed, during recovery, execution on this core is forwarded. The absence of re-executions in the recovery mechanism provides the compensation to the overhead involved in the architectural state and the L1 cache content copy from one core to the other.

Figure 3.42 describes an overview of the UnSync architecture. It shows two core-pairs of two way redundant UnSync architecture with their inter-core and intra-core communication links. Each core is configured with an on-core write-through L1 cache, and off-core L2 cache. While the shared L2 cache is protected with ECC, the L1 cache contains a parity bit on each cache line to detect single bit errors. Similarly, the core architecture blocks are equipped with error detection circuitry. The error detection blocks are connected to an Error Interrupt Handler (EIH) for each core-pair, to signal the recovery when an error is detected. Data committed into the L1 cache, from each core executing an identical thread of the program, is first written into a Communication Buffer (CB). From CB, one copy of the data is passed on to be written back in the protected L2 cache. Two identical cores in a core-pair execute the application where each core performs memory accesses on the shared L2 cache as independent cores. Data written on the L1 cache of a core, as it leaves the core as in a write-through cache, is written on a non-coalescing CB, one for each core in the core-pair, as plotted in Figure 3.42. In the CB, each updated entry is tagged with its corresponding instruction address. When the L1–L2 data bus is available for data transfer, the latest entry that has completed the execution on both is selected, and one copy of all the CB entries, earlier to this, is written into the L2 cache. This process ensures that, when processed data leaves the cores to be updated into the lower level

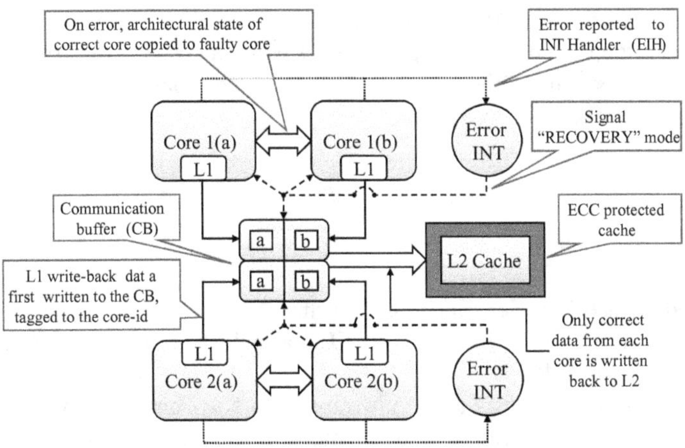


*Figure 3.42 UnSync architecture achieves area and power efficient protections by using hardware only error detection, un-synchronizing the comparisons from both cores, and developing always forward error recovery mechanism with the help of EIH (Error Interrupt Handler) and CB (Communication Buffer) [61]*

memory, both the cores have completed a particular state in the execution, and since no error was detected during this time, two copies are correct.

The main idea of UnSync about error protections is to apply power-, area-, and performance-efficient error detection mechanism rather than expensive error correction one. This error detection is effective with additional error recovery mechanism in UnSync that is significantly efficient at an infrequent error occurrence. Error detection in each core of the UnSync architecture is enforced by the use of hardware-only error detection blocks. The L1 cache, RF, and queuing structures are enabled with one bit parity-based detection owing to the fact that data write (parity generation) and read (parity verification) have a minimum of one-cycle time difference. On the other hand, for the architecture blocks like the program counter and pipeline registers, where data is read and written on every cycle, parity-based detection cannot be employed; therefore, dual mode redundancy-based error detection is enabled. If any of these detection blocks determines an error in the data on either core, an interrupt is transmitted to the EIH for that core-pair, which then performs the error recovery. The interconnect between the core and the EIH is described by the dotted arrows in Figure 3.42. Once the EIH receives an error interrupt, it signals "recovery" to both the cores and the CB of the corresponding core-pair. In this mode, the following procedure implements the "always forward execution" recovery mechanism in UnSync: (i) program execution on both the cores of the core-pair is stopped. (ii) The pipeline of the erroneous core is flushed so as to reset the pipeline registers. (iii) The architectural state (RF, PC, etc.) and the content of the L1 cache of the error-free core are copied onto those of the erroneous core, i.e., the core in which error was detected. This operation is performed by specific subroutines using the shared L2 cache. (iv) Data transfer from the CB to the L2 cache is stopped. Only the current data transfers in flight are completed. (v) The content of the CB, corresponding to the erroneous core, is overwritten by data from the error-free core. (vi) Once the architectural state, program counter, L1 cache contents, and the CB content of the erroneous core have been overwritten, both the cores resume execution of the program from the same program counter state as that was copied from the error-free core.

## References

[1] Naffziger SD, Colon-Bonet G, Fischer T, *et al.* The Implementation of the Itanium 2 Microprocessor. IEEE Journal of Solid-State Circuits. 2002; 37(11):1448–1460.

[2] Zhang W. Computing Cache Vulnerability to Transient Errors and Its Implication. In: 20th IEEE International Symposium on Defect and Fault Tolerance in VLSI Systems (DFT'05). Piscataway, NJ, USA: IEEE; 2005. p. 427–435.

[3] Naseer R, Boulghassoul Y, Draper J, *et al.* Critical Charge Characterization for Soft Error Rate Modeling in 90nm SRAM. In: 2007 IEEE International Symposium on Circuits and Systems. Piscataway, NJ, USA: IEEE; 2007. p. 1879–1882.

[4] Mitra S, Seifert N, Zhang M, *et al.* Robust System Design With Built-In Soft-Error Resilience. Computer. 2005;38(2):43–52.

[5] Shazli SZ, Abdul-Aziz M, Tahoori MB, *et al.* A Field Analysis of System-Level Effects of Soft Errors Occurring in Microprocessors Used in Information Systems. In: 2008 IEEE International Test Conference. Piscataway, NJ, USA: IEEE; 2008. p. 1–10.

[6] Li L, Degalahal V, Vijaykrishnan N, *et al.* Soft Error and Energy Consumption Interactions: A Data Cache Perspective. In: Proceedings of the 2004 International Symposium on Low Power Electronics and Design. ISLPED'04. New York, NY, USA: ACM; 2004. p. 132–137. Available from: http://doi.acm.org/10.1145/1013235.1013273.

[7] Hong J and Kim S. Flexible ECC Management for Low-Cost Transient Error Protection of Last-Level Caches. IEEE Transactions on Very Large Scale Integration (VLSI) Systems. 2016;24(6):2152–2164.

[8] Jeyapaul R and Shrivastava A. Smart Cache Cleaning: Energy Efficient Vulnerability Reduction in Embedded Processors. In: Proceedings of the 14th International Conference on Compilers, Architectures and Synthesis for Embedded Systems. CASES'11. New York, NY, USA: ACM; 2011. p. 105–114. Available from: http://doi.acm.org/10.1145/2038698.2038716.

[9] Ko Y, Jeyapaul R, Kim Y, *et al.* Protecting Caches From Soft Errors: A Microarchitect's Perspective. ACM Transactions on Embedded Computing Systems. 2017;16(4):93:1–93:28. Available from: http://doi.acm.org/10.1145/3063180.

[10] Zhang W, Gurumurthi S, Kandemir M, *et al.* ICR: In-Cache Replication for Enhancing Data Cache Reliability. In: 2003 International Conference on Dependable Systems and Networks, 2003. Proceedings. Piscataway, NJ, USA: IEEE; 2003. p. 291–300.

[11] Zhang W. Replication Cache: A Small Fully Associative Cache to Improve Data Cache Reliability. IEEE Transactions on Computers. 2005;54(12):1547–1555. Available from: https://doi.org/10.1109/TC.2005.202.

[12] Lee K, Shrivastava A, Issenin I, *et al.* Mitigating Soft Error Failures for Multimedia Applications by Selective Data Protection. In: Proceedings of the 2006 International Conference on Compilers, Architecture and Synthesis for Embedded Systems. CASES'06. New York, NY, USA: ACM; 2006. p. 411–420. Available from: http://doi.acm.org/10.1145/1176760.1176810.

[13] Mukherjee SS, Emer J, Fossum T, *et al.* Cache Scrubbing in Microprocessors: Myth or Necessity? In: Proceedings of the 10th IEEE Pacific Rim International Symposium on Dependable Computing (PRDC'04). PRDC'04. Washington, DC, USA: IEEE Computer Society; 2004. p. 37–42. Available from: http://dl.acm.org/citation.cfm?id=977407.978748.

[14] Farbeh H and Miremadi SG. PSP-Cache: A Low-Cost Fault-Tolerant Cache Memory Architecture. In: Proceedings of the Conference on Design, Automation & Test in Europe. DATE'14. Leuven, Belgium, Belgium: European Design and Automation Association; 2014. p. 164:1–164:4. Available from: http://dl.acm.org/citation.cfm?id=2616606.2616807.

[15] Chen L, Cao Y, and Zhang Z. Free ECC: An Efficient Error Protection for Compressed Last-Level Caches. In: 2013 IEEE 31st International Conference on Computer Design (ICCD). Piscataway, NJ, USA: IEEE; 2013. p. 278–285.
[16] Hong J and Kim S. Smart ECC Allocation Cache Utilizing Cache Data Space. IEEE Transactions on Computers. 2017;66(2):368–374.
[17] Manoochehri M, Annavaram M, and Dubois M. CPPC: Correctable Parity Protected Cache. In: Proceedings of the 38th Annual International Symposium on Computer Architecture. ISCA'11. New York, NY, USA: ACM; 2011. p. 223–234. Available from: http://doi.acm.org/10.1145/2000064.2000091.
[18] Yoon DH and Erez M. Memory Mapped ECC: Low-Cost Error Protection for Last Level Caches. In: Proceedings of the 36th Annual International Symposium on Computer Architecture. ISCA'09. New York, NY, USA: ACM; 2009. p. 116–127. Available from: http://doi.acm.org/10.1145/1555754.1555771.
[19] Hung LD, Irie H, Goshima M, *et al.* Utilization of SECDED for Soft Error and Variation-Induced Defect Tolerance in Caches. In: Proceedings of the Conference on Design, Automation and Test in Europe. DATE'07. San Jose, CA, USA: EDA Consortium; 2007. p. 1134–1139. Available from: http://dl.acm.org/citation.cfm?id=1266366.1266612.
[20] Sun H, Zheng N, and Zhang T. Leveraging Access Locality for the Efficient Use of Multibit Error-Correcting Codes in L2 Cache. IEEE Transactions on Computers. 2009;58(10):1297–1306.
[21] Alameldeen AR, Wagner I, Chishti Z, *et al.* Energy-Efficient Cache Design Using Variable-Strength Error-Correcting Codes. In: Proceedings of the 38th Annual International Symposium on Computer Architecture. ISCA'11. New York, NY, USA: ACM; 2011. p. 461–472. Available from: http://doi.acm.org/10.1145/2000064.2000118.
[22] Kim J, Hardavellas N, Mai K, *et al.* Multi-bit Error Tolerant Caches Using Two-Dimensional Error Coding. In: Proceedings of the 40th Annual IEEE/ACM International Symposium on Microarchitecture. MICRO 40. Washington, DC, USA: IEEE Computer Society; 2007. p. 197–209. Available from: http://dx.doi.org/10.1109/MICRO.2007.28.
[23] Shrivastava A, Lee J, and Jeyapaul R. Cache Vulnerability Equations for Protecting Data in Embedded Processor Caches from Soft Errors. In: Proceedings of the ACM SIGPLAN/SIGBED 2010 Conference on Languages, Compilers, and Tools for Embedded Systems. LCTES'10. New York, NY, USA: Association for Computing Machinery; 2010. p. 143–152. Available from: https://doi.org/10.1145/1755888.1755910.
[24] Kaxiras S, Hu Z, and Martonosi M. Cache Decay: Exploiting Generational Behavior to Reduce Cache Leakage Power. ACM SIGARCH Computer Architecture News. 2001 05;29.
[25] Hamming RW. Error Detecting and Error Correcting Codes. The Bell System Technical Journal. 1950;29(2):147–160.
[26] Li JF and Huang YJ. An Error Detection and Correction Scheme for RAMs with Partial-Write Function. In: Proceedings of the 2005 IEEE International

Workshop on Memory Technology, Design, and Testing. MTDT'05. Washington, DC, USA: IEEE Computer Society; 2005. p. 115–120. Available from: http://dx.doi.org/10.1109/MTDT.2005.16.

[27] Phelan R. Addressing Soft Errors in ARM Core-based Designs. White Paper. ARM; 2003.

[28] González A, Aliagas C, and Valero M. A Data Cache with Multiple Caching Strategies Tuned to Different Types of Locality. In: ACM International Conference on Supercomputing 25th Anniversary Volume. New York, NY, USA: Association for Computing Machinery; 1995. p. 217–226. Available from: https://doi.org/10.1145/2591635.2667170.

[29] Inc AMD. BIOS and Kernel Developer's Guide for AMD Athlon 64 and AMD Opteron Processors. 2006;26094.

[30] Tarjan D, Thoziyoor S, and Jouppi NP. CACTI 4.0. Technical Report HPL-2006-86. Palo Alto, CA, USA: HP Laboratories; 2006.

[31] Pekhimenko G, Seshadri V, Mutlu O, *et al.* Base-Delta-Immediate Compression: Practical Data Compression for On-Chip Caches. In: 2012 21st International Conference on Parallel Architectures and Compilation Techniques (PACT). Piscataway, NJ, USA: IEEE; 2012. p. 377–388.

[32] Benso A, Chiusano S, Di Natale G, *et al.* A Family of Self-Repair SRAM Cores. In: Proceedings of the 6th IEEE International On-Line Testing Workshop (Cat. No. PR00646). Piscataway, NJ, USA: IEEE; 2000. p. 214–218.

[33] Schober V, Paul S, and Picot O. Memory Built-In Self-Repair Using Redundant Words. In: Proceedings International Test Conference 2001 (Cat. No. 01CH37260). Piscataway, NJ, USA: IEEE; 2001. p. 995–1001.

[34] Stapper CH and Lee HS. Synergistic Fault-Tolerance for Memory Chips. IEEE Transactions on Computers. 1992;(9):1078–1087.

[35] Su CL, Yeh YT, and Wu CW. An Integrated ECC and Redundancy Repair Scheme for Memory Reliability Enhancement. In: 20th IEEE International Symposium on Defect and Fault Tolerance in VLSI Systems (DFT'05). Piscataway, NJ, USA: IEEE; 2005. p. 81–89.

[36] Chen CL and Hsiao M. Error-Correcting Codes for Semiconductor Memory Applications: A State-of-the-Art Review. IBM Journal of Research and Development. 1984;28(2):124–134.

[37] Chakraborty K and Mazumder P. Fault-Tolerance and Reliability Techniques for High-Density Random-Access Memories. Upper Saddle River, NJ: Prentice Hall PTR; 2002.

[38] Blome JA, Gupta S, Feng S, *et al.* Cost-Efficient Soft Error Protection for Embedded Microprocessors. In: Proceedings of the 2006 International Conference on Compilers, Architecture and Synthesis for Embedded Systems. CASES'06. New York, NY, USA: ACM; 2006. p. 421–431. Available from: http://doi.acm.org/10.1145/1176760.1176811.

[39] Dodd PE and Massengill LW. Basic Mechanisms and Modeling of Single-Event Upset in Digital Microelectronics. IEEE Transactions on Nuclear Science. 2003;50(3):583–602. Available from: http://ieeexplore.ieee.org/xpl/freeabs_all.jsp?arnumber=1208578.

[40] Gaisler J. Evaluation of a 32-Bit Microprocessor With Built-In Concurrent Error-Detection. In: Proceedings of IEEE 27th International Symposium on Fault Tolerant Computing. Piscataway, NJ, USA: IEEE; 1997. p. 42–46.

[41] McNairy C and Bhatia R. Montecito: A Dual-Core, Dual-Thread Itanium Processor. IEEE Micro. 2005;25(2):10–20. Available from: http://dx.doi.org/10.1109/MM.2005.34.

[42] Slegel TJ, Averill III RM, Check MA, *et al.* IBM's S/390 G5 Microprocessor Design. IEEE Micro. 1999;19(2):12–23. Available from: http://dx.doi.org/10.1109/40.755464.

[43] Tremblay M and Tamir Y. Support for Fault Tolerance in VLSI Processors. In: IEEE International Symposium on Circuits and Systems. vol. 1. Piscataway, NJ, USA: IEEE; 1989. p. 388–392.

[44] Montesinos P, Liu W, and Torrellas J. Using Register Lifetime Predictions to Protect Register Files against Soft Errors. In: 37th Annual IEEE/IFIP International Conference on Dependable Systems and Networks (DSN'07). Piscataway, NJ, USA: IEEE; 2007. p. 286–296.

[45] Kandala M, Zhang W, and Yang LT. An Area-Efficient Approach to Improving Register File Reliability against Transient Errors. In: 21st International Conference on Advanced Information Networking and Applications Workshops (AINAW'07). vol. 1. Piscataway, NJ, USA: IEEE; 2007. p. 798–803.

[46] Naseer R, Bhatti RZ, and Draper J. Analysis of Soft Error Mitigation Techniques for Register Files in IBM Cu-08 90nm Technology. In: 2006 49th IEEE International Midwest Symposium on Circuits and Systems. vol. 1. Piscataway, NJ, USA: IEEE; 2006. p. 515–519.

[47] Memik G, Kandemir MT, and Ozturk O. Increasing Register File Immunity to Transient Errors. In: Proceedings of the Conference on Design, Automation and Test in Europe – DATE'05. vol. 1. Washington, DC, USA: IEEE Computer Society; 2005. p. 586–591. Available from: http://dx.doi.org/10.1109/DATE.2005.181.

[48] Yan J and Zhang W. Compiler-Guided Register Reliability Improvement Against Soft Errors. In: Proceedings of the 5th ACM International Conference on Embedded Software. EMSOFT'05. New York, NY, USA: ACM; 2005. p. 203–209. Available from: http://doi.acm.org/10.1145/1086228.1086266.

[49] Lee J and Shrivastava A. A Compiler-Microarchitecture Hybrid Approach to Soft Error Reduction for Register Files. IEEE TCAD: IEEE Transactions on Computer Aided Design. 2010;29(7):1018–1027.

[50] Reinhardt SK and Mukherjee SS. Transient Fault Detection via Simultaneous Multithreading. vol. 28. New York, NY, USA: ACM; 2000.

[51] Mukherjee SS, Kontz M, and Reinhardt SK. Detailed Design and Evaluation of Redundant Multi-threading Alternatives. In: Computer Architecture, 2002. Proceedings of the 29th Annual International Symposium on. Piscataway, NJ, USA: IEEE; 2002. p. 99–110.

[52] Gomaa M, Scarbrough C, Vijaykumar T, *et al.* Transient-Fault Recovery for Chip Multiprocessors. In: Computer Architecture, 2003. Proceedings of the

30th Annual International Symposium on. Piscataway, NJ, USA: IEEE; 2003. p. 98–109.

[53] Rotenberg E. AR-SMT: A Microarchitectural Approach to Fault Tolerance in Microprocessors. In: Fault-Tolerant Computing, 1999. Digest of Papers. Twenty-Ninth Annual International Symposium on. Piscataway, NJ, USA: IEEE; 1999. p. 84–91.

[54] Vijaykumar T, Pomeranz I, and Cheng K. Transient-Fault Recovery Using Simultaneous Multithreading. In: ACM SIGARCH Computer Architecture News. vol. 30. Washington, DC, USA: IEEE Computer Society; 2002. p. 87–98.

[55] Schuchman E and Vijaykumar T. BlackJack: Hard Error Detection With Redundant Threads on SMT. In: Dependable Systems and Networks, 2007. DSN'07. 37th Annual IEEE/IFIP International Conference on. Piscataway, NJ, USA: IEEE; 2007. p. 327–337.

[56] Sundaramoorthy K, Purser Z, and Rotenberg E. Slipstream Processors: Improving Both Performance and Fault Tolerance. ACM SIGPLAN Notices. 2000;35(11):257–268.

[57] Smolens JC, Gold BT, Kim J, *et al.* Fingerprinting: Bounding Soft-Error Detection Latency and Bandwidth. In: ACM SIGPLAN Notices. vol. 39. New York, NY, USA: ACM; 2004. p. 224–234.

[58] Smolens JC, Gold BT, Falsafi B, *et al.* Reunion: Complexity-Effective Multicore Redundancy. In: Proceedings of the 39th Annual IEEE/ACM International Symposium on Microarchitecture. Washington, DC, USA: IEEE Computer Society; 2006. p. 223–234.

[59] LaFrieda C, Ipek E, Martinez JF, *et al.* Utilizing Dynamically Coupled Cores to Form a Resilient Chip Multiprocessor. In: 37th Annual IEEE/IFIP International Conference on Dependable Systems and Networks (DSN'07). Piscataway, NJ, USA: IEEE; 2007. p. 317–326.

[60] Austin TM. DIVA: A Reliable Substrate for Deep Submicron Microarchitecture Design. In: Microarchitecture, 1999. MICRO-32. Proceedings. 32nd Annual International Symposium on. Piscataway, NJ, USA: IEEE; 1999. p. 196–207.

[61] Jeyapaul R, Hong F, Shrivastava A, *et al.* UnSync: A Soft-Error Resilient Redundant Multicore Architecture. In: Proceedings of the International Conference on Parallel Processing (ICPP). New York, NY, USA: ACM; 2011.

[62] Jeyapaul R, Risheekesan A, Shrivastava A, *et al.* UnSync-CMP: Multicore CMP Architecture for Energy Efficient Soft Error Reliability. Transactions on Parallel and Distributed Systems. 2014;25:254–263.

[63] Martnez J, Renau J, Huang M, *et al.* Cherry: Checkpointed Early Resource Recycling in Out-of-Order Microprocessors. In: 35th Annual IEEE/ACM International Symposium on Microarchitecture, 2002. (MICRO-35). Proceedings. MICRO 35. Piscataway, NJ, USA: IEEE; 2002.

# Chapter 4
# Design techniques to improve the resilience of computing systems: software layer

*Alberto Bosio[1], Stefano Di Carlo[2], Giorgio Di Natale[3], Matteo Sonza Reorda[2], and Josie E. Rodriguez Condia[2]*

Hardware techniques to improve the robustness of a computing system can be very expensive, difficult to implement and validate. Moreover, they require long evaluation processes that could lead to the redesign of the hardware itself when reliability requirements are not satisfied. This chapter will cover the software techniques that allow improving the tolerance of the system to hardware faults by acting at software level only. We will cover the recently proposed approaches to detect and correct transient and permanent faults.

## 4.1 Introduction

This chapter presents the reliability issues and solutions targeting the software layer of a computing system. The software layer plays an important role from the system reliability point of view. Indeed, software can either mask or amplify errors thus improving or reducing the overall computing system reliability. This is the main idea behind the concept of Software-Implemented Fault Tolerance (SWIFT) techniques: how to write the software in order to maximize the error-masking effect.

Before moving to the details of software level fault-tolerant techniques, let us first introduce some basic concepts. Figure 4.1 depicts a simple view of a computing system divided into hardware and software layers. From the figure, it is possible to identify the "propagation" of the hardware faults (i.e., Physical faults) through the hardware layers composing the computing system. Some of these faults are masked by hardware layers, while some others reach the software layer. It is interesting to point out that at software level two more source of faults can be identified: the presence of **bugs** and the **misuse** of the software external User Interface (UI). These sources are completely independent of the hardware level.

[1]Lyon Institute of Nanotechnology, École Centrale de Lyon, Lyon, France
[2]Department of Control and Computer Engineering, Politecnico di Torino, Torino, Italy
[3]TIMA Laboratory, CNRS, Grenoble, France

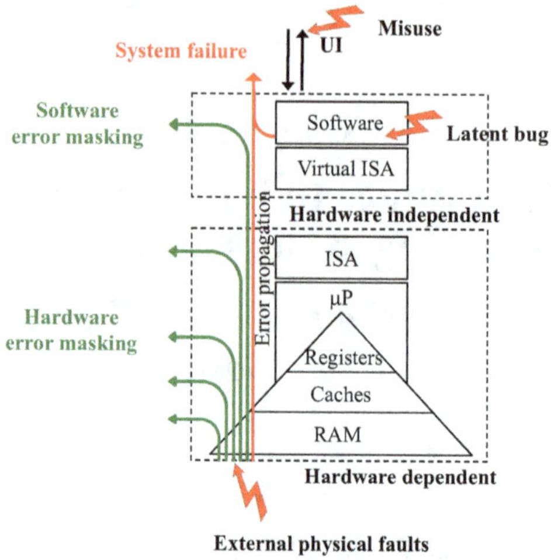


*Figure 4.1 System layers and fault propagation*

The chapter is structured as follows: Section 4.2 presents the taxonomy of the faults affecting the software layer. Section 4.3 reviews the existing Software-Implemented Hardware Fault Tolerance (SIHFT) solutions. Section 4.4 describes the techniques for Software-Based Self-Test (SBST), while Section 4.5 focuses the analysis of SBST solutions for GPUs.

## 4.2 Fault taxonomy

As already pointed out in the introduction, faults affecting software have different sources that can be classified using the following definitions:

- **Design faults**: these faults are introduced during the software implementation. Usually this kind of faults is referred to as *bugs*.
- **Physical faults**: these faults are originated at hardware level and reach the software level through propagation.
- **Interaction faults**: these faults are due by interaction between the software level and the external environment.

Independently from the earlier sources, faults can be further classified by the following characteristics:

- **Intent**: the fault can be intentionally or not introduced into the software. In the first case, the intent is to obtain a malfunction of the software, and in the literature the term *malicious* faults is usually adopted. In the second case, the term *non-malicious* fault is adopted [1].
- **Nature**: the fault is defined as *Permanent* if it is always present. The fault is defined as *Transient* if it appears at a certain time and then disappear.

- **Effect**: the fault effect can impact a single or multiple locations. The location can be either a variable or an instruction.
- **Impact**: the fault can have different impacts at application level:
  - *Hang*: the application does not terminate within a reasonable time interval. The time interval depends on the application itself.
  - *Silent Data Corruption (SDC)*: the output of application has been corrupted.
  - *Data Undetected Error (DUE)*: an unexpected exception, assertion, or segmentation fault, deadlock or interrupt occurred.
  - *Masked*: no mismatch at the application output.

Table 4.1 presents the fault taxonomy through a fault source/characteristic matrix. Each row corresponds to one fault characteristic, while each column corresponds to fault source. As it can be seen, independently from the fault source, all the fault characteristics have to be considered. For example, a design fault can be malicious if the software code has been intentionally modified to introduce the fault itself. The latter can be a Trojan or a virus [2,3]. In the same way, an interaction fault can be malicious too, and in this case we have to consider the case of intentional misuse that typically occurs during an *attack* [4]. Malicious Physical faults have been intentional introduced into the hardware level of the system (e.g., Trojan) [5].

The next subsection will present how faults are modeled at software level.

### *4.2.1 Software faults*

Table 4.2 reports a detailed list of software fault models induced by hardware faults. They can be grouped into three main categories:

- **Data fault models**: they enable to model faults corrupting data processed by a software application. They include (i) Wrong Data in an Operand, (ii) Not-accessible Operand and (iii) Operand Forced Switch;
- **Code fault models**: they enable one to model faults that corrupt the set of instructions composing a program. They include (i) Instruction Replacement, (ii) Faulty Instruction and (iii) Control Flow Error.
- **System fault models**: they enable one to model both timing faults and communication/synchronization faults during the software execution. They include (i) External Peripheral Communication Error, Signaling Error, Execution timing Error and Synchronization Error.

*Table 4.1 Software fault taxonomy/characteristic matrix*

| | **Design fault** | **Interaction fault** | **Physical fault** |
|---|---|---|---|
| Intent | | Malicious, non-malicious | |
| Nature | Permanent | Permanent, transient | Permanent |
| Effect | | Single, multiple | |
| Impact | | Hang, SDC, DUE, masked | |

Figure 4.2 shows an example of the earlier fault modeling. It represents the multiplication instruction as specified in the ARM Instruction Set Architecture (ISA) [6]. We will consider the case of three different faults (F1, F2 and F3) affecting different locations in different time. F1 affects the portion of the instruction responsible for the encoding of the destination register (Rd). Due to F1, it is possible that the Rd changes so that the result will be stored in a different register w.r.t. the fault-free one. This case is modeled by the Data fault model and more specifically by the Source Operand Forced Switch model. F2 affects the instruction opcode. This case may lead to a different opcode and thus the microprocessor decodes the faulty instruction as a different one w.r.t. the fault-free one. This case is modeled by the Code fault models

*Table 4.2 Software fault models*

| Software fault model | Description |
|---|---|
| Wrong Data in an Operand | An operand of the ISA instruction changes its value |
| Not-Accessible Operand | An operand of the ISA instruction cannot change its value |
| Source Operand Forced Switch | An operand is used in place of another |
| Instruction Replacement | An instruction is used in place of another |
| Faulty Instruction | The instruction is executed incorrectly |
| Control Flow Error | The control flow is not respected (control-flow faults) |
| External Peripheral Communication Error | An input value (from a peripheral) is corrupted or not arriving |
| Signaling Error | An internal signaling (exception, interrupt, etc.) is wrongly raised or suppressed |
| Execution Timing Error | An error in the timing management (e.g., PLL) interferes with the correct execution timing |
| Synchronization Error | An error in the scheduling processes causes an incoherent synchronization of processes/tasks |

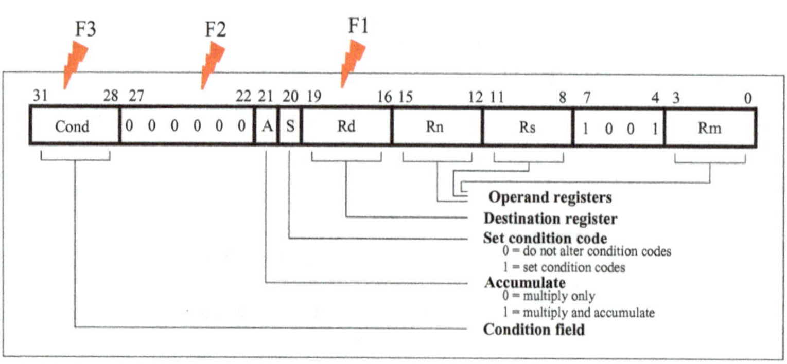


*Figure 4.2 Fault modeling example*

and more specifically by the Instruction Replacement. Finally, F3 affects the condition flags (cond) of the instruction. This case may lead to have an erroneous flag thus impacting the control flow of the program. This case is modeled by the Code fault models and more specifically by the Control Flow Error.

## 4.3 Software-Implemented Hardware Fault Tolerance

The concept of Commercial Off-The Shelf (COTS) hardware and software components has been introduced in safety-critical applications. These components guarantee high performance at the price of a low dependability. Since COTS hardware cannot be modified to introduce fault-tolerant mechanisms, the only possibility is to protect systems acting at the software layer. More in particular, the low-cost solution is to take advantage of SIHFT techniques that allow, by only using software, to detect and correct errors affecting the hardware.

SIHFT techniques are, in general, based on the addition, to the original target application, of software routines able to check the validity and correctness of the executed code and the managed data [7]. This section presents recent SIHFT techniques to guarantee the correct behavior of the system, even in the presence of hardware faults. Most of the existing solutions are inspired by equivalent solutions implemented in hardware but then adapted in software so that their cost is reduced. These techniques can be classified into two main categories:

1. techniques that modify the software in order to reduce the probability of fault occurrences;
2. techniques that allow detecting/tolerating the presence of an error: mainly based on redundancy, control flow integrity, checkpoints/rollbacks and the so-called Algorithmic-Based Fault Tolerance (ABFT).

The following subsections will detail each of the previous categories.

### *4.3.1 Modify the software in order to reduce the probability of fault occurrences*

These techniques mainly aim at modifying the code source in order to use in a more smart way the hardware resources. The ultimate goal is to reduce the probability that a fault-affecting hardware resources will propagate to the software.

Let us resort to an example depicted in the assembly code of Listing 4.1. The reader can notice that register *r*0 is written at line 2 and then read at line 5. This means that the *lifetime* of such a register corresponds to three cycles.* The point here is that higher the lifetime higher the exposure time and thus higher the probability to observe a single event upset. By simply changing the code source, it is possible to minimize the lifetime and thus reduce the probability of observing a fault at hardware

*For the sake of simplicity, we consider that each instruction needs one clock cycle to be executed.

level. The code shown in Listing 4.2 provides a lower fault probability because the lifetime of $r0$ has been reduced to 1 w.r.t. to the first code.

In [8–10] works, the main idea is to perform instruction rescheduling (after the performance-optimized scheduling) to reduce the vulnerable periods of registers. The main drawback of such approaches is that register file covers only a small portion of the processor layout. As a result, these techniques provide limited reliability improvement (from 2% to 9%) w.r.t. to a normal code.

### *4.3.2 Detecting/tolerating the presence of an error*

$N$-Version programing is probably the most applied software level fault diversity technique [11]. The idea behind $N$-version programing is the development of $N$ implementations of the same software application (with $N > 2$) by an independent development team. These versions are all functionally equivalent, i.e., they implement the same functionalities, but given the different instruction flow, they expose different failure characteristics that increase the likelihood that not all versions fail at the same time in a specific fault scenario. $N$-version programing can be coupled with redundancy techniques such as Dual Modular Redundancy and Triple Modular Redundancy.

When recovery from failures is a key point, software-based checkpoint recovery techniques are an interesting solution [12]. The overall idea when implementing checkpointing is to modify the software by inserting checkpoint instructions inside the code. A good practice to decide where checkpoint instructions must be inserted is to identify instructions with high error probability and to place checkpoints just before these instructions. Inserting a checkpoint means inserting calls to proper routines able to save the state of the program in a reliable storage area. In the case of failure, the program execution can be restarted from a safe state by restoring the latest

```
mov r0, @a
inc r0    ; r0 write
mov r1,@b
add r2, r1
mov @a, r0 ; r0 read
```

*Listing 4.1 Asm example*

```
mov r0, @a
inc r0    ; r0 write
mov @a, r0 ; r0 read
mov r1,@b
add r2, r1
```

*Listing 4.2 Asm example: reduced lifetime*

saved checkpoint. The literature is the reach of software-based checkpointing techniques. Compilers can be modified in order to assist the insertion of checkpoints as proposed in [13] that propose the use of an adaptive scheme to minimize the storage overhead required to save the checkpoints. Software libraries such as Libckpt [14] and libFT [15] have been released to the developers to facilitate the task of dumping the state of a program when performing checkpointing. However, full checkpointing automation is still not supported. At the Operating System (OS) level, [16] proposes a loadable kernel module for providing application-aware reliability and dynamically configuring reliability mechanisms. The module is implemented in Linux and supports the detection of application/OS failures and transparent application checkpointing.

Besides diversity and checkpointing, several software and compiler level techniques propose protection schemes based on data redundancy and control flow checking.

Data error detection using Duplicated Instructions [17] and SWIFT [18] and REliable Code COmpliler [10] represents the most famous software redundancy techniques based on duplicated instructions followed by checkpoint instructions able to compare the result of the two executions usually placed before store and/or conditional branches. These techniques generate a significant performance and memory overhead due to redundant instruction execution and shadow memory locations to store redundant data, respectively. Performance overhead can also be aggravated by the increased cache usage to hold redundant data for computation of original and duplicated instructions, generating additional memory traffic.

Control flow checking techniques instead aim at verifying that the control flow of the application is properly respected during the execution. The program is usually split into elementary blocks of instructions with a single entry and a single exit, usually referred to as basic blocks. A reference signature representing the correct execution flow in the blocks is calculated off-line and stored. At run-time the same signature is calculated again and compared with the golden one. Software-based control flow checking techniques insert appropriate instructions to compute the execution signature at run-time.

Different techniques such as Block Signature Self Checking [19], Control Checking with Assertions [20], Control-Flow Checking via regular expressions [21] and Control Flow Checking by Software Signatures [22] have been presented so far.

Similar to data redundancy techniques, also control-flow checking techniques may introduce significant overhead in the software execution associated with the tasks of computing and checking the software signatures.

A completely different approach is to implement fault tolerance techniques at the algorithm level by exploiting the characteristic of specific computations implementing the so-called ABFT [23]. Figure 4.3 shows a simple example of ABFT application. The algorithm is the matrix multiplication. Here it is possible to exploit the property of the algorithm in order to add an extra row and column of the two matrices. These extra elements contain a kind of code (in the example is the sum of the elements). After the multiplication, the extra row and column will satisfy the same property. It is thus possible to identify which element of the matrix has been affected by a fault.

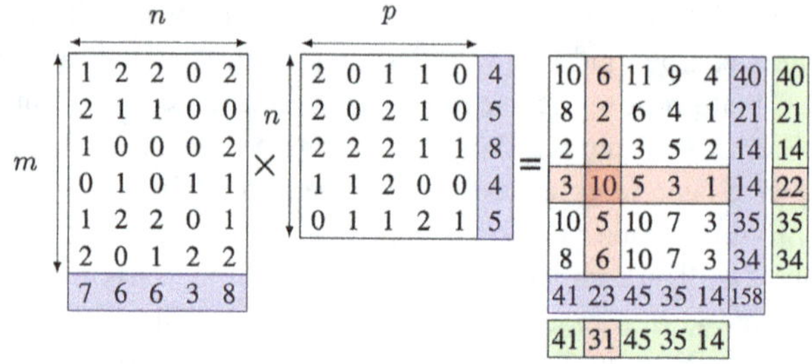


*Figure 4.3 ABFT example*

## 4.4 Software-Based Self-Test

In the last decades, electronic systems have been increasingly used in safety-critical applications, where the effects of possible faults affecting the hardware may have severe consequences. Several solutions were introduced to early detect the presence of permanent faults, or to mitigate their effects. The latter are based on (hardware, information, time) redundancy, the former involves in-field test, which allows detecting possible permanent faults arising during the operational phase (e.g., due to aging) before they cause serious consequences. In this way, the resulting failure probability can be decreased. In-field test of an embedded system can be performed in different ways depending on the considered scenario and the required reliability targets. In some cases, in-field test is performed at the system power-on, before the real application is started. In other cases, it is performed periodically, often exploiting the idle times of the application. Alternatively, the in-field test is performed concurrently to the application, e.g., by monitoring the produced results. In any case, in-field test must take the form of self-test, since no support from the outside can be provided, and must minimize the intrusiveness with respect to the resources used by the application. It is worth emphasizing that the activation frequency of in-field test and the fault coverage it must achieve are higher when semiconductor technologies with lower reliability are used. Since the latest technologies are known to be less reliable, their adoption in safety-critical systems makes the constraints on in-field test even harder. Different alternatives exist to implement effective in-field test solutions in a device used for a safety-critical application. If the device is specifically developed for that application, and hardware overhead constraints allow for that, solutions based on Design for Testability (DfT), e.g., Logic BIST, can be successfully exploited. This approach has several advantages, including a good support from commercial EDA tools, the ability to reach a high fault coverage (at least for static faults), and the possibility to reuse at least some of the hardware infrastructures already used for end-of-manufacturing test. On the other side, its main drawback lies in the fact that the required DfT hardware must be introduced early in the design flow, and the mechanism for its access from the outside must be agreed between the semiconductor company producing the device and the system company managing the in-field test. Moreover, when each activation of the in-field tests must fit into relatively short-time slots, this solution may not be suitable. As an alternative, in the last years several semiconductor and IP companies, including Infineon [24], STMicroelectronics [25], Renesas [26], Cypress [27], Microchip [28], ARM [29], started adopting a solution based on the so-called Self-Test Libraries (STLs). The idea is to use the CPU existing in most of the considered devices and to develop a set of procedures, whose execution can be easily triggered by the application software or by the OS (if any). When executed, these procedures perform a suitable sequence of operations, able to trigger possible permanent faults in the CPU or in other modules and to produce results that can reveal the existence of the faults. This approach is known in the literature as SBST [30]. Since STLs are developed by the semiconductor or IP companies, which know the structure of the hardware, the fault coverage (e.g., in terms of stuck-at faults) that STLs can achieve can be computed via fault simulation. This approach provides a nice compromise between the requirements of semiconductor and IP companies, which want to preserve the property of their hardware but must provide a flexible and effective solution for its test, and those of the system companies, which must test in-the-field the different devices composing their systems to achieve a given reliability or safety. As a further advantage, this approach performs a test of the whole device while it is operating in the same conditions of the application and can thus detect defects (e.g., delay or interconnection defects) that can hardly be caught by the DfT-based solutions. Finally, it is worth mentioning the fact that being based on the execution of a piece of code, a test based on SBST can be easily changed during the product life, e.g., to target new defects. On the other side, the major limitation of the solution based on STLs lies in the cost for their development, since this activity must be done manually without very limited support by EDA tools.

### *4.4.1 Basics on SBST*

The idea of using a piece of code to test a CPU was first proposed several decades ago [31] to face the scenario in which the CPU was a simple processor and its ISA and basic architecture were known only. More recently, the same idea was exploited to support end-of-production test of high-end processors, with the main goal of avoiding the usage of expensive high-frequency testers [30]. A similar approach found applications in industry to support silicon debug and speed binning [32]. A comprehensive overview about the usage of SBST for end-of-manufacturing CPU testing can be found in [33]. Similar solutions were also explored for testing communication peripheral components [34], system peripherals [35] and on-chip memories [36], including caches [37]. The growing interest toward in-field test pushed researchers to analyze how SBST could be effectively used in that domain. In principle, SBST has several nice properties, as mentioned earlier. However, some key points must be faced, which are not relevant when SBST is used for end-of-production test, such as (i) how to trigger the execution of each test procedure, (ii) how to retrieve the results, (iii) how to limit the invasiveness of each test procedure while still maintaining the final fault coverage, (iv) how to limit the duration of each test procedure to the maximum allowed time, (v) how to write the test code such that it complies with the coding stiles and rules that are valid for the application software. Some first analysis of the issues connected to the usage of SBST for on-line test, together with some first solutions are reported in [38]. In [39] some examples of solutions adopted on real test cases from industry are reported, while algorithms for automatically compacting existing test programs to reduce their size [40] or duration [41] have been recently developed. Finally, the work in [42] shows that formal techniques can be successfully used to automate the generation of test programs to be used for in-field test of pipelined processors. Current challenges in the area of SBST include the techniques for developing and optimizing STLs for multicore systems, and the solutions for addressing special categories of faults, such as the performance faults, i.e., those faults that only impact on the performance of a system, while still producing correct result values [43]. Extension of SBST techniques to special types of computing elements, such as VLIW processors [44] or GPGPUs [45] is also a hot topic at the moment.

## 4.5 SBST for GPGPUs

This section first summarizes the state of the art in terms of SBST solutions for permanent faults in GPGPUs. Then it shows that the potential effects of permanent faults in critical units of GPGPUs may become relevant. Finally, some SBST techniques are introduced to detect those faults.

### *4.5.1 Introduction*

GPGPUs are an effective solution to speed up massively parallel computation and are mainly employed as accelerators in highly data-intensive applications such as video, image and multi-signals processing, due to their powerful parallel architecture. Nowadays, these devices are promising solutions for new low-energy, real-time and high-performance applications with safety-critical requirements, such as autonomous automotive drivers and autonomous industrial machines [46]. As commented below, in order to match the requirements for these applications, these devices are designed using aggressive technology scaling techniques, thus increasing the fault-rate across the operational lifetime, mainly because these devices are prone to internal and external sensitive effects, such as aging and radiation [47–49]. Moreover, traditional end-of-manufacturing test solutions cannot guarantee the correct operation and unexpected misbehaviors could arise in the application. When considering system integration companies developing GPGPU-based solutions in safety-critical domains, a critical issue is that these companies often do not have detailed knowledge of the implementation of the adopted GPGPU devices. In this context, functional test techniques represent a viable solution to guarantee the correct in-field operation. This issue becomes critical when a product for safety-critical environments should follow industrial standards, such as the ISO 26262 for the automotive applications. The adoption of SBST techniques for GPGPU devices is feasible in such a scenario, although the cost of developing effective test solutions for such complex devices, including large numbers of parallel execution units, may be challenging. By principle, SBST techniques introduce zero hardware overhead. However, restrictions of execution cycles, resource overhead and power consumption should be considered. Moreover, GPGPUs are mainly special purpose processors and potentially all previous SBST solutions for single-core processor devices can be adapted to these parallel devices. Some SBST solutions for GPGPUs have been proposed in the past. Some of them focus on data-path modules, including the register files and the execution units [45] using adaptations of well-known SBST programs for single and multicore processors. Other works [50] employed internal thread identifiers to schedule tasks in the GPGPU, avoiding corrupted units and mitigating errors in the application. In [51], the authors proposed new mitigations strategies to face permanent faults in the processing core units, or Streaming Multiprocessors (SMs) (in Nvidia's terminology) employing a reverse engineering approach for the block scheduler policies and distributing the application blocks across the fault-free units. Finally, the work in [52] analyzed the fault sensibility and its relation with the employed sub-modules and program description.

### *4.5.2 Effects of permanent faults in GPGPU devices*

One important cause of permanent faults in GPGPU devices lies in the aging effects damaging the hardware integrity. In most multimedia applications, a fault located in the data-path can generate errors in the output. Nevertheless, some of these could be tolerated due to the graphical nature of the application (they only produce slight degradation of the image quality, which is often even difficult to detect). On the other hand, a fault located in a control-unit can generate severe consequences for the running application. When hitting a sensitive location, a fault could generate execution hanging or thread execution missing. In order to present an example showing the effects of permanent faults affecting the control logic of a GPGPU device, we considered an image preprocessing application (edge-detection) and performed some experiments resorting to the GPGPU-SIM simulator [53]. In this case study, a permanent fault is injected in a memory cell of the scheduler. The fault prevents the execution of a particular thread in the program kernel. During the execution of the GPGPU program, the affected thread can partially damage the neighbor thread results introducing errors. Results are graphically visible in the produced image (Figure 4.4), as the reader can see, the effects are far from being negligible. The previous application shows the impact of one permanent fault in a sensitive location. One or a few faults in a more complex and critical application could produce critical misbehaviors compromising the entire execution. SBST solutions can be applied to detect permanent faults in special purpose units of a GPGPU. The next subsection introduces some strategies applied to control units in GPGPU devices.

### *4.5.3 SBST techniques for testing the GPGPU scheduler*

Developing effective SBST procedures for control-path modules in processor-based systems is not a trivial task. This is true also for GPGPU-based systems. The warp

*Figure 4.4 Effects of one permanent fault in the control unit of a GPGPU. Original fault-free gray-scale image (left), fault-free edge-detection output image (top-right) and faulty edge-detection output image (bottom-right).*

scheduler is one of the control-path modules, and it is a critical unit for the GPGPU operation. This unit manages the parallel execution of multiple threads inside the SM, and the detection of permanent faults in this unit is crucial to avoid the application collapsing. The basic functions of this unit are (i) warp submission, (ii) warp execution checking (this process is done after finishing each instruction by the warp and updating the related information) and (iii) warp termination. A fault in this unit is able to generate critical issues, such as execution hanging, performance degradation and SDC effects. This module includes some sub-modules, such as warp generators, dispatchers and checkers. Additionally, some special purpose memories are included. One of these memories is the status warp pool memory. This memory stores the status information of each warp dispatched to the SM in an entry line. Each entry line is composed of a Thread-Mask (ThMk) field, indicating the number of active threads per warp, the warp program-counter (WPc) field and some other fields. In [54], some approaches to detect faults affecting the warp scheduler based on the SBST solution have been proposed. The authors used the available instructions to design SBST programs targeting permanent faults in the warp pool memory of a GPGPU. These techniques are mainly based on combinations of multiple instructions and clever algorithmic mechanisms to generate the input sequences to the targeted unit in order to make the faulty effects visible. In this work, architectural information about the targeted GPGPU was available, and it was possible to use it to develop a suitable test for each field of each entry line inside the memory. The proposed algorithms are based on a sequence of subroutines to generate stimuli able to write and to read from a specific field inside the warp entry line. The targeted fields in the entry line were the ThMk and WPc. These fields depend on control-flow instructions. ThMk can be written by adding multiple combinations of conditional control-flow instructions in the program kernel. The WPc field can be modified through unconditional control-flow instructions. A major difference with respect to other strategies is the observability mechanism, based on signatures. The method that presents better results implements the test by means of a subroutine, which computes one signature per thread to check its correct execution and hence detect possible faults affecting the ThMk and WPc fields. The subroutine execution generates thread divergence. Moreover, this changes the program counter location. Then, on each path (taken and not-taken) each thread modifies its signature, allowing the fault detection. At the end, the signature is stored in global memory and checked by the host. This strategy takes the advantage of supported instructions and includes zero hardware overhead in the system. A moderate memory overhead is required and the total number of required memory locations is equal to the number of threads per block to be executed by the SM. A comparison between a reference application, a typical embarrassing parallel application (denoted as Basic) and the proposed approach is shown in Figure 4.5. Results show that the proposed approach increases the percentage

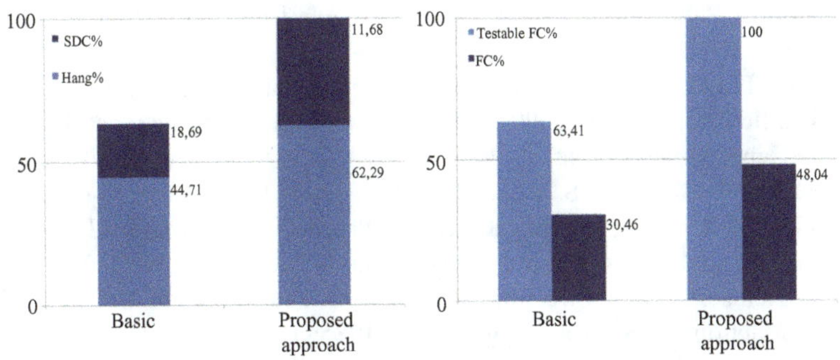


*Figure 4.5 Comparison of fault coverage between a typical parallel application and the proposed method to detect permanent faults in the scheduler warp pool memory of a GPGPU (detailed testable FC (left), testable FC and FC (right))*

of permanent faults detection. The method is able to reach the 100% of detectable fault coverage. The detectable fault coverage corresponds to the total number of faults that can be detected using SBST strategies. On the other hand, the total fault coverage is lower. This difference occurs due to the presence of faults that cannot be tested, e.g., because they relate to unused memory bits.

## References


[1] Avizienis A, Laprie JC, Randell B, *et al.* Basic Concepts and Taxonomy of Dependable and Secure Computing. IEEE Transactions on Dependable and Secure Computing. 2004;1(1):11–33.

[2] Avizienis A and He Y. Microprocessor entomology: A taxonomy of design faults in COTS microprocessors. In: Dependable Computing for Critical Applications 7; 1999. p. 3–23.

[3] Xiao G, Zheng Z, Yin B, *et al.* Experience report: Fault triggers in Linux operating system: From evolution perspective. In: 2017 IEEE 28th International Symposium on Software Reliability Engineering (ISSRE); 2017. p. 101–111.

[4] Singh YN and Singh SK. A Taxonomy of Biometric System Vulnerabilities and Defences. International Journal of Biometrics. 2013;5(2):137–159. Available from: http://dx.doi.org/10.1504/IJBM.2013.052964.

[5] Xiao K, Forte D, Jin Y, *et al.* Hardware Trojans: Lessons Learned After One Decade of Research. ACM Transactions on Design Automation of Electronic Systems. 2016;22(1):6:1–6:23. Available from: http://doi.acm.org/10.1145/2906147.

[6] ARM ISA. Accessed: 2019-06-27. Available from: https://www.arm.com.

[7] Benso A, Di Carlo S, Di Natale G, *et al.* Data criticality estimation in software applications. In: International Test Conference, 2003. Proceedings. ITC 2003. vol. 1; 2003. p. 802–810.

[8] Rehman S, Shafique M, and Henkel J. In: Introduction. Cham: Springer International Publishing; 2016. p. 1–21. Available from: https://doi.org/10.1007/978-3-319-25772-3_1.

[9] Xu J, Tan Q, and Shen R. The instruction scheduling for soft errors based on data flow analysis. In: 2009 15th IEEE Pacific Rim International Symposium on Dependable Computing; 2009. p. 372–378.

[10] Benso A, Chiusano S, Prinetto P, *et al.* A C/C++ source-to-source compiler for dependable applications. In: Proceeding International Conference on Dependable Systems and Networks. DSN 2000; 2000. p. 71–78.

[11] Avizienis A. The N-Version Approach to Fault-Tolerant Software. IEEE Transactions on Software Engineering. 1985;(12):1491–1501.

[12] Koo R and Toueg S. Checkpointing and Rollback-Recovery for Distributed Systems. IEEE Transactions on Software Engineering. 1987;(1):23–31.

[13] Li CC and Fuchs WK. Catch-compiler-assisted techniques for checkpointing. In: Digest of Papers. Fault-Tolerant Computing: 20th International Symposium. IEEE; 1990. p. 74–81.

[14] Plank JS, Beck M, Kingsley G, *et al.* Libckpt: Transparent Checkpointing Under Unix. Computer Science Department; 1994.

[15] Huang Y and Kintala C. Software implemented fault tolerance: Technologies and experience. In: FTCS. vol. 23. IEEE Computer Society Press; 1993. p. 2–9.

[16] Wang L, Kalbarczyk Z, Gu W, *et al.* An OS-level framework for providing application-aware reliability. In: 2006 12th Pacific Rim International Symposium on Dependable Computing (PRDC'06). IEEE; 2006. p. 55–62.

[17] Oh N, Shirvani PP, and McCluskey EJ. Error Detection by Duplicated Instructions in Super-Scalar Processors. IEEE Transactions on Reliability. 2002;51(1):63–75.

[18] Reis GA, Chang J, Vachharajani N, *et al.* Software-Controlled Fault Tolerance. ACM Transactions on Architecture and Code Optimization. 2005;2(4): 366–396. Available from: http://doi.acm.org/10.1145/1113841.1113843.

[19] Miremadi G, Harlsson J, Gunneflo U, *et al.* Two software techniques for on-line error detection. In: Digest of Papers. FTCS-22: The Twenty-Second International Symposium on Fault-Tolerant Computing. IEEE; 1992. p. 328–335.

[20] Alkhalifa Z, Nair VS, Krishnamurthy N, *et al.* Design and Evaluation of System-Level Checks for On-Line Control Flow Error Detection. IEEE Transactions on Parallel and Distributed Systems. 1999;10(6):627–641.

[21] Benso A, Di Carlo S, Di Natale G, *et al.* Control-flow checking via regular expressions. In: Proceedings 10th Asian Test Symposium. IEEE; 2001. p. 299–303.

[22] Oh N, Shirvani PP, and McCluskey EJ. Control-Flow Checking by Software Signatures. IEEE Transactions on Reliability. 2002;51(1):111–122.

[23] Huang K-H and Abraham JA. Algorithm-Based Fault Tolerance for Matrix Operations. IEEE Transactions on Computers. 1984;C-33(6):518–528.

[24] Infineon. 2018. Available from: https://www.hitex.com/software-components/selftest-libraries-safety-libs/pro-sil-safetcore-safetlib/.

[25] STMicroelectronics. AN3307 – Application note. In Guidelines for Obtaining IEC 60335 Class B Certification for any STM32 Application; 2016.

[26] Renesas. 2018. Available from: https://www.renesas.com/en-eu/products/synergy/software/add-ons.html#read.

[27] Cypress. AN204377 FM3 and FM4 Family, IEC61508 SIL2 Self-Test Library; 2017.

[28] Microchip. DS52076A 16-bit CPU Self-Test Library User's Guide; 2012.

[29] ARM. 2018. Available from: https://developer.arm.com/technologies/functional-safety.

[30] Chen L and Dey S. Software-Based Self-Testing Methodology for Processor Cores. IEEE Transactions on Computer-Aided Design of Integrated Circuits and Systems. 2001;20(3):369–380.

[31] Thatte SM and Abraham JA. Test Generation for Microprocessors. IEEE Transactions on Computers. 1980;C-29(6):429–441.

[32] Parvathala P, Maneparambil K, and Lindsay W. FRITS – A microprocessor functional BIST method. In: Proceedings. International Test Conference; 2002. p. 590–598.

[33] Psarakis M, Gizopoulos D, Sanchez E, *et al.* Microprocessor Software-Based Self-Testing. IEEE Design Test of Computers. 2010;27(3):4–19.

[34] Apostolakis A, Gizopoulos D, Psarakis M, *et al.* Test Program Generation for Communication Peripherals in Processor-Based SoC Devices. IEEE Design Test of Computers. 2009;26(2):52–63.

[35] Grosso M, Perez WJH, Ravotto D, *et al.* A software-based self-test methodology for system peripherals. In: 2010 15th IEEE European Test Symposium; 2010. p. 195–200.

[36] van de Goor A, Gaydadjiev G, and Hamdioui S. Memory testing with a RISC microcontroller. In: 2010 Design, Automation Test in Europe Conference Exhibition (DATE 2010); 2010. p. 214–219.

[37] Di Carlo S, Prinetto P, and Savino A. Software-Based Self-Test of Set-Associative Cache Memories. IEEE Transactions on Computers. 2011;60(7):1030–1044.

[38] Paschalis A and Gizopoulos D. Effective Software-Based Self-Test Strategies for On-Line Periodic Testing of Embedded Processors. IEEE Transactions on Computer-Aided Design of Integrated Circuits and Systems. 2005;24(1): 88–99.

[39] Bernardi P, Cantoro R, De Luca S, *et al.* Development Flow for On-Line Core Self-Test of Automotive Microcontrollers. IEEE Transactions on Computers. 2016;65(3):744–754.

[40] Gaudesi M, Reorda MS, and Pomeranz I. On test program compaction. In: 2015 20th IEEE European Test Symposium (ETS); 2015. p. 1–6.

[41] Gaudesi M, Pomeranz I, Reorda MS, *et al.* New Techniques to Reduce the Execution Time of Functional Test Programs. IEEE Transactions on Computers. 2017;66(7):1268–1273.

[42] Riefert A, Cantoro R, Sauer M, *et al.* A Flexible Framework for the Automatic Generation of SBST Programs. IEEE Transactions on Very Large Scale Integration (VLSI) Systems. 2016;24(10):3055–3066.

[43] Sabena D, Reorda MS, and Sterpone L. On the Automatic Generation of Optimized Software-Based Self-Test Programs for VLIW Processors. IEEE Transactions on Very Large Scale Integration (VLSI) Systems. 2014;22(4):813–823.

[44] Hatzimihail M, Psarakis M, Gizopoulos D, *et al.* A methodology for detecting performance faults in microprocessors via performance monitoring hardware. In: 2007 IEEE International Test Conference; 2007. p. 1–10.

[45] Di Carlo S, Gambardella G, Indaco M, *et al.* A software-based self test of CUDA Fermi GPUs. In: 2013 18th IEEE European Test Symposium (ETS); 2013. p. 1–6.

[46] Shi W, Alawieh MB, Li X, *et al.* Algorithm and Hardware Implementation for Visual Perception System in Autonomous Vehicle: A Survey. Integration. 2017;59:148–156. Available from: http://www.sciencedirect.com/science/article/pii/S0167926017303218.

[47] Hamdioui S, Gizopoulos D, Guido G, *et al.* Reliability challenges of real-time systems in forthcoming technology nodes. In: 2013 Design, Automation Test in Europe Conference Exhibition (DATE); 2013. p. 129–134.

[48] Agbo I, Taouil M, Hamdioui S, *et al.* Read path degradation analysis in SRAM. In: 2016 21th IEEE European Test Symposium (ETS); 2016. p. 1–2.

[49] Baumann RC. Radiation-Induced Soft Errors in Advanced Semiconductor Technologies. IEEE Transactions on Device and Materials Reliability. 2005;5(3):305–316.

[50] Defour D and Petit E. A Software Scheduling Solution to Avoid Corrupted Units on GPUs. Journal of Parallel and Distributed Computing. 2016;90–91:1–8. Available from: http://www.sciencedirect.com/science/article/pii/S0743731516000022.

[51] Di Carlo S, Gambardella G, Martella I, *et al.* An improved fault mitigation strategy for CUDA Fermi GPUs. In: Dependable GPU Computing workshop, Dresden; 2014.

[52] Farazmand N, Ubal R, and Kaeli D. Statistical fault injection-based AVF analysis of a GPU architecture. In: IEEE Workshop on Silicon Errors in Logic; 2012.

[53] Bakhoda A, Yuan GL, Fung W, *et al.* Analyzing CUDA workloads using a detailed GPU simulator; 2009. p. 163–174.

[54] Du B, Condia JER, Reorda MS, *et al.* About the functional test of the GPGPU scheduler. In: 2018 IEEE 24th International Symposium on On-Line Testing And Robust System Design (IOLTS); 2018. p. 85–90.

# Chapter 5
# Cross-layer resilience

*Eric Cheng*[1] *and Subhasish Mitra*[1,2]

Resilience to errors in the underlying hardware is a key design objective for a large class of computing systems, from embedded systems all the way to the cloud. Sources of hardware errors include radiation, circuit aging, variability induced by manufacturing and operating conditions, manufacturing test escapes, and early-life failures. Many publications have suggested that *cross-layer resilience*, where multiple error resilience techniques from different layers of the system stack cooperate to achieve cost-effective resilience, is essential for designing cost-effective resilient digital systems. This chapter presents a unique framework to address fundamental cross-layer resilience questions: achieve desired resilience targets at minimal costs (energy, power, execution time, and area) by combining resilience techniques across various layers of the system stack (circuit, logic, architecture, software, and algorithm). This framework systematically explores the large space of comprehensive resilience techniques and their combinations across various layers of the system stack, derives cost-effective solutions that achieve resilience targets at minimal costs, and provides guidelines for the design of new resilience techniques.

## 5.1 Introduction

This chapter addresses the *cross-layer resilience challenge* for designing resilient systems: given a set of resilience techniques at various abstraction layers (circuit, logic, architecture, software, and algorithm), how does one protect a given design from radiation-induced soft errors using (perhaps) a ***combination*** of these techniques, ***across multiple abstraction layers***, such that overall soft error resilience targets are met at minimal costs (energy, power, execution time, and area)? Specific soft error resilience targets include: *Silent Data Corruption* (*SDC*), where an error causes the system to output an incorrect result without error indication; and *Detected but Uncorrected Error* (*DUE*), where an error is detected (e.g., by a resilience technique or a system crash or hang) but is not recovered automatically without user intervention.

[1]Department of Electrical Engineering, Stanford University, Stanford, CA, USA
[2]Department of Computer Science, Stanford University, Stanford, CA, USA

The need for *cross-layer resilience*, where multiple error resilience techniques from different layers of the system stack cooperate to achieve cost-effective error resilience, is articulated in several publications (e.g., [1–7]).

There are numerous publications on error resilience techniques, many of which span multiple abstraction layers. These publications mostly describe specific implementations. Examples include structural integrity checking [8] and its derivatives (mostly spanning architecture and software layers) or the combined use of circuit hardening, Error Detection (ED) (e.g., using logic parity checking and residue codes), and instruction-level retry [9–11] (spanning circuit, logic, and architecture layers). Cross-layer resilience implementations in commercial systems are often based on "designer experience" or "historical practice." There exists no comprehensive framework to systematically address the cross-layer resilience challenge. Creating such a framework is difficult. It must encompass the entire design flow end-to-end, from comprehensive and thorough analysis of various combinations of error resilience techniques all the way to layout-level implementations, such that one can (automatically) determine which resilience technique or combination of techniques (either at the same abstraction layer or across different abstraction layers) should be chosen. However, such a framework is essential in order to answer the following important cross-layer resilience questions:

1. Is cross-layer resilience the best approach for achieving a given resilience target at low cost?
2. Are all cross-layer solutions equally cost-effective? If not, which cross-layer solutions are the best?
3. How do cross-layer choices change depending on application-level energy, latency, and area constraints?
4. How can one create a cross-layer resilience solution that is cost-effective across a wide variety of application workloads?
5. Are there general guidelines for new error resilience techniques to be cost-effective?

CLEAR (Cross-Layer Exploration for Architecting Resilience) is a first of its kind framework [12–16] that addresses the cross-layer resilience challenge. In this chapter, the focus is on the use of CLEAR for resilient systems that operate in the presence of radiation-induced soft errors in terrestrial settings; however, other error sources (e.g., voltage noise) are discussed in [13,14,16].

Although the soft error rate of an Static Random Access Memory (SRAM) cell or a flip-flop stays roughly constant or even decreases over technology generations, the system-level soft error rate increases with increased integration [17–19]. Moreover, soft error rates can increase when lower supply voltages are used to improve energy efficiency [20,21]. This chapter focuses on *flip-flop soft errors* because design techniques to protect them are generally expensive. Coding techniques are routinely used for protecting on-chip memories. Combinational logic circuits are significantly less susceptible to soft errors and do not pose a concern [19,22]. Both Single-Event Upsets (*SEUs*) and Single-Event Multiple Upsets (*SEMUs*) [21,23] are considered. While CLEAR can address soft errors in various digital components of a complex System-on-a-Chip (SoC) (including uncore components [24] and hardware accelerators), a detailed analysis of soft errors in all these components is beyond the scope of this chapter. Hence, this chapter will focus on soft errors in processor cores.

To demonstrate the effectiveness and practicality of CLEAR, an exploration of 586 cross-layer combinations using ten representative ED/correction techniques and four hardware error recovery techniques is conducted. These techniques span various layers of the system stack: circuit, logic, architecture, software, and algorithm (Figure 5.1). This extensive cross-layer exploration encompasses over 9 million flip-flop soft error injections into two diverse processor core architectures (Table 5.1): a simple in-order SPARC Leon3 core (*InO-core*) and a complex super-scalar out-of-order Alpha IVM core (*OoO-core*), across 18 benchmarks: SPECINT2000 [25] and DARPA PERFECT [26]. Such extensive exploration enables conclusive answers to the previous cross-layer resilience questions:

1. For a wide range of error resilience targets, optimized cross-layer combinations can provide low-cost solutions for soft errors.
2. Not all cross-layer solutions are cost-effective.
    i For general-purpose processor cores, a carefully optimized combination of selective circuit-level hardening, logic-level parity checking, and micro-architectural recovery provides a highly effective cross-layer resilience solution. For example, a 50 × SDC improvement (defined in Section 5.2.1) is achieved at 2.1% and 6.1% energy costs for the OoO-cores and InO-cores, respectively. The use of selective circuit-level hardening and logic-level parity checking is guided by a thorough analysis of the effects of soft errors on application benchmarks.
    ii When the application space can be restricted to matrix operations, a cross-layer combination of Algorithm-Based Fault Tolerance (ABFT) correction, selective circuit-level hardening, logic-level parity checking, and micro-architectural recovery can be highly effective. For example, a 50 × SDC improvement is achieved at 1.9% and 3.1% energy costs for the OoO-cores and InO-cores, respectively. But, this approach may not be practical for general-purpose processor cores targeting general applications.
    iii Selective circuit-level hardening, guided by a thorough analysis of the effects of soft errors on application benchmarks, provides a highly effective soft error resilience approach. For example, a 50 × SDC improvement is achieved at 3.1% and 7.3% energy costs for the OoO-cores and InO-cores, respectively.
3. The previous conclusions about cost-effective soft error resilience techniques largely hold across various application characteristics (e.g., latency constraints despite errors in soft real-time applications).
4. Selective circuit-level hardening (and logic-level parity checking) techniques are guided by the analysis of the effects of soft errors on application benchmarks. Hence, one must address the challenge of potential mismatch between application benchmarks vs. applications in the field, especially when targeting high degrees of resilience (e.g., 10× or more SDC improvement). This challenge is overcome using various flavors of circuit-level hardening techniques (details in Section 5.4).
5. Cost-effective resilience approaches discussed previously provide bounds that new soft error resilience techniques must achieve to be competitive. It is, however,

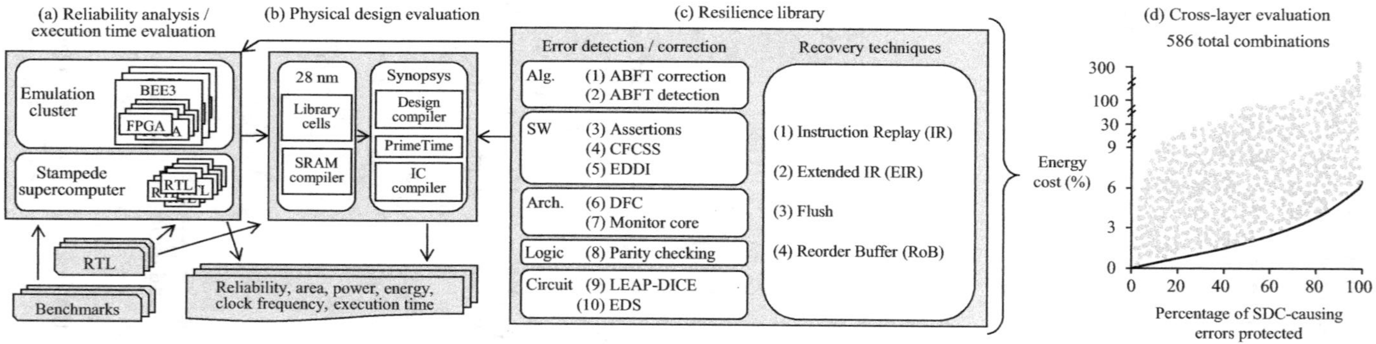


*Figure 5.1* *CLEAR framework: (a) BEE3 emulation cluster/stampede supercomputer injects over 9 million errors into two diverse processor architectures running 18 full-length application benchmarks. (b) Accurate physical design evaluation accounts for resilience overheads. (c) Comprehensive resilience library consisting of ten error detection/correction techniques+four hardware error recovery techniques. (d) Example illustrating thorough exploration of 586 cross-layer combinations with varying energy costs vs. percentage of SDC-causing errors protected.*

*Table 5.1 Processor designs studied*

| Core | Design | Description | Clk. freq. | Error injections | Instructions per cycle |
|---|---|---|---|---|---|
| InO | Leon3 [27] | Simple, in-order (1,250 flip-flops) | 2.0 GHz | 5.9 million | 0.4 |
| OoO | IVM [28] | Complex, super-scalar, out-of-order (13,819 flip-flops) | 600 MHz | 3.5 million | 1.3 |

crucial that the benefits and costs of new techniques are evaluated thoroughly and correctly.

## 5.2 CLEAR framework

Figure 5.1 gives an overview of the CLEAR framework. Individual components of the framework are discussed next.

### *5.2.1 Reliability analysis*

CLEAR is not merely an error-rate-projection tool; rather, reliability analysis is a component of the overall CLEAR framework that helps one to enable the design of resilient systems.

Flip-flop soft error injections are used for reliability analysis with respect to radiation-induced soft errors. This is because radiation test results confirm that injection of single bit-flips into flip-flops closely models soft error behaviors in actual systems [29,30]. Furthermore, flip-flop-level error injection is crucial, since naïve high-level error injections can be highly inaccurate [31]. For individual flip-flops, both SEUs and SEMUs manifest as single-bit errors. To ensure that this assumption holds for both the baseline and resilient designs, it is crucial to make the use of SEMU-tolerant circuit hardening and layout implementations, as is demonstrated here.

The following demonstration is based on injection of over 9 million flip-flop soft errors into the Register-Transfer Level (RTL) of the processor designs using three BEE3 Field-Programmable Gate Array (FPGA) emulation systems and also using mixed-mode simulations on the Stampede supercomputer (TACC at The University of Texas at Austin) (similar to [28,31–33]). This ensures that error injection results have less than a 0.1% margin of error with a 95% confidence interval per benchmark. Errors are injected uniformly into all flip-flops and application regions, to mimic real-world scenarios.

The SPECINT2000 [25] and DARPA PERFECT [26] benchmark suites are used for evaluation.* The PERFECT suite complements SPEC by adding applications targeting signal and image processing domains. Benchmarks are run in their entirety.

*11 SPEC/7 PERFECT benchmarks for InO-cores and 8 SPEC/3 PERFECT for OoO-cores (missing benchmarks contain floating-point instructions not executable by the OoO-core RTL model).

Flip-flop soft errors can result in the following outcomes [28,30,31,34,35]: **Vanished**—normal termination and output files match error-free runs; **Output Mismatch (OMM)**—normal termination, but output files are different from error-free runs; **Unexpected Termination (UT)**—program terminates abnormally; **Hang**—no termination or output within $2\times$ the nominal execution time; **ED**—an employed resilience technique flags an error, but the error is not recovered using a hardware recovery mechanism.

Using the previous outcomes, any error that results in OMM causes SDC (i.e., an *SDC-causing error*). Any error that results in UT, Hang, or ED causes DUE (i.e., a DUE-*causing error*). Note that there are no ED outcomes if no ED technique is employed. The resilience of a protected (new) design compared to an unprotected (original, baseline) design can be defined in terms of *SDC improvement* (5.1) or *DUE improvement* (5.2). The susceptibility of flip-flops to soft errors is assumed to be uniform across all flip-flops in the design (but this parameter is adjustable in our framework).

Resilience techniques that increase the execution time of an application or add additional hardware also increase the susceptibility of the design to soft errors. To accurately account for this situation, one needs to calculate, based on [36], a correction factor $\gamma$ (where $\gamma \geq 1$) that is applied to ensure a fair and accurate comparison for all techniques. It is common in research literature to consider $\gamma = 1$ (which is optimistic). Although all the conclusions in this chapter would hold for $\gamma = 1$, this chapter provides a more accurate analysis by considering and reporting results using true $\gamma$ values. Take for instance the monitor core technique; in a representative implementation, the technique increases the number of flip-flops in a resilient OoO-core by 38%. These extra flip-flops become additional locations for soft errors to occur. This results in a $\gamma$ correction of 1.38 in order to account for the increased susceptibility of the design to soft errors. Techniques that increase execution time have a similar impact. For example, Control Flow Checking by Software Signatures (CFCSS) incurs a 40.6% execution time impact; a corresponding $\gamma$ correction of 1.41. A technique such as Data Flow Checking (DFC), which increases flip-flop count (20%) and execution time (6.2%), would need a $\gamma$ correction of 1.28 ($1.2\times1.062$), since the impact is multiplicative (increased flip-flop count over an increased duration). The $\gamma$ correction factor accounts for these increased susceptibilities for fair and accurate comparisons of all resilience techniques considered [36]. SDC and DUE improvements with $\gamma = 1$ can be back calculated by multiplying the reported $\gamma$ value in Table 5.3 and do not change the conclusions.

$$\text{SDC improvement} = \frac{(\text{original OMM count})}{(\text{new OMM count})} \times \gamma^{-1} \tag{5.1}$$

$$\text{DUE improvement} = \frac{(\text{original (UT + Hang) count})}{(\text{new (UT + Hang + ED) count})} \times \gamma^{-1} \tag{5.2}$$

Reporting SDC and DUE improvements allows results to be agnostic to absolute error rates. Although this chapter describes the use of error-injection-driven reliability analysis, the modular nature of CLEAR allows other approaches to be swapped in as appropriate (e.g., error injection analysis could be substituted with techniques like [37], once they are properly validated).

As shown in Table 5.2, across the set of applications, not all flip-flops will have errors that result in SDC or DUE (errors in 19% of flip-flops in the InO-core and 39% of flip-flops in the OoO-core always vanish regardless of the application). This phenomenon has been documented in the literature [38] and is due to the fact that errors that impact certain structures (e.g., branch predictor and trap status registers) have no effect on program execution or correctness. Additionally, this means that resilience techniques would not normally need to be applied to these flip-flops. However, for completeness, this chapter also reports design points that would achieve the maximum improvement possible, where resilience is added to every single flip-flop (including those with errors that always vanish). This maximum improvement point provides an upper bound for cost (given the possibility that for a future application, a flip-flop that currently has errors that always vanish may encounter an SDC-causing or DUE-causing error).

*ED Latency* (the time elapsed from when an error occurs in the processor to when a resilience technique detects the error) is also an important aspect to consider. An end-to-end reliable system must not only detect errors but also recover from these detected errors. Long detection latencies impact the amount of computation that needs to be recovered and can also limit the types of recovery that are capable of recovering the detected error (Section 5.2.4).

### *5.2.2 Execution time*

Execution time is estimated using FPGA emulation and RTL simulation. Applications are run to completion to accurately capture the execution time of an unprotected design. The error-free execution time impact associated with resilience techniques at the architecture, software, and algorithm levels is reported. For resilience techniques at the circuit and logic levels, the CLEAR design methodology maintains the same clock speed as the unprotected design.

### *5.2.3 Physical design*

Synopsys design tools (Design Compiler, IC compiler, and Primetime) [39] with a commercial 28 nm technology library (with corresponding SRAM compiler) are used to perform Synthesis, Place-and-Route (SP&R), and power analysis. *SP&R* was run for all configurations of the design (before and after adding resilience techniques) to ensure that all constraints of the original design (e.g., timing and physical design) were met for the resilient designs. Design tools often introduce artifacts (e.g., slight

*Table 5.2 Distribution of flip-flops with errors resulting in SDC and/or DUE over all benchmarks studied*

| Core | Percentage of FFs with SDC-causing errors | Percentage of FFs with DUE-causing errors | Percentage of FFs with both SDC-causing and DUE-causing errors |
|---|---|---|---|
| InO | 60.1 | 78.3 | 81.2 |
| OoO | 35.7 | 52.1 | 61 |

variations in the final design over multiple SP&R runs) that impact the final design characteristics (e.g., area and power). These artifacts can be caused by small variations in the RTL or optimization heuristics, for example. To account for these artifacts, separate resilient designs based on error injection results for each individual application benchmark are generated. SP&R was then performed for each of these designs, and the reported design characteristics were averaged to minimize the artifacts. For example, for each of the 18 application benchmarks, a separate resilient design that achieves a 50× SDC improvement using LEAP-Dual Interlocked storage Cell (DICE) only is created. The costs to achieve this improvement are reported by averaging across the 18 designs. Relative standard deviation (i.e., standard deviation/mean) across all experiments ranges from 0.6% to 3.1%. Finally, it is important to note that all layouts created during physical design are carefully generated in order to mitigate the impact of SEMUs (as explained in Section 5.2.4).

### *5.2.4 Resilience library*

This chapter considers ten carefully chosen ED and correction techniques together with four hardware error recovery techniques. These techniques largely cover the space of existing soft error resilience techniques. The characteristics (e.g., costs and resilience improvement) of each resilience technique when used as a *stand-alone solution* (e.g., an ED/correction technique by itself or, optionally, in conjunction with a recovery technique) are presented in Table 5.3.

**Circuit**: The *hardened flip-flops* (LEAP-DICE, Light Hardened LEAP (LHL), LEAP-ctrl) in Table 5.4 are designed to tolerate both SEUs and SEMUs at both nominal and near-threshold operating voltages [23,40]. SEMUs especially impact circuit techniques since a single strike affects multiple nodes within a flip-flop. Thus, these specially designed hardened flip-flops, which tolerate SEMUs through charge cancellation, are required. These hardened flip-flops have been experimentally validated using radiation experiments on test chips fabricated in 90, 45, 40, 32, 28, 20, and 14 nm nodes in both bulk and SOI technologies and can be incorporated into standard cell libraries (i.e., standard cell design guidelines are satisfied) [23,40–44]. The LEAP-ctrl flip-flop is a special design that can operate in resilient (high resilience, high power) and economy (low resilience, low power) modes. It is useful in situations where a software or algorithm technique only provides protection when running specific applications and thus, selectively enabling low-level hardware resilience when the former techniques are unavailable may be beneficial. While *ED Sequential* (*EDS*) [45,46] was originally designed to detect timing errors, it can be used to detect flip-flop soft errors as well. While EDS incurs less overhead at the individual flip-flop level vs. LEAP-DICE, for example, EDS requires delay buffers to ensure minimum hold constraints, aggregation, and routing of ED signals to an output (or recovery module), and a recovery mechanism to correct detected errors. These factors can significantly increase the overall costs for implementing a resilient design utilizing EDS (Table 5.17).

**Logic**: *Parity checking* provides ED by checking flip-flop inputs and outputs [49]. The design heuristics presented in this chapter reduce the cost of parity while also

*Table 5.3 Individual resilience techniques: costs and improvements when implemented as a standalone solution*

| Layer | Technique | | Area cost (%) | Power cost (%) | Energy cost (%) | Exec. time impact (%) | Avg. SDC improve | Avg. DUE improve | False positive (%) | Detection latency | $\gamma$ |
|---|---|---|---|---|---|---|---|---|---|---|---|
| Circuit[a] | LEAP-DICE (no additional recovery needed) | InO | 0–9.3 | 0–22.4 | 0–22.4 | 0 | 1×–5,000×[b] | 1×–5,000×[b] | 0 | n/a | 0 |
| | | OoO | 0–6.5 | 0–9.4 | 0–9.4 | | | | | | |
| | EDS (without recovery–unconstrained) | InO | 0–10.7 | 0–22.9 | 0–22.9 | 0 | 1×–100,000×[b] | 0.1×[b]–1× | 0 | 1 cycle | 0 |
| | | OoO | 0–12.2 | 0–11.5 | 0–11.5 | | | | | | |
| | EDS (with IR recovery) | InO | 0–16.7 | 0–43.9 | 0–43.9 | 0 | 1×–100,000×[b] | 1×–100,000×[b] | 0 | 1 cycle | 1.4 |
| | | OoO | 0–12.3 | 0–11.6 | 0–11.6 | | | | | | 1.06 |
| Logic[a] | Parity (without recovery –unconstrained) | InO | 0–10.9 | 0–23.1 | 0–23.1 | 0 | 1×–100,000×[b] | 0.1×[e]–1× | 0 | 1 cycle | 0 |
| | | OoO | 0–14.1 | 0–13.6 | 0–13.6 | | | | | | |
| | Parity (with IR recovery) | InO | 0–26.9 | 0–44 | 0–44 | 0 | 1×–100,000×[b] | 1×–100,000×[b] | 0 | 1 cycle | 1.4 |
| | | OoO | 0–14.2 | 0–13.7 | 0–13.7 | | | | | | 1.06 |
| Arch. | DFC (without recovery–unconstrained) | InO | 3 | 1 | 7.3 | 6.2 | 1.2× | 0.5× | 0 | 15 cycles | 1.28 |
| | | OoO | 0.2 | 0.1 | 7.2 | 7.1 | | | | | 1.09 |
| | DFC (with EIR recovery) | InO | 37 | 33 | 41.2 | 6.2 | 1.2× | 1.4× | 0 | 15 cycles | 1.48 |
| | | OoO | 0.4 | 0.2 | 7.3 | 7.1 | | | | | 1.14 |
| | Monitor core (with RoB recovery) | OoO | 9 | 16.3 | 16.3 | 0 | 19× | 15× | 0 | 128 cycles | 1.38 |
| Software[c] | Software assertions for general-purpose processors (without recovery—unconstrained) | InO | 0 | 0 | 15.6 | 15.6[d] | 1.5×[e] | 0.6× | 0.003 | 9.3M cycles[f] | 1.16 |
| | CFCSS (without recovery–unconstrained) | InO | 0 | 0 | 40.6 | 40.6 | 1.5× | 0.5× | 0 | 6.2M cycles[f] | 1.41 |
| | EDDI (without recovery–unconstrained) | InO | 0 | 0 | 110 | 110 | 37.8×[g] | 0.3× | 0 | 287k cycles[f] | 2.1 |

*Table 5.3 (Continued)*

| Layer | Technique | | Area cost (%) | Power cost (%) | Energy cost (%) | Exec. time impact (%) | Avg. SDC improve | Avg. DUE improve | False positive (%) | Detection latency | $\gamma$ |
|---|---|---|---|---|---|---|---|---|---|---|---|
| Alg. | ABFT correction (no additional recovery needed) | InO<br>OoO | 0 | 0 | 1.4 | 1.4 | 4.3× | 1.2× | 0 | n/a | 1.01 |
| | ABFT detection (without recovery–unconstrained) | InO<br>OoO | 0 | 0 | 24 | 1–56.9[h] | 3.5× | 0.5× | 0 | 9.6M cycles[f] | 1.24 |

[a]Circuit and logic techniques have tunable costs/resilience (e.g., for InO-cores, 5× SDC improvement using LEAP-DICE is achieved at 4.3% energy cost, while 50× SDC improvement is achieved at 7.3% energy cost). This is achievable through selective insertion guided by error injection using application benchmarks.

[b]Maximum improvement reported is achieved by protecting every single flip-flop in the design.

[c]Software techniques are generated for InO-cores only since the LLVM compiler no longer supports the alpha architecture.

[d]Some software assertions for general-purpose processors (e.g., [47]) suffer from false positives (i.e., an error is reported during an error-free run). The execution time impact reported discounts the impact of false positives.

[e]Improvements differ from previous publications that injected errors into architectural registers. Reference [31] demonstrated that such injections can be highly inaccurate; the CLEAR methodology presented in this chapter used highly accurate flip-flop-level error injections.

[f]Actual detection latency for software and algorithm techniques may be shorter in practice. On the emulation platforms, measured detection latencies include the time to trap and cleanly exit execution (on the order of a few thousand cycles).

[g]Results for EDDI with store-readback [48] are reported. Without this enhancement, EDDI provides 3.3× SDC/0.4× DUE improvement.

[h]Execution time impact for ABFT detection can be high, since error detection checks may require computationally expensive calculations.

ensuring that clock frequency is maintained as in the original design (by varying the number of flip-flops checked together, grouping flip-flops by timing slack, pipelining parity checker logic, etc.). Naïve implementations of parity checking can otherwise degrade design frequency by up to 200 MHz (20%) or increase energy cost by 80% on the InO-core. SEMUs are minimized through layouts that ensure a minimum spacing (the size of one flip-flop) between flip-flops checked by the same parity checker. This ensures that only one flip-flop, in a group of flip-flops checked by the same parity checker, will encounter an upset due to a single strike in our 28 nm technology in terrestrial environments [50]. Although a single strike could impact multiple flip-flops, since these flip-flops are checked by different checkers, the upsets will be detected. Since this absolute minimum spacing will remain constant, the relative spacing required between flip-flops will increase at smaller technology nodes, which may exacerbate the difficulty of implementation. Minimum spacing is enforced by applying design constraints during the layout stage. This constraint is important because even in large designs, flip-flops will still tend to be placed very close to one another. Table 5.5 shows the distribution of distances that each flip-flop has to its next nearest neighbor in a baseline design (this does not correspond to the spacing between flip-flops checked by the same logic parity checker). As shown, the majority of flip-flops are actually placed such that they would be susceptible to a SEMU. After applying parity checking, it is evident that no flip-flop, within a group checked by the same parity checker, is placed such that it will be vulnerable to a SEMU (Table 5.6).

*Table 5.4 Resilient flip-flops*

| **Type** | **Soft Error Rate (SER)** | **Area** | **Power** | **Delay** | **Energy** |
|---|---|---|---|---|---|
| Baseline | 1 | 1 | 1 | 1 | 1 |
| Light Hardened LEAP (LHL) | $2.5\times10^{-1}$ | 1.2 | 1.1 | 1.2 | 1.3 |
| LEAP-DICE | $2.0\times10^{-4}$ | 2.0 | 1.8 | 1 | 1.8 |
| LEAP-ctrl (economy mode) | 1 | 3.1 | 1.2 | 1 | 1.2 |
| LEAP-ctrl (resilient mode) | $2.0\times10^{-4}$ | 3.1 | 2.2 | 1 | 2.2 |
| Error Detection Sequential (EDS)[a] | ~100% detect | 1.5 | 1.4 | 1 | 1.4 |

[a]For EDS, the costs are listed for the flip-flop only. Error signal routing and delay buffers (included in Table 5.3) increase overall cost [46].

*Table 5.5 Distribution of spacing between a flip-flop and its nearest neighbor in a baseline (original, unprotected) design*

| **Distance** | **InO-core (%)** | **OoO-core (%)** |
|---|---|---|
| <1 flip-flop length away (i.e., flip-flops are adjacent and vulnerable to a SEMU) | 65.2 | 42.2 |
| 1–2 flip-flop lengths away | 30 | 30.6 |
| 2–3 flip-flop lengths away | 3.7 | 18.4 |
| 3–4 flip-flop lengths away | 0.6 | 3.5 |
| >4 flip-flop lengths away | 0.5 | 5.3 |

Logic parity is implemented using an XOR-tree-based predictor and checker, which detects flip-flops soft errors. This implementation differs from logic parity prediction, which also targets errors inside combinational logic [51]. XOR-tree logic parity is sufficient for detecting flip-flop soft errors (with the minimum spacing constraint applied). "Pipelining" in the predictor tree (Figure 5.2) may be required to ensure 0% clock period impact. The following heuristics for forming *parity groups* (the specific flip-flops that are checked together) to minimize cost of parity (cost comparisons in Table 5.7) were evaluated:

1. Parity group size: Flip-flops are clustered into a constant power of two-sized group that amortizes the parity logic cost by allowing the use of full binary trees at the predictor and checker. The last set of flip-flops will consist of modulo "group size" of flip-flops.
2. Vulnerability: Flip-flops are sorted by decreasing susceptibility to errors causing SDC or DUE and grouped into a constant power of two-sized group. The last set of flip-flops will consist of modulo "group size" of flip-flops.
3. Locality: Flip-flops are grouped by their location in the layout, in which flip-flops in the same functional unit are grouped together to help one to reduce wire routing for the predictor and checker logic. A constant power of two-sized groups is formed with the last group consisting of modulo "group size" of flip-flops.

*Table 5.6 Distribution of spacing between a flip-flop and its nearest neighbor in the same parity group (i.e., minimum distance between flip-flops checked by the same parity checker)*

| Distance | InO-core (%) | OoO-core (%) |
|---|---|---|
| <1 flip-flop length away (i.e., flip-flops are adjacent and vulnerable to a SEMU) | 0 | 0 |
| 1–2 flip-flop lengths away | 7.8 | 8.8 |
| 2–3 flip-flop lengths away | 5.3 | 10.6 |
| 3–4 flip-flop lengths away | 3.4 | 18.3 |
| >4 flip-flop lengths away | 83.3 | 62.2 |
| Average distance | 4.4 flip-flops | 12.8 flip-flops |

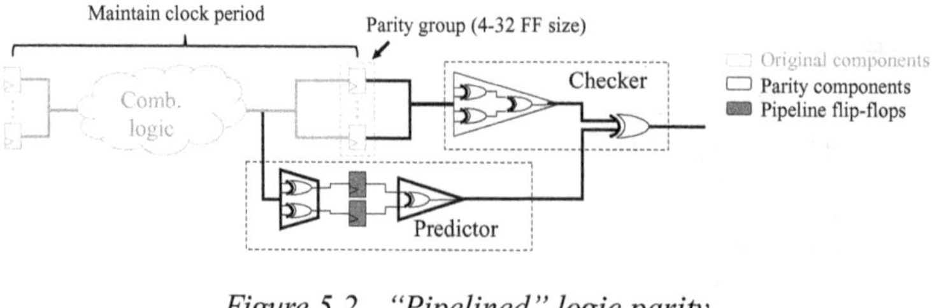


*Figure 5.2 "Pipelined" logic parity*

4. Timing: Flip-flops are sorted based on their available timing path slack and grouped into a constant power of two-sized group. The last set of flip-flops will consist of modulo "group size" of flip-flops.
5. Optimized: Figure 5.3 describes our heuristic. This solution is the most optimized and is the configuration used to report overhead values.

When unpipelined parity can be used, it is better to use larger sized groups (e.g., 32-bit groups) in order to amortize the additional predictor/checker logic to the number of flip-flops protected. However, when pipelined parity is required, the use of 16-bit groups was found to be a good option. This is because beyond 16-bits, additional pipeline flip-flops begin to dominate costs. These factors have driven the implementation of the previously described heuristics.

**Architecture**: The implementation of *DFC*, which checks static dataflow graphs, presented in this chapter includes *Control Flow Checking*, which checks static control-flow graphs. This combination checker resembles that of [52], which is also similar to the checker in [8].

Compiler optimization allows for static signatures required by the checkers to be embedded into unused delay slots in the software, thereby reducing execution time overhead by 13%.

Table 5.8 helps one to explain why DFC is unable to provide high SDC and DUE improvement. Of flip-flops that have errors that result in SDCs and DUEs (Section 5.2.1), DFC checkers detect SDCs and DUEs in less than 68% of these

*Table 5.7 Comparison of heuristics for "pipelined" logic parity implementations to protect all flip-flops on the InO-core*

| Heuristic | Area cost (%) | Power cost (%) | Energy cost (%) |
|---|---|---|---|
| Vulnerability (4-bit parity group) | 15.2 | 42 | 42 |
| Vulnerability (8-bit parity group) | 13.4 | 29.8 | 29.8 |
| Vulnerability (16-bit parity group) | 13.3 | 27.9 | 27.9 |
| Vulnerability (32-bit parity group) | 14.6 | 35.3 | 35.3 |
| Locality (16-bit parity group) | 13.4 | 29.4 | 29.4 |
| Timing (16-bit parity group) | 11.5 | 26.8 | 26.8 |
| Optimized (16-bit/32-bit groups) | 10.9 | 23.1 | 23.1 |

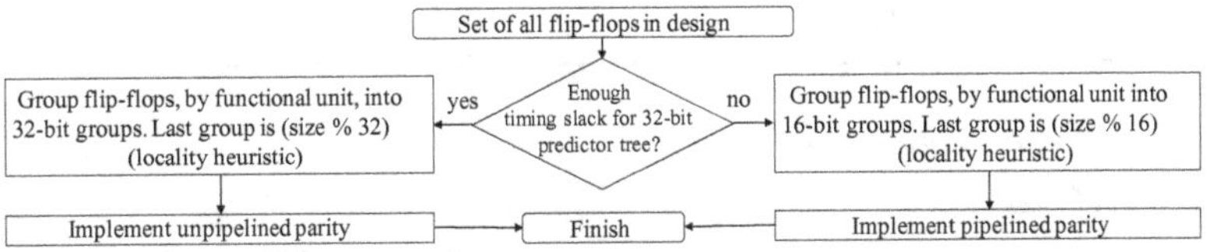


*Figure 5.3 Logic parity heuristic for low-cost parity implementation. 32-bit unpipelined parity and 16-bit pipelined parity were experimentally determined to be the lowest cost configurations.*

flip-flops (these 68% of flip-flops are distributed across all pipeline stages). For these 68% of flip-flops, on average, DFC detects less than 40% of the errors that result in SDCs or DUEs. This is because not all errors that result in an SDC or DUE will corrupt the dataflow or control flow signatures checked by the technique (e.g., register contents are corrupted and written out to a file, but the executed instructions remain unchanged). The combination of these factors means DFC is only detecting ~30% of SDCs or DUEs; thus, the technique provides low resilience improvement. These results are consistent with previously published data (detection of ~16% of non-vanished errors) on the effectiveness of DFC checkers in simple cores [52].

*Monitor cores* are checker cores that validate instructions executed by the main core (e.g., [8,53]). In this chapter, monitor cores similar to [53] are analyzed. For InO-cores, the size of the monitor core is of the same order as the main core, and hence, excluded from this study. For OoO-cores, the simpler monitor core can have lower throughput compared to the main core and thus stall the main core. It was confirmed (via Instructions Per Cycle estimation) that the monitor core implementation used does not stall the main core (Table 5.9).

**Software**: *Software assertions for general-purpose processors* check program variables to detect errors. This chapter combines assertions from [47,54] to check both data and control variables to maximize error coverage. Checks for data variables (e.g., end result) are added via compiler transformations using training inputs to determine the valid range of values for these variables (e.g., likely program invariants). Since such assertion checks are added based on training inputs, it is possible to encounter false positives, where an error is reported in an error-free run. This false-positive rate was determined by training the assertions using representative inputs. However,

*Table 5.8 DFC error coverage*

| | **InO** | | **OoO** | |
|---|---|---|---|---|
| | **SDC** | **DUE** | **SDC** | **DUE** |
| Percentage of flip-flops with an SDC-causing/ DUE-causing errors that are detected by DFC | 57% | 68% | 65% | 66% |
| Percentage of SDC-causing/DUE-causing errors detected (average per FF that is protected by DFC) | 30% | 30% | 29% | 40% |
| Overall percentage of SDC-causing/DUE-causing errors detected (for all flip-flops in the design) | 15.9% | 27% | 19.3% | 30% |
| Resulting improvement (5.1) | 1.2× | 1.4× | 1.2× | 1.4× |

*Table 5.9 Monitor core vs. main core*

| **Design** | **Clk. freq.** | **Average Instructions Per Cycle (IPC)** |
|---|---|---|
| OoO-core | 600 MHz | 1.3 |
| Monitor core | 2 GHz | 0.7 |

final analysis is performed by incorporating the input data used during evaluation into the training step in order to give the technique the best possible benefit and to eliminate the occurrence of false positives. Checks for control variables (e.g., loop index, stack pointer, and array address) are determined using application profiling and are manually added in the assembly code.

Table 5.10 provides the breakdown for the contribution to cost, improvement, and false positives resulting from assertions checking data variables [47] vs. those checking control variables [54]. Table 5.11 demonstrates the importance of evaluating resilience techniques using accurate injection [31].[†] Depending on the particular error injection model used, SDC improvement could be overestimated for one benchmark and underestimated for another. For instance, using inaccurate architecture register error injection (regU), one would be led to believe that software assertions provide 3× the SDC improvement than they do in reality (e.g., when evaluated using flip-flop-level error injection).

In order to pinpoint the sources of inaccuracy between the actual improvement rates that were determined using accurate flip-flop-level error injection vs. those published in the literature, error injection campaigns at other levels of abstraction (architecture register and program variable) were conducted. However, even

*Table 5.10 Comparison of assertions checking data (e.g., end result) vs. control (e.g., loop index) variables*

| | **Data variable check** | **Control variable check** | **Combined check** |
|---|---|---|---|
| Execution time impact | 12.1% | 3.5% | 15.6% |
| SDC improvement | 1.5× | 1.1× | 1.5× |
| DUE improvement | 0.7× | 0.9× | 0.6× |
| False[†]-positive rate | 0.003% | 0% | 0.003% |

*Table 5.11 Comparison of SDC improvement and detection for assertions when injecting errors at various levels*

| **App.**[a] | **Flip-flop (ground truth)** | **Register uniform (regU)** | **Register write regW** | **Program variable uniform (varU)** | **Program variable write (varW)** |
|---|---|---|---|---|---|
| bzip2 | 1.8× | 1.6× | 1.1× | 1.9× | 1.5× |
| crafty | 0.5× | 0.3× | 0.5× | 0.7× | 1.1× |
| gzip | 2× | 19.3× | 1× | 1.6× | 1.1× |
| mcf | 1.1× | 1.3× | 0.9× | 1× | 1.8× |
| parser | 2.4× | 1.7× | 1× | 2.4× | 2× |
| avg. | 1.6× | 4.8× | 0.9× | 1.5× | 1.5× |

[a]See footnote[†].

[†]The same SPEC applications evaluated in [47] were studied in this chapter.

then, there was an inability to exactly reproduce previously published improvement rates. Some additional differences in the architecture and program variable injection methodology used in this chapter compared to the published methodology may account for this discrepancy:

1. The architecture register and program variable evaluations used in this chapter were conducted on a SPARCv8 in-order design rather than a SPARCv9 out-of-order design.
2. The architecture register and program variable methodology used in this chapter injects errors uniformly into all program instructions, while previous publications chose to only inject into integer instructions of floating-point benchmarks.
3. The architecture register and program variable methodology used in this chapter injects errors uniformly over the full application rather than injecting only into the core of the application during computation.
4. Since the architecture register and program variable methodology used in this chapter injects errors uniformly into all possible error candidates (e.g., all cycles and targets), the calculated improvement covers the entire design. Previous publications calculated improvement over the limited subset of error candidates (out of all possible error candidates) that were injected into and thus only covers a subset of the design.

*CFCSS* checks static control flow graphs and is implemented via compiler modification similar to [55]. It is possible to analyze CFCSS in further detail to gain deeper understanding as to why improvement for the technique is relatively low (Table 5.12). Compared to DFC (a technique with a similar concept), it can be seen that CFCSS offers slightly better SDC improvement. However, since CFCSS only checks control flow signatures, many SDCs will still escape (e.g., the result of an add is corrupted and written to file). Additionally, certain DUEs, such as those that may cause a program crash, will not be detectable by CFCSS, or other software techniques, since execution may abort before a corresponding software check can be triggered. The relatively low resilience improvement using CFCSS has been corroborated in actual systems as well [56].

*ED by Duplicated Instructions* (*EDDI*) provides instruction redundant execution via compiler modification [57]. This chapter utilizes EDDI with store-readback [48] to maximize coverage by ensuring that values are written correctly. From Table 5.13,

*Table 5.12 CFCSS error coverage*

| | SDC | DUE |
|---|---|---|
| Percentage of flip-flops with an SDC-causing/DUE-causing error that is detected by CFCSS | 55% | 66% |
| Percentage of SDC-causing/DUE-causing errors that are detected per FF that is protected by CFCSS | 61% | 14% |
| Resulting improvement (5.1) | 1.5× | 0.5× |

it is clear why store-readback is important for EDDI. In order to achieve high SDC improvements, nearly all SDC-causing errors need to be detected. By detecting an additional 12% of SDCs, store-readback increases SDC improvement of EDDI by an order of magnitude. Virtually, all escaped SDCs are caught by ensuring that the values being written to the output are indeed correct (by reading back the written value). However, given that some SDC-causing or DUE-causing errors are still not detected by the technique, the results show that using naïve high-level injections will still yield incorrect conclusions (Table 5.14). Enhancements to EDDI such as Error detectors [58] and reliability-aware transforms [59] are intended to reduce the number of EDDI checks (i.e., selective insertion of checks) in order to minimize execution time impact while maintaining high overall error coverage. An evaluation of the Error detectors technique using flip-flop-level error injection found that the technique provides an SDC improvement of 2.6× improvement (a 21% reduction in SDC improvement as compared to EDDI without store-readback). However, Error detectors require software path tracking to recalculate important variables, which introduced a 3.9× execution time impact, greater than that of the original EDDI technique. The overhead corresponding to software path tracking can be reduced by implementing path tracking in hardware (as was done in the original work), but doing so eliminates the benefits of EDDI as a software-only technique.

**Algorithm**: *ABFT* can detect (*ABFT detection*) or detect and correct errors (*ABFT correction*) through algorithm modifications [60–63]. Although ABFT correction

*Table 5.13 EDDI: importance of store-readback*

| | **SDC improvement** | **SDC errors detected (%)** | **SDC errors escaped** | **DUE improvement** | **DUE errors detected (%)** | **DUE errors escaped** |
|---|---|---|---|---|---|---|
| Without store-readback | 3.3× | 86.1 | 49 | 0.4× | 19 | 3,090 |
| With store-readback | 37.8× | 98.7 | 6 | 0.3× | 19.8 | 3,006 |

*Table 5.14 Comparison of SDC improvement and detection for EDDI when injecting errors at various levels (without store-readback)*

| **Injection location** | **SDC improvement** | **SDC detected (%)** |
|---|---|---|
| Flip-flop (ground truth) | 3.3× | 86.1 |
| Register Uniform (regU) | 2.0× | 48.8 |
| Register Write (regW) | 6.6× | 84.8 |
| Program Variable Uniform (varU) | 12.6× | 92.1 |
| Program Variable Write (varU) | 100,000× | 100 |

algorithms can be used for detection-only (with minimally reduced execution time impact), ABFT detection algorithms cannot be used for correction. There is often a large difference in execution time impact between ABFT algorithms as well depending on the complexity of check calculation required. An ABFT correction technique for matrix inner product, for example, requires simple modular checksums (e.g., generated by adding all elements in a matrix row)—an inexpensive computation. On the other hand, ABFT detection for FFT, for example, requires expensive calculations using Parseval's theorem [64]. For the particular applications studied, the algorithms that were protected using ABFT detection often required more computationally expensive checks than algorithms that were protected using ABFT correction; therefore, the former generally had greater execution time impact (relative to each of their own original baseline execution times). An additional complication arises when an ABFT detection-only algorithm is implemented. Due to the long ED latencies imposed by ABFT detection (9.6 million cycles, on average), hardware recovery techniques are not feasible and higher level recovery mechanisms will impose significant overheads.

**Recovery**: Two recovery scenarios are considered: *bounded latency*, i.e., an error must be recovered within a fixed period of time after its occurrence, and *unconstrained*, i.e., where no latency constraints exist and errors are recovered externally once detected (no hardware recovery is required). Bounded latency recovery is achieved using one of the following hardware recovery techniques (Table 5.15): *flush* or *Reorder Buffer* (*RoB*) recovery (both of which rely on flushing noncommitted instructions followed by re-execution) [65,66]; *Instruction Replay* (*IR*) or *Extended IR* (*EIR*) recovery (both of which rely on instruction checkpointing to rollback and replay instructions) [10]. EIR is an extension of IR with additional buffers required by DFC for recovery. Flush and RoB are unable to recover from errors detected after the memory write stage of InO-cores or after the RoB of OoO-cores, respectively (these errors will have propagated to architecture visible states). Hence, LEAP-DICE is used to protect flip-flops in these pipeline stages when using flush/RoB recovery. IR and EIR can recover detected errors in any pipeline flip-flop. IR recovery is shown in Figure 5.4 and flush recovery is shown in Figure 5.5. Since recovery hardware serves as single points of failure, flip-flops in the recovery hardware itself needs to be capable of error correction (e.g., protected using hardened flip-flops when considering soft errors).

**Additional techniques**: Many additional resilience techniques have been published in literature; but, these techniques are closely related to our evaluated techniques. Therefore, the results presented in this chapter are believed to be representative and largely cover the cross-layer design space.

At the circuit-level, hardened flip-flops like DICE [67], BCDMR (Bistable Cross-coupled Dual Modular Redundancy) [68], and BISER (Built In Soft Error Resilience) [69] are similar in cost to LEAP-DICE, the most resilient hardened flip-flop studied. The DICE technique suffers from an inability to tolerate SEMUs, unlike LEAP-DICE. BISER is capable of operating in both economy and resilient modes. This enhancement is provided by LEAP-ctrl. Hardened flip-flops like RCC (Reinforcing Charge Collection) [18] offer around 5× soft error rate improvement at around 1.2× area, power, and energy cost. LHL provides slightly more soft error tolerance at roughly the same cost as RCC. Circuit-level detection techniques such as [70–72] are similar to EDS. Like EDS, these techniques can detect soft errors while offering minor differences in actual implementation. Stability checking [73] works on a similar principle of time sampling to detect errors.

*Table 5.15 Hardware error recovery costs*

| Core | Type | Area (%) | Power (%) | Energy (%) | Recovery latency (cycles) | Unrecoverable flip-flop errors |
|---|---|---|---|---|---|---|
| InO | Instruction Replay (IR) recovery | 16 | 21 | 21 | 47 | None (all pipeline FFs recoverable) |
| | EIR recovery | 34 | 32 | 32 | 47 | |
| | Flush recovery | 0.6 | 0.9 | 1.8 | 7 | FFs after memory write stage |
| OoO | Instruction Replay (IR) recovery | 0.1 | 0.1 | 0.1 | 104 | None (all pipeline FFs recoverable) |
| | EIR recovery | 0.2 | 0.1 | 0.1 | 104 | |
| | Reorder Buffer (ROB) recovery | 0.01 | 0.01 | 0.01 | 64 | FFs after reorder buffer stage |

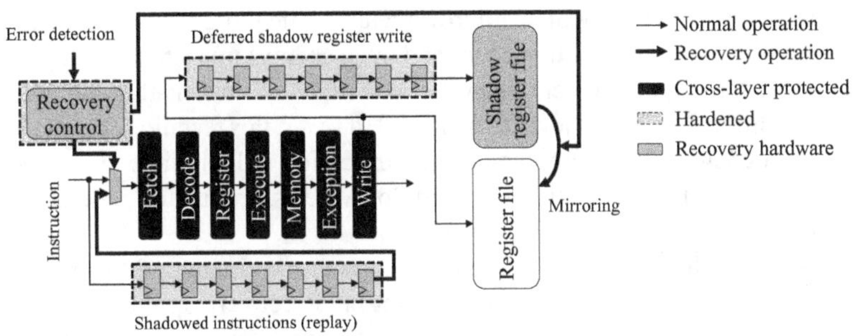


*Figure 5.4 Instruction Replay (IR) recovery*

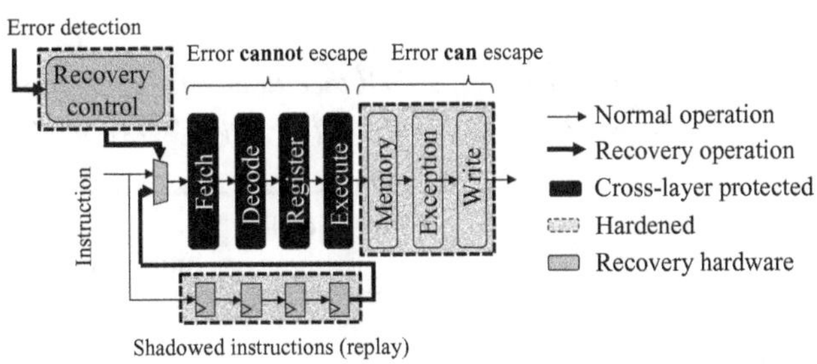


*Figure 5.5 Flush recovery*

Logic-level techniques like residue codes [9] can be effective for specific functional units like multipliers but are costlier to implement than the simple XOR-trees used in logic parity. Additional logic-level coding techniques like Berger codes [74] and Bose-Lin codes [75] are costlier to implement than logic parity. Like logic parity checking, residue, Berger, and Bose-Lin codes only detect errors.

Techniques like DMR (Dual Modular Redundancy) and TMR (Triple Modular Redundancy) at the architecture level can be easily ruled out since these techniques will incur more than 100% area, power, and energy costs. RMT (Redundant Multi-Threading) [76] has been shown to have high (>40%) energy costs (which can increase due to recovery since RMT only serves to detect errors). Additionally, RMT is highly architecture dependent, which limits its applicability.

Software techniques like Shoestring [77], Error detectors [58], Reliability-driven transforms [59], and SWIFT (Software Implemented Fault Tolerance) [78] are similar to EDDI but offer variations to the technique by reducing the number of checks added. As a result, EDDI can be used as a bound on the maximum ED possible. An enhancement to SWIFT, known as CompileR Assisted Fault Tolerance (CRAFT) [79], uses HW acceleration to improve reliability, but doing so eliminates the benefit of EDDI as a software-only technique. Although it is difficult to faithfully compare these "selective" EDDI techniques as published (since the original authors evaluated improvements using high-level error injection at the architecture register level which are generally inaccurate), the published results for these "selective" EDDI techniques show insufficient benefit (Table 5.16). Enhancements that reduce the execution time impact provide very low SDC improvements, while those that provide moderate improvement incur high execution time (and thus, energy) impact (much higher than providing the same improvement using LEAP-DICE, for instance). Fault screening [65] is an additional software-level technique. However, this technique also checks to ensure intermediate values computed during execution fall within expected bounds, which is similar to the mechanisms behind Software assertions for general-purpose processors, and thus, is covered by the latter.

**Low-level techniques**: Resilience techniques at the circuit and logic layer (i.e., *low-level techniques*) are tunable as they can be selectively applied to individual flip-flops. As a result, a range of SDC/DUE improvements can be achieved for varying costs (Table 5.17). These techniques offer the ability to finely tune the specific flip-flops to protect in order to achieve the degree of resilience improvements required.

*Table 5.16 Comparison of "selective" EDDI techniques as reported in literature compared to EDDI evaluated using flip-flop-level error injection*

| | Error-injection | SDC improvement | Exec. time impact |
|---|---|---|---|
| EDDI with store-readback (implemented) | Flip-flop | 37.8× | 2.1× |
| Reliability-aware transforms (published) | Arch. reg. | 1.8× | 1.05× |
| Shoestring (published) | Arch. reg. | 5.1× | 1.15× |

Table 5.17 [a] *Costs vs. SDC and DUE improvements for tunable resilience techniques*

| | | | Bounded latency recovery | | | | | | | | | | Unconstrained recovery[b] | | | | | | | | | | Exec. time impact (%) |
|---|---|---|---|---|---|---|---|---|---|---|---|---|---|---|---|---|---|---|---|---|---|---|---|
| | | | SDC improvement | | | | | DUE improvement | | | | | SDC improvement | | | | | DUE improvement | | | | | |
| | | | 2 | 5 | 50 | 500 | max | 2 | 5 | 50 | 500 | max | 2 | 5 | 50 | 500 | max | 2 | 5 | 50 | 500 | max | |
| InO | Selective hardening | *A* | 0.8 | 1.8 | 2.9 | 3.3 | 9.3 | 0.7 | 1.7 | 3.8 | 5.1 | 9.3 | 0.8 | 1.8 | 2.9 | 3.3 | 9.3 | 0.7 | 1.7 | 3.8 | 5.1 | 9.3 | 0 |
| | using LEAP-DICE | *E* | 2 | 4.3 | 7.3 | 8.2 | 22.4 | 1.5 | 3.8 | 9.5 | 12.5 | 22.4 | 2 | 4.3 | 7.3 | 8.2 | 22.4 | 1.5 | 3.8 | 9.5 | 12.5 | 22.4 | |
| | Logic parity only | *A* | 17.3 | 18.6 | 20.3 | 20.7 | 26.9 | 16.9 | 18.3 | 21.5 | 22.8 | 23.3 | 1.3 | 2.6 | 4.3 | 4.7 | 10.9 | – | – | – | – | – | 0 |
| | (+IR recovery) | *E* | 23.4 | 26 | 29.4 | 30.5 | 44.1 | 22.5 | 25.4 | 31.9 | 35 | 35.9 | 2.4 | 5 | 8.4 | 9.5 | 23.1 | | | | | | |
| | EDS-only | *A* | 17.1 | 18.1 | 19.7 | 20.5 | 26.7 | 16.8 | 18 | 20.3 | 22.5 | 26.2 | 1.1 | 2.1 | 3.7 | 4.5 | 10.7 | – | – | – | – | – | 0 |
| | (+IR recovery) | *E* | 23.1 | 25.4 | 28.5 | 29.6 | 43.9 | 22.1 | 25.2 | 31.5 | 39.2 | 43.7 | 2.1 | 4.4 | 7.5 | 8.6 | 22.9 | | | | | | |
| OoO | Selective hardening | *A* | 1.1 | 1.3 | 2.2 | 2.4 | 6.5 | 1.3 | 1.6 | 3.1 | 3.6 | 6.5 | 1.1 | 1.3 | 2.2 | 2.4 | 6.5 | 1.3 | 1.6 | 3.1 | 3.6 | 6.5 | 0 |
| | using LEAP-DICE | *E* | 1.5 | 1.7 | 3.1 | 3.5 | 9.4 | 2 | 2.3 | 4.2 | 5.1 | 9.4 | 1.5 | 1.7 | 3.1 | 3.5 | 9.4 | 2 | 2.3 | 4.2 | 5.1 | 9.4 | |
| | Logic parity only | *A* | 1.9 | 2.1 | 6.1 | 6.3 | 14.2 | 1.7 | 2.6 | 4.5 | 5 | 13.8 | 1.8 | 2 | 5.9 | 6.2 | 14.1 | – | – | – | – | – | 0 |
| | (+IR recovery) | *E* | 1.6 | 2.4 | 4.1 | 5.1 | 13.7 | 2.4 | 3 | 4.4 | 5.4 | 13.6 | 1.5 | 2.3 | 4 | 5 | 13.6 | | | | | | |
| | EDS-only | *A* | 1.4 | 1.8 | 3.3 | 4 | 12.3 | 1.3 | 2 | 3.6 | 4 | 11.8 | 1.3 | 1.7 | 3.2 | 3.9 | 12.2 | – | – | – | – | – | 0 |
| | (+IR recovery) | *E* | 1.7 | 2.1 | 3.5 | 4 | 11.6 | 2.1 | 2.5 | 4.4 | 5.3 | 11.4 | 1.6 | 2 | 3.4 | 3.9 | 11.5 | | | | | | |

*A*, area cost %; *P*, power cost %; *E*, energy cost %; $P = E$ for these combinations—no clock/execution time impact.
[a]Costs are generated per benchmark. The average cost over all benchmarks is reported. Relative standard deviation is 0.6%–3.1%.
[b] DUE improvements are not possible with detection-only techniques given unconstrained recovery.

**High-level techniques**: In general, techniques at the architecture, software, and algorithm layers (i.e., *high-level techniques*) are less tunable as there is little control of the exact subset of flip-flops a high-level technique will protect. From Table 5.3, it can be seen that no high-level technique provides more than 38× improvement (while most offer far less improvement). As a result, to achieve a 50× improvement, for example, augmentation from low-level techniques at the circuit-level and logic-level is required, regardless.

## 5.3 Cross-layer combinations

CLEAR uses a top-down approach to explore the cost-effectiveness of various cross-layer combinations. For example, resilience techniques at the upper layers of the system stack (e.g., ABFT correction) are applied before incrementally moving down the stack to apply techniques from lower layers (e.g., an optimized combination of logic parity checking, circuit-level LEAP-DICE, and micro-architectural recovery). This approach (example shown in Figure 5.6) ensures that resilience techniques from various layers of the stack effectively interact with one another. Resilience techniques from the algorithm, software, and architecture layers of the stack generally protect multiple flip-flops (determined using error injection); however, a designer typically has little control over the specific subset protected. Using multiple resilience techniques from these layers can lead to situations where a given flip-flop may be protected (sometimes unnecessarily) by multiple techniques. At the logic and circuit layers, fine-grained protection is available, since these techniques can be applied selectively to individual flip-flops (those not sufficiently protected by higher level techniques).

A total of 586 cross-layer combinations are explored using CLEAR (Table 5.18). Not all combinations of the ten resilience techniques and four recovery techniques are valid (e.g., it is unnecessary to combine ABFT correction and ABFT detection, since the techniques are mutually exclusive or to explore combinations of monitor cores to protect an InO-core due to the high cost). Accurate flip-flop-level injection and layout evaluation reveal that many individual techniques provide minimal (less than 1.5×) SDC/DUE improvement (contrary to conclusions reported in the literature that were derived using inaccurate architecture-level or software-level injection), have high costs, or both. The consequence of this revelation is that most cross-layer combinations have high cost (detailed results for these costly combinations are omitted for brevity but are shown in Figure 5.1).

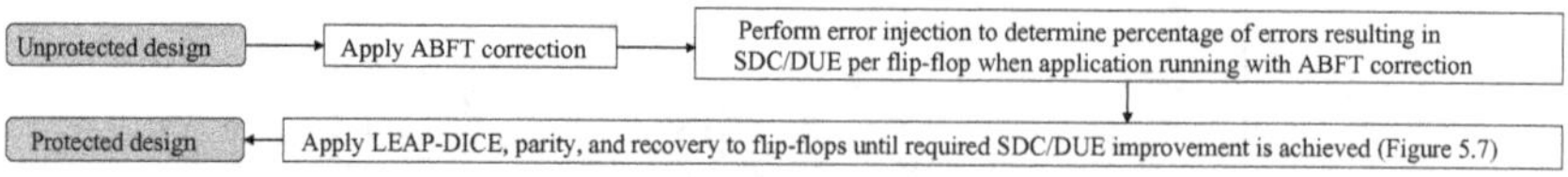


*Figure 5.6 Cross-layer methodology example for combining ABFT correction, LEAP-DICE, logic parity, and micro-architectural recovery*

### *5.3.1 Combinations for general-purpose processors*

Among the 586 cross-layer combinations explored using CLEAR, a highly promising approach combines selective circuit-level hardening using LEAP-DICE, logic parity, and micro-architectural recovery (flush recovery for InO-cores, RoB recovery for OoO-cores). Thorough error injection using application benchmarks plays a critical role in selecting the flip-flops protected using these techniques. Figure 5.7 and Heuristic 1 detail the methodology for creating this combination. If recovery is not needed (e.g., for unconstrained recovery), the "Harden" procedure in Heuristic 1 can be modified to always return false.

For example, to achieve a 50× SDC improvement, the combination of LEAP-DICE, logic parity, and micro-architectural recovery provides a 1.5× and 1.2× energy savings for the OoO-cores and InO-cores, respectively, compared to selective circuit hardening using LEAP-DICE (Table 5.19). The relative benefits are consistent across benchmarks and over the range of SDC/DUE improvements. The overheads in

*Table 5.18 Creating 586 cross-layer combinations*

| | | No recovery | Flush/RoB recovery | IR/EIR recovery | Total |
|---|---|---|---|---|---|
| InO | Combinations of LEAP-DICE, EDS, parity, DFC, Assertions, CFCSS, EDDI | 127 | 3 | 14 | 144 |
| | ABFT correction/detection alone | 2 | 0 | 0 | 2 |
| | ABFT correction+previous combinations | 127 | 3 | 14 | 144 |
| | ABFT detection+previous combinations | 127 | 0 | 0 | 127 |
| | InO-core total | – | – | – | 417 |
| OoO | Combinations of LEAP-DICE, EDS, parity, DFC, monitor cores | 31 | 7 | 30 | 68 |
| | ABFT correction/detection alone | 2 | 0 | 0 | 2 |
| | ABFT correction+previous combinations | 31 | 7 | 30 | 68 |
| | ABFT detection+previous combinations | 31 | 0 | 0 | 31 |
| | OoO-core total | – | – | – | 169 |
| Combined total | | | | | 586 |

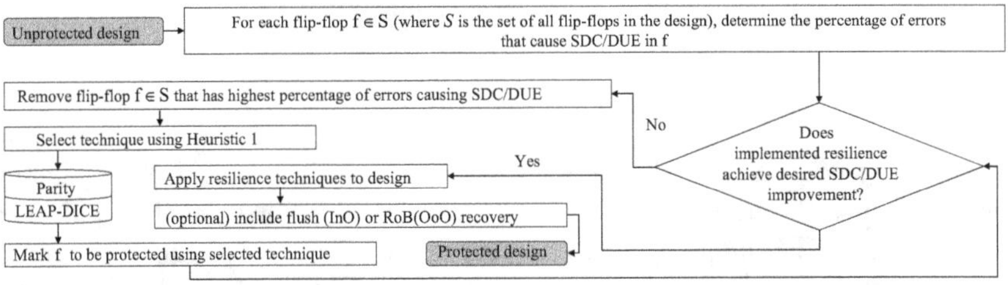


*Figure 5.7 Cross-layer resilience methodology for combining LEAP-DICE, parity, and micro-architectural recovery*

Table 5.19 are small because only the most energy-efficient resilience solutions are reported. Most of the 586 combinations are far costlier.

The scenario where recovery hardware is not needed (e.g., unconstrained recovery) can also be considered. In this case, a minimal (<0.2% energy) savings can be achieved when targeting SDC improvement. However, without recovery hardware, DUEs increase, since detected errors are now uncorrectable; thus, no DUE improvement is achievable.

Finally, one may suppose that the inclusion of EDS into cross-layer optimization may yield further savings, since EDS costs ~25% less area, power, energy than LEAP-DICE. However, a significant portion of EDS overhead is not captured solely by cell overhead. In fact, the additional cost of aggregating and routing the EDS ED signals and the cost of adding delay buffers to satisfy minimum delay constraints posed by EDS dominate cost and prevent cross-layer combinations using EDS from yielding benefits (Table 5.19). Various additional cross-layer combinations spanning circuit, logic, architecture, and software layers are presented in Table 5.19.

---

**Heuristic 1:** Choose LEAP-DICE or parity technique

---

**Input:** `f`: flip-flop to be protected

**Output:** Technique to apply to `f` (LEAP-DICE, parity)

**if** HARDEN(`f`) **then return** LEAP-DICE

**if** PARITY(`f`) **then return** parity

**return** LEAP-DICE

---

**procedure** HARDEN(`f`)

**if** an error in `f` cannot be flushed (i.e., `f` is in the memory, exception, writeback stages of InO or after the RoB of OoO)

**then return** TRUE; **else return** FALSE

**end procedure**

---

**procedure** PARITY(`f`)

**if** `f` has timing path slack greater than delay imposed by 32-bit XOR-tree (this implements low-cost parity checking as explained in Section 5.2.4)

**then return** TRUE, **else return** FALSE

**end procedure**

---

Up to this point, SDC and DUE improvements have been considered separately. However, it may be useful to achieve a specific improvement in SDC and DUE simultaneously. When targeting SDC improvement, DUE improvement also improves (and vice versa); however, it is unlikely that the two improvements will be the same, since flip-flops with high SDC vulnerability will not necessarily be the same flip-flops that have high DUE vulnerability. A simple method for targeting joint SDC/DUE

*Table 5.19 Costs vs. SDC and DUE improvements for various combinations in general-purpose processors*

| | | Bounded latency recovery | | | | | | | | | | Unconstrained recovery | | | | | | | | | | Exec. time impact (%) |
|---|---|---|---|---|---|---|---|---|---|---|---|---|---|---|---|---|---|---|---|---|---|---|
| | | SDC improvement | | | | | DUE improvement | | | | | SDC improvement | | | | | DUE improvement | | | | | |
| | | 2 | 5 | 50 | 500 | max | 2 | 5 | 50 | 500 | max | 2 | 5 | 50 | 500 | max | 2 | 5 | 50 | 500 | max | |
| InO LEAP-DICE+logic | *A* | 0.7 | 1.7 | 2.5 | 3 | 8 | 0.6 | 1.5 | 3.6 | 4.4 | 8 | 0.7 | 1.6 | 2.4 | 2.8 | 7.6 | – | – | – | – | – | 0 |
| parity (+flush | *P* | 1.9 | 3.9 | 6.1 | 6.7 | 17.9 | 1.5 | 3.4 | 8.4 | 10.4 | 17.9 | 1.9 | 3.8 | 5.9 | 6.5 | 17.2 | | | | | | |
| recovery) | *E* | 1.9 | 3.9 | 6.1 | 6.7 | 17.9 | 1.5 | 3.4 | 8.4 | 10.4 | 17.9 | 1.9 | 3.8 | 5.9 | 6.5 | 17.2 | | | | | | |
| EDS+LEAP- | *A* | 0.9 | 2.3 | 2.7 | 3.3 | 8.4 | 0.8 | 2.1 | 3.8 | 4.8 | 8.4 | 0.9 | 2.2 | 2.5 | 3.2 | 8.1 | – | – | – | – | – | 0 |
| DICE+logic parity | *P* | 1.9 | 4.3 | 6.6 | 7.2 | 19.3 | 1.7 | 3.8 | 8.5 | 11 | 19.3 | 1.9 | 4.2 | 6.3 | 7.1 | 19 | | | | | | |
| (+flush recovery) | *E* | 1.9 | 4.3 | 6.6 | 7.2 | 19.3 | 1.7 | 3.8 | 8.5 | 11 | 19.3 | 1.9 | 4.2 | 6.3 | 7.1 | 19 | | | | | | |
| DFC+LEAP- | *A* | 39.3 | 41.1 | 41.5 | 43.1 | 45 | 39.3 | 39.9 | 41.9 | 42.5 | 45 | 3.3 | 5.1 | 5.6 | 7.1 | 10.6 | – | – | – | – | – | 6.2 |
| DICE+logic parity | *P* | 32.4 | 35.5 | 38.7 | 41 | 50.9 | 32.5 | 33.9 | 38.4 | 40.7 | 50.9 | 1.4 | 4.8 | 8.1 | 10 | 18.2 | | | | | | |
| (+EIR recovery) | *E* | 44.2 | 56.7 | 60.2 | 62.7 | 60.3 | 45.8 | 48.9 | 58.3 | 63 | 60.3 | 10.6 | 13.9 | 17.4 | 19.9 | 25.5 | | | | | | |
| Assertions+LEAP- | *A* | – | – | – | – | – | – | – | – | – | – | 0.7 | 0. | 1 | 1.1 | 7.6 | – | – | – | – | – | 15.6 |
| DICE+logic parity | *P* | | | | | | | | | | | 1.4 | 1.8 | 2.2 | 2.2 | 17.2 | | | | | | |
| (no recovery) | *E* | | | | | | | | | | | 17.1 | 17.5 | 18 | 18 | 24.5 | | | | | | |
| CFCSS+LEAP- | *A* | – | – | – | – | – | – | – | – | – | – | 0.3 | 1 | 1.4 | 1.3 | 7.6 | – | – | – | – | – | 40.6 |
| DICE+logic parity | *P* | | | | | | | | | | | 0.8 | 1.8 | 2.9 | 3.1 | 17.2 | | | | | | |
| (no recovery) | *E* | | | | | | | | | | | 41.5 | 43 | 44.6 | 44.9 | 64.8 | | | | | | |
| EDDI+LEAP- | *A* | – | – | – | – | – | – | – | – | – | – | 0 | 0 | 0.7 | 0.9 | 7.6 | – | – | – | – | – | 110 |
| DICE+logic parity | *P* | | | | | | | | | | | 0 | 0 | 0.6 | 0.8 | 17.2 | | | | | | |
| (no recovery) | *E* | | | | | | | | | | | 110 | 110 | 111 | 111 | 146 | | | | | | |

*Table 5.19 (Continued)*

| | | Bounded latency recovery | | | | | | | | | | Unconstrained recovery | | | | | | | | | | Exec. time impact (%) |
|---|---|---|---|---|---|---|---|---|---|---|---|---|---|---|---|---|---|---|---|---|---|---|
| | | SDC improvement | | | | | DUE improvement | | | | | SDC improvement | | | | | DUE improvement | | | | | |
| | | **2** | **5** | **50** | **500** | **max** | **2** | **5** | **50** | **500** | **max** | **2** | **5** | **50** | **500** | **max** | **2** | **5** | **50** | **500** | **max** | |
| OoO LEAP-DICE+logic | *A* | 0.06 | 0.1 | 1.4 | 2.2 | 4.9 | 0.5 | 0.7 | 2.6 | 3 | 4.9 | 0.06 | 0.1 | 1.4 | 2.2 | 4.9 | – | – | – | – | – | 0 |
| parity (+ROB | *P* | 0.1 | 0.2 | 2.1 | 2.4 | 7 | 0.1 | 0.1 | 2 | 1.8 | 7 | 0.1 | 0.2 | 2.1 | 2.4 | 7 | | | | | | |
| recovery) | *E* | 0.1 | 0.2 | 2.0 | 2.4 | 7 | 0.1 | 0.1 | 2 | 1.8 | 7 | 0.1 | 0.2 | 2.1 | 2.4 | 7 | | | | | | |
| EDS+LEAP- | *A* | 0.07 | 0.1 | 1.6 | 2.2 | 5.4 | 0.6 | 0.8 | 2.6 | 3 | 5.4 | 0.07 | 0.1 | 1.6 | 2.2 | 5.4 | – | – | – | – | – | 0 |
| DICE+logic parity | *P* | 0.1 | 0.2 | 2.3 | 2.5 | 8.1 | 0.1 | 0.1 | 2 | 1.8 | 8.1 | 0.1 | 0.2 | 2.3 | 2.5 | 8.1 | | | | | | |
| (+ROB recovery) | *E* | 0.1 | 0.2 | 2.3 | 2.5 | 8.1 | 0.1 | 0.1 | 2 | 1.8 | 8.1 | 0.1 | 0.2 | 2.3 | 2.5 | 8.1 | | | | | | |
| DFC+LEAP- | *A* | 0.2 | 1 | 1.8 | 2 | 5.3 | 0.2 | 0.4 | 1.7 | 3.9 | 5.3 | 0.1 | 0.8 | 1.6 | 1.8 | 5.1 | – | – | – | – | – | 7.1 |
| DICE+logic parity | *P* | 1.1 | 1.4 | 2 | 2.8 | 7.2 | 0.2 | 0.2 | 2.6 | 3.3 | 7.2 | 1 | 1.3 | 1.9 | 2.7 | 7.1 | | | | | | |
| (+EIR recovery) | *E* | 21.2 | 21.5 | 22.2 | 23 | 14.8 | 20 | 20.1 | 22.9 | 23.6 | 14.8 | 10 | 11.4 | 12.1 | 12.9 | 14.7 | | | | | | |
| Monitor core+ | *A* | 9 | 9 | 9.8 | 10.5 | 13.9 | 9 | 9 | 10.1 | 11.2 | 13.9 | 9 | 9 | 9.8 | 10.5 | 13.9 | – | – | – | – | – | 0 |
| LEAP-DICE+logic | *P* | 16.3 | 16.3 | 20 | 20.2 | 23.3 | 16.3 | 16.3 | 20.1 | 21.5 | 22.3 | 16.3 | 16.3 | 20 | 20.2 | 23.3 | | | | | | |
| parity (+ROB rec.) | *E* | 16.3 | 16.3 | 20 | 20.2 | 23.3 | 16.3 | 16.3 | 20.1 | 21.5 | 22.3 | 16.3 | 16.3 | 20 | 20.2 | 23.3 | | | | | | |

*A*, area cost %; *P*, power cost %; *E*, energy cost %.

*Table 5.20 Cost to achieve joint SDC/DUE improvement with a combination of LEAP-DICE, parity, and flush/RoB recovery*

| **Joint SDC/DUE improvement** | **InO** | | | **OoO** | | |
|---|---|---|---|---|---|---|
| | **Area (%)** | **Power (%)** | **Energy (%)** | **Area (%)** | **Power (%)** | **Energy (%)** |
| 2× | 0.7 | 2 | 2 | 0.6 | 0.1 | 0.1 |
| 5× | 1.9 | 4.2 | 4.2 | 0.9 | 0.4 | 0.4 |
| 50× | 4.1 | 9 | 9 | 2.8 | 2.2 | 2.2 |
| 500× | 4.6 | 10.8 | 10.8 | 3.1 | 2.8 | 2.8 |
| max | 8 | 17.9 | 17.9 | 4.9 | 7 | 7 |

improvement is to implement resilience until SDC (DUE) improvement is reached and then continue implementing resilience to unprotected flip-flops until DUE (SDC) improvement is also achieved. This ensures that both SDC and DUE improvement meet (or exceed) the targeted minimum required improvement. Table 5.20 details the costs required to achieve joint SDC/DUE improvement using this methodology when considering a combination of LEAP-DICE, parity, and flush/RoB recovery.

### 5.3.2 *Targeting specific applications*

When the application space targets specific algorithms (e.g., matrix operations), a cross-layer combination of LEAP-DICE, parity, ABFT correction, and micro-architectural error recovery (flush/RoB) provides additional energy savings (compared to the general-purpose cross-layer combinations presented in Section 5.3.1). Since ABFT correction performs in-place error correction, no separate recovery mechanism is required for ABFT correction. For this study, it is possible to apply ABFT correction to three of the PERFECT benchmarks: 2d_convolution, debayer_filter, and inner_product (the rest were protected using ABFT detection).

The results in Table 5.21 confirm that combinations of ABFT correction, LEAP-DICE, parity, and micro-architectural recovery provide up to 1.1× and 2× energy savings over the previously presented combination of LEAP-DICE, parity, and recovery when targeting SDC improvement for the OoO-cores and InO-cores, respectively. However, as will be discussed in Section 5.3.2.1, the practicality of ABFT is limited when considering general-purpose processors.

When targeting DUE improvement, including ABFT correction provides no energy savings for the OoO-core. This is because ABFT correction (along with most architecture and software techniques like DFC, CFCSS, and Assertions) performs checks at set locations in the program. For example, a DUE resulting from an invalid pointer access can cause an immediate program termination before a check is invoked. As a result, this DUE would not be detected by the resilience technique.

Although ABFT correction is useful for general-purpose processors limited to specific applications, the same cannot be said for ABFT detection (Table 5.21).

Table 5.21 *Costs vs. SDC and DUE improvement for various cross-layer combinations involving ABFT*

| | | | Bounded latency recovery | | | | | | | | | | Unconstrained recovery | | | | | | | | | | Exec. time impact (%) |
|---|---|---|---|---|---|---|---|---|---|---|---|---|---|---|---|---|---|---|---|---|---|---|---|
| | | | SDC improvement | | | | | DUE improvement | | | | | SDC improvement | | | | | DUE improvement | | | | | |
| | | | **2** | **5** | **50** | **500** | **max** | **2** | **5** | **50** | **500** | **max** | **2** | **5** | **50** | **500** | **max** | **2** | **5** | **50** | **500** | **max** | |
| InO | ABFT correction+ LEAP-DICE+logic parity (+flush recovery) | *A* | 0 | 0.4 | 1.0 | 1.2 | 8 | 0.3 | 0.4 | 1.5 | 2.7 | 8 | 0 | 0.4 | 0.9 | 1.1 | 7.6 | – | – | – | – | – | 1.4 |
| | | *P* | 0 | 0.7 | 1.7 | 1.8 | 17.9 | 1 | 1 | 3.3 | 5.7 | 17.9 | 0 | 0.7 | 1.6 | 1.8 | 17.2 | | | | | | |
| | | *E* | 1.4 | 2.2 | 3.1 | 3.2 | 19.6 | 2.4 | 2.4 | 4.8 | 7.2 | 19.6 | 1.4 | 2.2 | 3 | 3.2 | 18.8 | | | | | | |
| | ABFT detection+ LEAP-DICE+logic parity (no recovery) | *A* | – | – | – | – | – | – | – | – | – | – | 0 | 1.2 | 2 | 2.5 | 7.6 | – | – | – | – | – | 24 |
| | | *P* | | | | | | | | | | | 0 | 2.4 | 4.8 | 5.7 | 17.2 | | | | | | |
| | | *E* | | | | | | | | | | | 1.4 | 27 | 30 | 31.1 | 45.3 | | | | | | |
| | ABFT correction+ LEAP-ctrl+LEAP-DICE+logic parity (+flush recovery) | *A* | 1.5 | 2.5 | 3.8 | 4.1 | 8 | 1 | 1 | 4.1 | 5 | 8 | 1.5 | 2.3 | 3.4 | 4 | 7.6 | – | – | – | – | – | 1.4 |
| | | *P* | 0.6 | 1.3 | 2.6 | 2.8 | 17.9 | 1.3 | 1.3 | 4.6 | 7 | 17.9 | 0.6 | 1.2 | 2.6 | 2.7 | 17.2 | | | | | | |
| | | *E* | 1.9 | 2.7 | 4.0 | 4.2 | 19.6 | 2.8 | 2.8 | 6.1 | 8.5 | 19.6 | 1.9 | 2.6 | 4 | 4.1 | 18.8 | | | | | | |
| OoO | ABFT correction+ LEAP-DICE+logic parity (+ROB recovery) | *A* | 0 | 0.01 | 0.3 | 0.5 | 4.9 | 0.4 | 0.6 | 2.1 | 3 | 4.9 | 0 | 0.01 | 0.3 | 0.5 | 4.8 | – | – | – | – | – | 1.4 |
| | | *P* | 0 | 0.01 | 0.5 | 0.8 | 7 | 0.1 | 0.1 | 3 | 1.6 | 7 | 0 | 0.01 | 0.5 | 0.8 | 6.9 | | | | | | |
| | | *E* | 1.4 | 1.5 | 1.9 | 2.2 | 8.5 | 1.5 | 1.5 | 4.2 | 3 | 8.5 | 1.4 | 1.5 | 1.9 | 2.2 | 8.4 | | | | | | |
| | ABFT detection+ LEAP-DICE+logic parity (no recovery) | *A* | – | – | – | – | – | – | – | – | – | – | 0 | 0.1 | 0.7 | 1.2 | 4.8 | – | – | – | – | – | 24 |
| | | *P* | | | | | | | | | | | 0 | 0.2 | 1.2 | 1.6 | 6.9 | | | | | | |
| | | *E* | | | | | | | | | | | 24 | 24.2 | 25.5 | 26 | 32.6 | | | | | | |
| | ABFT correction+ LEAP-ctrl+LEAP-DICE+logic parity (+ROB recovery) | *A* | 1.5 | 1.8 | 2.9 | 3.2 | 4.9 | 0.6 | 0.9 | 2.8 | 3.6 | 4.9 | 1.5 | 1.8 | 2.9 | 3.2 | 4.9 | – | – | – | – | – | 1.4 |
| | | *P* | 0.3 | 0.3 | 1.0 | 1.3 | 7 | 0.1 | 0.1 | 3 | 1.6 | 7 | 0.3 | 0.3 | 1 | 1.3 | 6.9 | | | | | | |
| | | *E* | 1.7 | 1.7 | 2.5 | 2.7 | 8.5 | 1.5 | 1.5 | 4.3 | 3.1 | 8.5 | 1.7 | 1.7 | 2.5 | 2.7 | 8.4 | | | | | | |

*A*, area cost %; *P*, power cost %; *E*, energy cost %.

Figure 5.8 shows that since ABFT detection cannot perform in-place correction, ABFT detection benchmarks cannot provide DUE improvement (any detected error necessarily increases the number of DUEs). Additionally, given the lower average SDC improvement and generally higher execution time impact for ABFT detection algorithms, combinations with ABFT detection do not yield low-cost solutions.

#### 5.3.2.1 Additional considerations for ABFT

Since most applications are not amenable to ABFT correction, the flip-flops protected by ABFT correction must also be protected by techniques such as LEAP-DICE or parity (or combinations thereof) for processors targeting general-purpose applications. This requires circuit hardening techniques (e.g., [69,80]) with the ability to selectively operate in an error-resilient mode (high resilience, high energy) when ABFT is unavailable, or in an economy mode (low resilience, low power mode) when ABFT is available. The LEAP-ctrl flip-flop accomplishes this task. The addition of LEAP-ctrl can incur an additional ~1% energy cost and ~3% area cost (Table 5.21).

Although 44% (22% for OoO-cores) of flip-flops would need to be implemented using LEAP-ctrl, only 5% (2% for OoO-cores) would be operating in economy mode at any given time (Table 5.22). Unfortunately, this requirement of fine-grained operating mode control is difficult to implement in practice, since it would require some firmware or software control to determine and pass information to a hardware controller indicating whether or not an ABFT application were running and which flip-flops to place in resilient mode and which to place in economy mode (rather than a simple switch setting all such flip-flops into the same operating mode). Therefore, cross-layer combinations using ABFT correction may not be practical or useful in general-purpose processors targeting general applications.

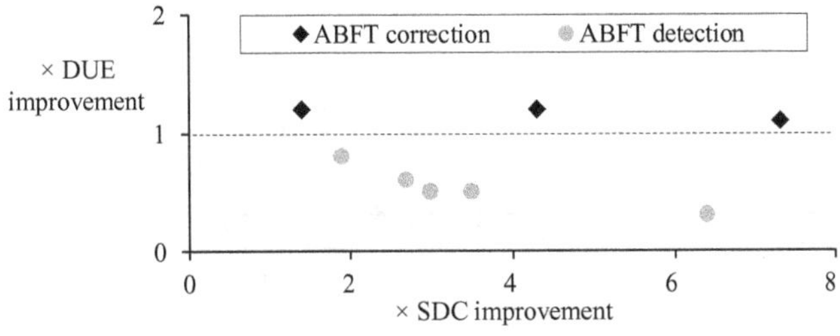


*Figure 5.8 ABFT correction and ABFT detection benchmark comparison*

*Table 5.22 Impact of ABFT correction on flip-flops*

| Core | Percentage of FFs with an error corrected by any ABFT algorithm (∪) | Percentage of FFs with an error corrected by every ABFT algorithm (∩) |
|---|---|---|
| InO | 44 | 5 |
| OoO | 22 | 2 |

## 5.4 Application benchmark dependence

The most cost-effective resilience techniques rely on selective circuit hardening/parity checking guided by error injection using application benchmarks. This raises the question: what happens when the applications in the field do not match application benchmarks? This situation is referred to as *application benchmark dependence.*

To quantify this dependence, one can randomly select 4 (of 11) SPEC benchmarks as a *training set* and use the remaining 7 as a *validation set.* Resilience is implemented using the training set and the resulting design's resilience is determined using the validation set. Therefore, the training set tells which flip-flops to protect and the validation set allows for the determination of what the actual improvement would be when this same set of flip-flops is protected. This chapter evaluated 50 training/validation pairs.

Since high-level techniques cannot be tuned to achieve a given resilience improvement, each technique is analyzed as a stand-alone technique to better understand how they perform individually. For stand-alone resilience techniques, the average inaccuracy between the results of trained and validated resilience is generally very low (Table 5.23) and is likely due to the fact that the improvements that the techniques themselves provided are already very low. *P*-values [81], which provide a measure of how likely the validated improvement and trained improvement would match, are also reported.

Table 5.24 indicates that validated SDC and DUE improvements are generally underestimated. Fortunately, when targeting <10× SDC improvement, the underestimation is minimal. This is due to the fact that the most *vulnerable* 10% of flip-flops (i.e., the flip-flops that result in the most SDCs or DUEs) are consistent across benchmarks. Since the number of errors resulting in SDC or DUE are not uniformly distributed among flip-flops, protecting these top 10% of flip-flops will result in the

*Table 5.23 Trained vs. validated improvement for high-level techniques*

| Core | Technique | SDC | | | | DUE | | | |
|---|---|---|---|---|---|---|---|---|---|
| | | Train | Validate | Underestimate[a] (%) | *P*-value | Train | Validate | Underestimate[a] (%) | *P*-value |
| InO | DFC | 1.3× | 1.2× | −7.7 | $3.8\times10^{-9}$ | 1.4× | 1.3× | −7.1 | $3.9\times10^{-17}$ |
| | Assertions | 1.5× | 1.4× | −6.7 | $2.4\times10^{-1}$ | 0.6× | 0.6× | 0 | $8\times10^{-2}$ |
| | CFCSS | 1.6× | 1.5× | −6.3 | $5.7\times10^{-1}$ | 0.6× | 0.6× | 0 | $9.2\times10^{-1}$ |
| | EDDI | 37.8× | 30.4× | −19.6 | $6.9\times10^{-1}$ | 0.4× | 0.4× | 0 | $2.2\times10^{-1}$ |
| | ABFT correction | 4.3× | 3.9× | −9.3 | $6.7\times10^{-1}$ | 1.2× | 1.2× | 0 | $1.8\times10^{-1}$ |
| OoO | DFC | 1.3× | 1.2× | −7.7 | $1.9\times10^{-5}$ | 1.4× | 1.3× | −7.1 | $1.4\times10^{-10}$ |
| | Monitor core | 19.6× | 17.5× | −5.6 | $8.3\times10^{-3}$ | 15.2× | 13.9× | −8.6 | $3.5\times10^{-7}$ |
| | ABFT correction | 4.3× | 3.7× | −14 | $7.2\times10^{-1}$ | 1.1× | 1.1× | 0 | $1.5\times10^{-1}$ |

[a]Underestimation is low because improvements are already low.

Table 5.24 *Improvement, cost before and after applying LHL to otherwise unprotected flip-flops*

| Core | SDC improvement | | | Cost before LHL insertion | | Cost after LHL insertion | | DUE improvement | | | Cost before LHL insertion | | Cost after LHL insertion | |
|---|---|---|---|---|---|---|---|---|---|---|---|---|---|---|
| | Train | Validate | After LHL | Area (%) | Power/energy (%) | Area (%) | Power/energy (%) | Train | Validate | After LHL | Area (%) | Power/energy (%) | Area (%) | Power/energy (%) |
| InO | 5× | 4.8× | 19.3× | 1.6 | 3.6 | 3.1 | 5.7 | 5× | 4.7× | 18.7× | 1.5 | 3.4 | 3.3 | 5.9 |
| | 10× | 9.6× | 38.2× | 1.7 | 3.9 | 3.1 | 5.7 | 10× | 8.7× | 34.6× | 1.9 | 4.2 | 3.5 | 6.5 |
| | 20× | 19.1× | 75.8× | 1.9 | 4.4 | 3.2 | 6.1 | 20× | 16.3× | 64.5× | 2.4 | 5.3 | 3.7 | 7 |
| | 30× | 26.8× | 105.6× | 2.2 | 4.8 | 3.2 | 6.3 | 30× | 23.5× | 92.7× | 2.8 | 6.6 | 3.7 | 8.1 |
| | 40× | 32.9× | 129.4× | 2.3 | 5.3 | 3.3 | 6.7 | 40× | 29.9× | 117.6× | 3.3 | 7.5 | 4.1 | 8.7 |
| | 50× | 38.9× | 152.3× | 2.4 | 5.7 | 3.3 | 6.9 | 50× | 35.9× | 140.6× | 3.6 | 8.4 | 4.2 | 9.4 |
| | 500× | 433.1× | 1,326.1× | 2.9 | 6.3 | 3.4 | 7.1 | 500× | 243.5× | 840.3× | 4.4 | 10.4 | 4.8 | 10.9 |
| | Max | 5,568.9× | 5,568.9× | 8 | 17.9 | 8 | 17.9 | Max | 5,524.7× | 5,524.7× | 8 | 17.9 | 8 | 17.9 |
| OoO | 5× | 4.8× | 35.1× | 0.1 | 0.2 | 0.9 | 1.8 | 5× | 4.4× | 28.7× | 0.7 | 0.1 | 1.8 | 1.7 |
| | 10× | 8.8× | 40.7× | 0.4 | 0.6 | 1.1 | 2.1 | 10× | 8.7× | 36.6× | 1.1 | 0.5 | 2.1 | 2 |
| | 20× | 18.8× | 65.6× | 0.7 | 1 | 1.3 | 2.3 | 20× | 17.3× | 70.2× | 1.5 | 0.9 | 2.5 | 2 |
| | 30× | 21.3× | 82.3× | 0.9 | 1.4 | 1.4 | 2.4 | 30× | 22.2× | 81.5× | 1.8 | 1.3 | 2.6 | 2.1 |
| | 40× | 26.4× | 130.2× | 1.2 | 1.7 | 1.7 | 2.5 | 40× | 26.1× | 115.1× | 2.1 | 1.6 | 2.8 | 2.4 |
| | 50× | 32.1× | 204.3× | 1.4 | 2.1 | 1.9 | 2.7 | 50× | 29.8× | 121.3× | 2.5 | 2 | 3.1 | 2.6 |
| | 500× | 301.4× | 1,084.1× | 2.2 | 2.4 | 2.4 | 2.8 | 500× | 153.2× | 625.1× | 2.9 | 1.9 | 3.4 | 2.7 |
| | Max | 6,625.8× | 6,625.8× | 4.9 | 7 | 4.9 | 7 | Max | 6,802.6× | 6,802.6× | 4.9 | 7 | 4.9 | 7 |

~10 × SDC improvement regardless of the benchmark considered. The vulnerabilities of the remaining 90% of flip-flops are more benchmark dependent. Concretely, it is possible to analyze benchmark similarity by examining the vulnerable flip-flops indicated by each application benchmark. Per benchmark, one can group the most vulnerable 10% of flip-flops into a subset (e.g., subset 1). The next 10% of vulnerable flip-flops (e.g., 10%–20%) are grouped into subset 2 (and so on up to subset 10). Therefore, given the 18 benchmarks used in this chapter, create 18 distinct subset 1s, 18 distinct subset 2s, and so on, are created. Each group of 18 subsets (e.g., all subset 1s) can then be assigned a similarity as given in (5.3). The similarity of subset $x$ is the number of flip-flops that exist in all subset $x$ (e.g., subset intersection) divided by the number of unique flip-flops in every subset $x$ (e.g., subset union). From Table 5.25, it is clear that only the top 10% most vulnerable flip-flops have very high commonality across all benchmarks (the last two subsets have high similarity because these are the flip-flops that have errors that always vanish). All other flip-flops are relatively distributed across the spectrum depending on the specific benchmark being run.

$$\text{Similarity (subset } x) = \frac{|\cap \text{(all flip-flops in every subset } x)|}{|\cup \text{(all flip-flops in every subset } x)|} \tag{5.3}$$

It is clear that for highly resilient designs, one must develop methods to combat this sensitivity to benchmarks. Benchmark sensitivity may be minimized by training using additional benchmarks or through better benchmarks (e.g., [82]). An alternative approach is to apply the CLEAR framework using available benchmarks, and then replace all remaining unprotected flip-flops using LHL (Table 5.4). This enables the resilient designs to meet (or exceed) resilience targets at ~1% additional cost for SDC and DUE improvements >10×.

*Table 5.25 Subset similarity across all 18 benchmarks for the InO-core (subsets broken into groups consisting of 10% of all flip-flops)*

| Subset (ranked by decreasing SDC+DUE vulnerability) (%) | Similarity (5.2) |
|---|---|
| 1: 0–10 | 0.83 |
| 2: 10–20 | 0.05 |
| 3: 20–30 | 0 |
| 4: 30–40 | 0 |
| 5: 40–50 | 0 |
| 6: 50–60 | 0 |
| 7: 60–70 | 0 |
| 8: 70–80 | 0 |
| 9: 80–90 | 0.71 |
| 10: 90–100 | 1 |

## 5.5 The design of new resilience techniques

CLEAR has been used to comprehensively analyze the design space of existing resilience techniques (and their combinations). As new resilience techniques are proposed, CLEAR can incorporate and analyze these techniques as well. However, CLEAR can also be used today to guide the design of new resilience techniques.

All resilience techniques will lie on a two-dimensional plane of energy cost vs. SDC improvement (Figure 5.9). The range of designs formed using combinations of LEAP-DICE, parity, and micro-architectural recovery forms the lowest cost cross-layer combination available using today's resilience techniques. In order for new resilience techniques to be able to create competitive cross-layer combinations, they must have energy and improvement trade-offs that place the technique under the region bounded by our LEAP-DICE, parity, and micro-architectural recovery solution. Since certain stand-alone techniques, like LEAP-DICE, can also provide highly competitive solutions, it is useful to understand the cost vs. improvement trade-offs for new techniques in relation to this best stand-alone technique as well (Figure 5.10).

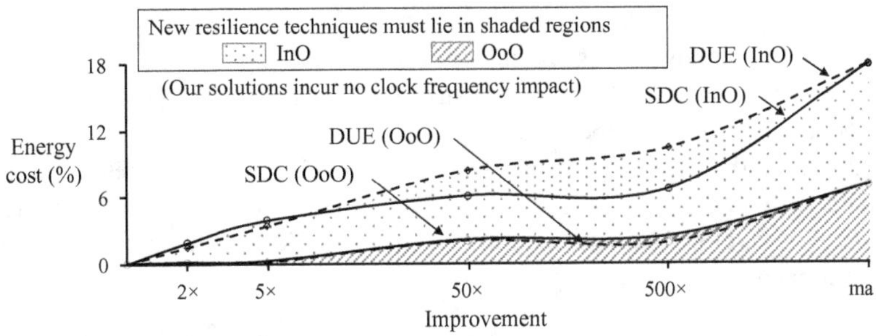


*Figure 5.9 New resilience techniques must have cost and improvement trade-offs that lie within the shaded regions bounded by LEAP-DICE+parity+micro-architectural recovery*

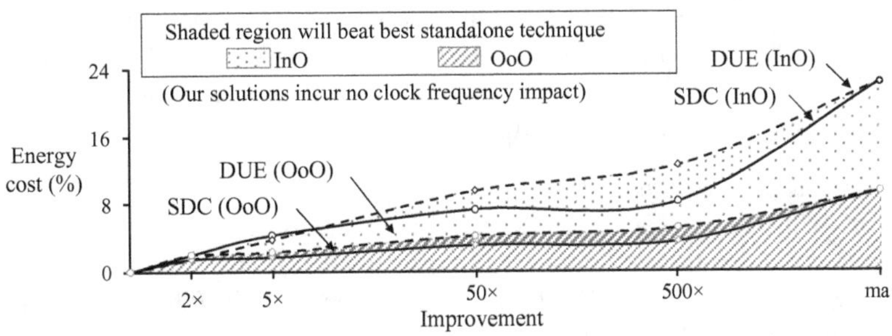


*Figure 5.10 To be competitive standalone solutions, new resilience techniques must have cost and improvement trade-offs that lie within the shaded regions bounded by LEAP-DICE*

## 5.6 Conclusions

CLEAR is a first of its kind cross-layer resilience framework that enables effective exploration of a wide variety of resilience techniques and their combinations across several layers of the system stack. Extensive cross-layer resilience studies using CLEAR demonstrate the following:

1. A *carefully optimized* combination of selective circuit-level hardening, logic-level parity checking, and micro-architectural recovery provides a highly cost-effective soft error resilience solution for general-purpose processors.
2. Selective circuit-level hardening alone, guided by thorough analysis of the effects of soft errors on application benchmarks, also provides a cost-effective soft error resilience solution (with ~1% additional energy cost for a 50× SDC improvement compared to the previous approach).
3. ABFT correction combined with selective circuit-level hardening (and logic-level parity checking and micro-architectural recovery) can further improve soft error resilience costs. However, existing ABFT correction techniques can only be used for a few applications; this limits the applicability of this approach in the context of general-purpose processors.
4. Based on this analysis, it is possible to derive bounds on energy costs vs. degree of resilience (SDC or DUE improvements) that new soft error resilience techniques must achieve to be competitive.
5. It is crucial that the benefits and costs of new resilience techniques are evaluated thoroughly and correctly. Detailed analysis (e.g., flip-flop-level error injection or layout-level cost quantification) identifies hidden weaknesses that are often overlooked.

While this chapter focuses on soft errors in processor cores, CLEAR has also been applied to a wide range of diverse hardware designs (general-purpose processor cores, hardware accelerators, uncore components such as cache controllers, memory controllers and I/O controllers) used in SoCs [16]. CLEAR results have also been validated for industrial designs. For example, CLEAR soft error explorations demonstrate a 50-fold improvement in SDC rate at 1.6% and 5.7% energy costs for uncore components and hardware accelerators, respectively. The voltage noise design space is explored to derive a solution for in-order processor cores at 2.1% energy cost.

Comprehensive insights derived from CLEAR are as follows:

1. *Circuit-level techniques are crucial* for cost-effective error resilience (e.g., LEAP-DICE error correction [23,40] for soft errors or circuit failure prediction using critical path monitors [83] for voltage noise). For example, application-guided selective LEAP-DICE insertion achieves a 50-fold improvement in soft-error-induced SDC rate for out-of-order processor cores at 3% energy cost (no speed impact). For accelerators, similar improvement is achieved at 7% energy cost.
2. *Application-level guidance is essential* (e.g., to identify vulnerable design blocks) for cost-effective resilience. However, some application-level knowledge may

be unavailable at design time. CLEAR demonstrates ways to overcome this challenge, but more research is required moving forward.

3. *Existing architecture and software-level resilience techniques are generally expensive* (16%–110% energy costs) and their low-cost variants provide too little resilience (e.g., less than 2-fold SDC rate improvement). Hence, circuit-level techniques are crucial (insight 1). In some cases, application-level techniques can be cost-effective, e.g., output timing monitoring using a watchdog timer reduces unsafe application behaviors 63-fold for embedded control [84].
4. *Some of the previously published resilience techniques, in particular those implemented at the architecture and software layers, suffer from incorrect evaluation.* CLEAR uses accurate evaluation methods to correct some of these incorrect conclusions and optimistic projections from the past. For example, injecting random errors into software-visible registers might lead to an optimistic estimation of 19-fold (vs. actual 2-fold) SDC rate improvement (for soft errors) using certain software-level resilience techniques. Optimistic projections may also result from overlooking hidden costs in physical designs.
5. *Cost-effective protection from multiple error sources (e.g., soft errors and voltage noise) is achieved by combining resilience techniques that specifically target each error source.* For example, an in-order processor core can achieve combined soft error and voltage noise resilience at 8.2% energy cost vs. 28% or more using resilience techniques that are agnostic to error sources.
6. *CLEAR derives highly effective and low-cost resilience techniques by combining existing techniques*. Hence, new resilience techniques must be extremely efficient in terms of cost vs. degree of resilience achieved to be competitive.

## Acknowledgments

This work is supported in part by DARPA MTO (contract no. HR0011-13-C-0022), DTRA, NSF, and SRC. The views expressed are those of the authors and do not reflect the official policy or position of the Department of Defense or the US Government. Distribution Statement "A" (Approved for Public Release, Distribution Unlimited).

## References

[1] Borkar, S., "Designing reliable systems from unreliable components: the challenges of transistor variability and degradation," *IEEE Micro*, vol. 25, no. 6, pp. 10–16, 2005.

[2] Cappello, F., A. Geist, W. Gropp, S. Kale, B. Kramer, and M. Snir, "Toward Exascale Resilience: 2014 update," *Supercomputing Frontiers and Innovations*, vol. 1, no. 1, pp. 5–28, 2014.

[3] Carter, N. P., H. Naeimi, and D. S. Gardner, "Design techniques for cross-layer resilience," *Design, Automation Test in Europe Conference Exhibition*, pp. 1023–1028, 2014.

[4] DeHon, A., H. M. Quinn, and N. P. Carter, "Vision for cross-layer optimization to address the dual challenges of energy and reliability," *Design, Automation, & Test in Europe*, pp. 1017–1022, 2010.

[5] Gupta, M. S., J. A. Rivers, L. Wang, and P. Bose, "Cross-layer system resilience at affordable power," *IEEE Intl. Reliability Physics Symp.*, pp. 2B.1.1–2B.1.8, 2014.

[6] Henkel, J., L. Bauer, H. Zhang, S. Rehman, and M. Shafique, "Multi-layer dependability: from microarchitecture to application level," *ACM/EDAC/IEEE Design Automation Conf.*, pp. 1–6, 2014.

[7] Pedram, M., D. Brooks, and T. Pinkston, "Report for the NSF workshop on cross-layer power optimization and management," *NSF*, 2012.

[8] Lu, D. J., "Watchdog processor and structural integrity checking," *IEEE Transactions on Computers*, vol. C-31, no. 7, pp. 681–685, 1982.

[9] Ando, H., Y. Yoshida, and A. Inoue, "A 1.3-GHz fifth-generation SPARC64 microprocessor," *Solid-State Circuits Conference. Digest of Technical Papers. ISSCC. 2003 IEEE International*, vol. 1, pp. 246–491, 2003.

[10] Meaney, P. J., S. B. Swaney, P. N. Sanda, and L. Spainhower, "IBM z990 soft error detection and recovery," *IEEE Transactions on Device and Materials Reliability*, vol. 5, no. 3, pp. 419–427, 2005.

[11] Sinharoy, B., R. Kalla, W. J. Starke, *et al.*, "IBM POWER7 multicore server processor," *IBM Journal of Research and Development*, vol. 55, no. 3, pp. 1:1–1:29, 2011.

[12] Cheng, E., S. Mirkhani, L. G. Szafaryn, *et al.*, "CLEAR: cross-layer exploration for architecting resilience – combining hardware and software techniques to tolerate soft errors in processor cores," in *Proceedings of the 53rd Annual Design Automation Conference*, ACM, New York, NY, USA.

[13] Cheng, E., J. Abraham, P. Bose, *et al.*, "Cross-layer resilience in low-voltage digital systems: key insights," *ICCD*, pp. 593–596, 2017.

[14] Cheng, E., *Cross-layer resilience to tolerate hardware errors in digital systems*, Ph.D. dissertation, Stanford University, 2018.

[15] Cheng, E., S. Mirkhani, L. G. Szafaryn, *et al.*, "Tolerating soft errors in processor cores using CLEAR (cross-layer exploration for architecting resilience)," *IEEE Transactions on Computer-Aided Design of Integrated Circuits and Systems*, vol. 37, no. 9, pp. 1839–1852, 2018.

[16] Cheng, E., D. Mueller-Gritschneder, J. Abraham, *et al.*, "Cross-layer resilience: challenges, insights and the road ahead," *DAC*, pp. 1–4, 2019.

[17] Mitra, S., P. Bose, E. Cheng, *et al.*, "The resilience wall: cross-layer solution strategies," *Intl. Symp. VLSI Technology, Systems and Applications*, pp. 1–11, 2014.

[18] Seifert, N., V. Ambrose, B. Gill, *et al.*, "On the radiation-induced soft error performance of hardened sequential element in advanced bulk CMOS technologies," *IEEE Intl. Reliability Physics Symp.*, pp. 188–197, 2010.

[19] Seifert N., B. Gill, S. Jahinuzzaman, *et al.*, "Soft error susceptibilities of 22 nm tri-gate devices," *IEEE Transactions on Nuclear Science*, vol. 59, no. 6, pp. 2666–2673.

[20] Mahatme, N. N., N. J. Gaspard, S. Jagannathan, *et al.*, "Impact of supply voltage and frequency on the soft error rates of logic circuits," *IEEE Transactions on Nuclear Science*, vol. 60, no. 6, pp. 4200–4206,

[21] Pawlowski, R., J. Crop, M. Cho, *et al.*, "Characterization of radiation-induced SRAM and logic soft errors from 0.33 V to 1.0 V in 65 nm CMOS," *IEEE Custom Integrated Circuits Conf.*, pp. 1–4, 2014.

[22] Gill, B., N. Seifert, and V. Zia, "Comparison of alpha-particle and neutron-induced combinational and sequential logic error rates at the 32 nm technology node," *IEEE Intl. Reliability Physics Symp.*, pp. 199–205, 2009.

[23] Lee, H.-H. K., K. Lilja, M. Bounasser, *et al.*, "LEAP: layout design through error-aware transistor positioning for soft-error resilient sequential cell design," *IEEE Intl. Reliability Physics Symp.*, pp. 203–212, 2010.

[24] Cho, H., C. Y. Cher, T. Shepherd, and S. Mitra, "Understanding soft errors in uncore components," *ACM/EDAC/IEEE Design Automation Conf.*, pp. 1–6, 2015.

[25] Henning, J. L., "SPEC CPU2000: measuring CPU performance in the new millennium," *IEEE Computer*, vol. 33, no. 7, pp. 28–45, 2000.

[26] K. Barker, T. Benson, D. Campbell *et al.*, "PERFECT: power efficiency revolution for embedded computing technologies. Benchmark Suite Manual, Pacific Northwest National Laboratory and Georgia Tech Research Institute, December 2013, http: //hpc.pnnl.gov/projects/PERFECT.

[27] Aeroflex G., "Leon3 processor," http://www.gaisler.com.

[28] Wang N. J., J. Quek, T. M. Rafacz, and S. J. Patel, "Characterizing the effects of transient faults on a high-performance processor pipeline," *IEEE/IFIP Intl. Conf. Dependable Systems & Networks*, pp. 61–70, 2004.

[29] Bottoni, C., M. Glorieux, J. M. Daveau, *et al.*, "Heavy ions test result on a 65 nm SPARC-V8 radiation-hard microprocessor," *IEEE Intl. Reliability Physics Symp.*, pp. 5F.5.1–5F.5.6, 2014.

[30] Sanda, P. N., J. W. Kellington, and P. Kudva, "Soft-error resilience of the IBM POWER6 processor," *IBM J. of Res. and Dev.*, vol. 52, no. 3, pp. 275–284, 2008.

[31] Cho, H., S. Mirkhani, C. Y. Cher, J. A. Abraham and S. Mitra, "Quantitative evaluation of soft error injection techniques for robust system design," *50th ACM/EDAC/IEEE Design Automation Conf.*, pp. 1–10, 2013.

[32] Davis, J, and C. Thacker, "BEE3: revitalizing computer architecture research," *Microsoft*, 2009.

[33] Ramachandran, P., P. Kudva, J. Kellington, J. Schumann, and P. Sanda, "Statistical fault injection," *IEEE International Conference on Dependable Systems and Networks With FTCS and DCC (DSN)*, pp. 122–127, 2008.

[34] Michalak, S. E., A. J. DuBois, C. B. Storlie *et al.*, "Assessment of the impact of cosmic-ray-induced neutrons on hardware in the roadrunner supercomputer," *IEEE Trans. on Dev. Mat. Rel.*, vol. 12, no. 2, pp. 445–454, 2012.

[35] Wang, N. J., A. Mahesri, and S. J. Patel, "Examining ACE analysis reliability estimates using fault-injection," *Proceedings of the 34th Annual International Symposium on Computer Architecture*, ACM, New York, NY, USA, 2007.

[36] Schirmeier, H., C. Borchert, and O. Spinczyk, "Avoiding pitfalls in fault-injection based comparison of program susceptibility to soft errors," *45th Annual IEEE/IFIP International Conference on Dependable Systems and Networks*, pp. 445–454, 2015.

[37] Mirkhani, S., S. Mitra, C. Y. Cher, and J. Abraham, "Efficient soft error vulnerability estimation of complex designs," *Design, Automation Test in Europe Conference Exhibition*, pp. 103–108, 2015.

[38] Sullivan, M., B. Zimmer, S. Hari, T. Tsai, and S. W. Keckler, "An analytical model for hardened latch selection and exploration," *Silicon Errors in Logic – System Effects*, 2016.

[39] Synopsys, Inc., "Synopsys design suite," https://www.synopsys.co.

[40] Lilja, K., M. Bounasser, and S. J. Wen, "Single-event performance and layout optimization of flip-flops in a 28-nm bulk technology," *IEEE Transactions on Nuclear Science*, vol. 60, no. 4, pp. 2782–2788, 2013.

[41] Lilja, K., M. Bounasser, and T. R. Assis, "SER prediction in advanced finFET and SOI finFET technologies; challenges and comparisons to measurements," *Single Events Effects Symp.*, 2016.

[42] Quinn, R. C., J. S. Kauppila, T. D. Loveless *et al.*, "Frequency trends observed in 32 nm SOI flip-flops and combinational logic," *IEEE Nuclear and Space Radiation Effects Conf.*, 2015.

[43] Quinn, R. C., J. S. Kauppila, T. D. Loveless *et al.*, "Heavy ion SEU test data for 32 nm SOI flip-flops," *IEEE Radiation Effects Data Workshop*, pp. 1–5, 2015.

[44] Turowski, M., "32 nm SOI SRAM and latch SEU cross-sections measured (heavy ion data) and determined with simulations," *Single Event Effects Symp.*, 2015.

[45] Bowman, K. A., J. W. Tschanz, N. S. Kim *et al.*, "Energy-efficient and metastability-immune resilient circuits for dynamic variation tolerance," *IEEE J. Solid-State Circuits*, vol. 44, no. 1, pp. 49–63, 2009.

[46] Bowman, K. A., J. W. Tschanz, S. L. L. Lu *et al.*, "A 45 nm resilient microprocessor core for dynamic variation tolerance," *IEEE J. Solid-State Circuits*, vol. 46, no. 1, pp. 194–208, 2011.

[47] Sahoo S. K., M. L. Li, P. Ramachandran *et al.*, "Using likely program invariants to detect hardware errors," *IEEE Intl. Conf. Dependable Systems and Networks with FTCS and DCC (DSN)*, pp. 70–79, 2008.

[48] Lin, D., T. Hong, Y. Li *et al.*, "Effective post-silicon validation of system-on-chips using quick error detection," *IEEE Transactions on Computer-Aided Design of Integrated Circuits and Systems*, vol. 33, no. 10, pp. 1573–1590, 2014.

[49] Spainhower, L. and T. A. Gregg, "IBM S/390 parallel enterprise server G5 fault tolerance: a historical perspective," *IBM J. Res. & Dev.*, vol. 43, no. 5.6, pp. 863–873, 1999.

[50] Amusan, O. A., *Effects of single-event-induced charge sharing in sub-100 nm bulk CMOS technologies*, Vanderbilt University Dissertation, pp. 985–994, 2009.

[51] Mitra S. and E. J. McCluskey, "Which concurrent error detection scheme to choose?," *IEEE Intl. Test Conf.*, 2000.

[52] Meixner, A., M. E. Bauer, and D. Sorin, "Argus: low-cost comprehensive error detection in simple cores," *40th Annual IEEE/ACM Intl. Symp. Microarchitecture*, pp. 210–222, 2007.

[53] Austin, T. M., "DIVA: a reliable substrate for deep submicron microarchitecture design," *32nd Annual Intl. Symp. Microarchitecture, MICRO-32*, pp. 196–207, 1999.

[54] Hari S. K. S., S. V. Adve, and H. Naeimi, "Low-cost program-level detectors for reducing silent data corruptions," *IEEE/IFIP Intl. Conf. Dependable Systems & Networks*, pp. 1–12, 2012.

[55] Oh, N., P. P. Shirvani, and E. J. McCluskey, "Control flow checking by software signatures," *IEEE Transactions on Reliability*, vol. 51, no. 1, pp. 111–122, 2002.

[56] Lovellette, M. N., K. S. Wood, D. L. Wood *et al.*, "Strategies for fault-tolerant, space-based computing: lessons learned from the ARGOS testbed," *IEEE Aerospace Conf.*, vol. 5, pp. 5-2109–5-2119, 2002.

[57] Oh, N., P. P. Shirvani, and E. J. McCluskey, "Error detection by duplicated instructions in super-scalar processors," *IEEE Transactions on Reliability*, vol. 51, no. 1, pp. 63–75, 2002.

[58] Pattabiraman K., Z. T. Kalbarczyk, and R. K. Iyer, "Automated derivation of application-aware error detectors using static analysis: the trusted ILLIAC approach," *IEEE Transactions on Dependable and Secure Computing*, vol. 8, no. 1, pp. 44–57, 2011.

[59] Rehman, S., F. Kriebel, M. Shafique, and J. Henkel, "Reliability-driven software transformations for unreliable hardware," *IEEE Transactions on Computer-Aided Design of Integrated Circuits and Systems*, vol. 33, no. 11, pp. 1597–1610, 2014.

[60] Bosilca, G., R. Delmas, J. Dongarra, and J. Langou, "Algorithmic based fault tolerance applied to high performance computing," *Journal of Parallel and Distributed Computing*, vol. 69, no. 4, pp. 410–416, 2009.

[61] Chen, Z. and J. Dongarra, "Numerically stable real number codes based on random matrices," *Computational Science – ICCS 2005: 5th International Conference*, Berlin, Springer Berlin Heidelberg, pp. 115–122, 2005.

[62] Huang, K.-H. and J. A. Abraham, "Algorithm-based fault tolerance for matrix operations," *IEEE Transactions on Computers*, vol. C-33, no. 6, pp. 518–528, 1984.

[63] Nair, V. S. S. and J. A. Abraham, "Real-number codes for fault-tolerant matrix operations on processor arrays," *IEEE Transactions on Computers*, vol. 39, no. 4, pp. 426–435, 1990.

[64] Reddy, A. L. N. and P. Banerjee, "Algorithm-based fault detection for signal processing applications," *IEEE Transactions on Computers*, vol. 39, no. 10, pp. 1304–1308, 1990.

[65] Racunas P., K. Constantinides, S. Manne, and S. S. Mukherjee, "Perturbation-based fault screening," *13th IEEE Intl. Symp. High Performance Computer Architecture*, pp. 169–180, 2007.

[66] Wang, N. J. and S. J. Patel, "ReStore: symptom based soft error detection in microprocessors," *Intl. Conf. Dependable Systems and Networks (DSN)*, pp. 30–39, 2005.

[67] Calin, T., M. Nicolaidis, and R. Velazco, "Upset hardened memory design for submicron CMOS technology," *IEEE Transactions on Nuclear Science*, vol. 43, no. 6, pp. 2874–2878, 1996.

[68] Furuta, J., C. Hamanaka, K. Kobayashi, and H. Onodera, "A 65 nm bistable cross-coupled dual modular redundancy flip-flop capable of protecting soft errors on the C-element," *Symp. VLSI Circuits*, pp. 123–124, 2010.

[69] Mitra S., N. Seifert, M. Zhang, Q. Shi, and K. S. Kim, "Robust system design with built-in soft error resilience," *IEEE Computer*, vol. 38, no. 2, pp. 43–52, 2005.

[70] Blaauw, D., S. Kalaiselvan, K. Lai *et al.*, "Razor II: in situ error detection and correction for PVT and SER tolerance," *IEEE Intl. Solid-State Circuits Conf. – Digest of Technical Papers*, pp. 400–622, 2008.

[71] Fojtik, M., D. Fick, Y. Kim *et al.*, "Bubble razor: eliminating timing margins in an ARM cortex-M3 processor in 45 nm CMOS using architecturally independent error detection and correction," *IEEE J. Solid-State Circuits*, vol. 48, no. 1, pp. 66–81, 2013.

[72] Nicolaidis, M., "Time redundancy based soft-error tolerance to rescue nanometer technologies," *IEEE VLSI Test Symp.*, pp. 86–94, 1999.

[73] Franco, P., and E. J. McCluskey, "On-line delay testing of digital circuits," *IEEE VLSI Test Symp.*, pp. 167–173, 1994.

[74] Berger, J. M., "A note on error detection codes for asymmetric channels," *Information and Control*, vol. 4, no. 1, pp. 68–73, 1961.

[75] Bose, B., and D. J. Lin, "Systematic unidirectional error-detecting codes," *IEEE Transactions on Computers*, 1985.

[76] Mukherjee, S. S., M. Kontz, and S. K. Reinhardt, "Detailed design and evaluation of redundant multi-threading alternatives," *29th Annual Intl. Symp. on Computer Architecture*, pp. 99–110, 2002.

[77] Feng, S., S. Gupta, A. Ansari, and S. Mahlke, "Shoestring: probabilistic soft error reliability on the cheap," *Proceedings of the 15th Edition of ASPLOS on Architectural Support for Programming Languages and Operating Systems*, ACM, New York, NY, USA, 2010

[78] Reis, G. A., J. Chang, N. Vachharajani, R. Rangan, and D. I. August, "SWIFT: software implemented fault tolerance," *IEEE/ACM Intl. Symp. Code Generation and Optimization*, pp. 243–254, 2005.

[79] Reis, G. A., J. Chang, N. Vachharajani, S. S. Mukherjee, R. Rangan, and D. I. August, "Design and evaluation of hybrid fault-detection systems," *32nd Intl. Symp. on Computer Architecture*, pp. 148–159, 2005.

[80] Zhang, M., S. Mitra, T. M. Mak, *et al.*, "Sequential element design with built-in soft error resilience," *IEEE Transactions on Very Large Scale Integration*, vol. 14, no. 12, pp. 1368–1378, 2006.

[81] Wasserstein, R. L. and N. A. Lazar, "The ASA's statement on p-values: context, process, and purpose," *The American Statistician*, vol. 70, no. 2, pp. 129–133, 2016.

[82] Mirkhani, S., B. Samynathan, and J. A. Abraham, "In-depth soft error vulnerability analysis using synthetic benchmarks," *IEEE 33rd VLSI Test Symp.*, pp. 1–16, 2015.

[83] Lefurgy, C. R., A. J. Drake, M. S. Floyd, *et al.*, "Active management of timing guardband to save energy in POWER7," *IEEE/ACM Intl. Symp. on Microarchitecture*, pp. 1–11, 2011.

[84] Mueller-Gritschneder, D., M. Dittrich, J. Weinzierl *et al.*, "ETISS-ML: a multi-level instruction set simulator with RTL-level fault injection support for the evaluation of cross-layer resiliency techniques," *DATE,* pp. 609–612, 2018.

*Part II*

# Reliability assessment

# Chapter 6
# Physical stress

*Fernando Fernandes dos Santos*[1], *Fabio Benevenuti, Gennaro Rodrigues*[2], *Fernanda Kastensmidt*[2], *and Paolo Rech*[1]

## 6.1 Introduction

The physical fault injection techniques apply external perturbations on the circuit or code under test to evaluate its reliability. This chapter will present the most important techniques, challenges, and benefits of the different available physical fault injection. In particular, we will cover field test and accelerated particle (neutrons, protons, or heavy ions) beams and laser injection as techniques for artificially reproducing realistic faults. It is important to notice that the choice of which physical fault source to use strongly depends on the application for which the device reliability needs to be qualified. Neutrons are used for terrestrial or avionic applications. Protons, x-rays, and gamma rays are used for low orbit or space applications, while heavy ions normally are necessary for deep space applications. Laser techniques, as we will discuss in Section 6.7, are used to identify the circuit vulnerable regions.

The main advantage of most physical fault injections (laser is an exception discussed later) is that they provide a realistic evaluation of the device error rate in a given application. By injecting faults in all the processor resources, physical fault injection enables us to assess the actual reliability for different benchmarks, which typically cannot be achieved by Software Fault Injection (SFI) only. In fact, SFI is, in most of the cases, limited to a subset of user-accessible resources. Even if Register-Transfer Level (RTL) level description is available, in principle allowing the user to inject faults in all the possible sites, the full device characterization may take too long to be feasible. Moreover, without beam experiments it is nearly impossible to find the precise fault generation probability and the realistic fault model. As such, pure SFI could lead to rough error rate evaluations. Still, SFI is fundamental to fully understand a system reliability because during radiation experiments faults are observed only at the code output, limiting the observability and insights that can be directly applied for hardening solutions. In other words, beam experiments do not allow one to correlate

[1]PPGC, Instituto de Informática/UFRGS, Porto Alegre, Brazil
[2]PGMICRO, Instituto de Informática/UFRGS, Porto Alegre, Brazil

the observed effect with the cause. When an error is observed at the output, there is no indication of where/when the beam generated the fault.

Field experiments are a very expensive and time-consuming physical stress test during which a high number of devices are exposed to the natural flux of particles. As the flux (i.e., number of particles/cm$^2$/h) in the natural environment is very low, to have sufficient statistic it is necessary to expose a very high number of devices for a long time. The prohibitive costs of field experiments limit the data available to some industrial data [1] and some supercomputer data [2].

The main advantage of physical fault injection using accelerated particle beams is that, by increasing the number of particles that hit the device, the time needed to gather a statistically significant amount of data is drastically reduced. Given the high particle flux compared to the device realistic application and the consequent high probability of faults, it is essential to tune the experiment setup not to introduce artifacts in the reliability evaluation, as explained in the following sections.

In this chapter we will, in Section 6.2, clearly discuss the sources of faults in modern computing devices. In Section 6.3 we list the metrics adopted to quantify the reliability of electronic devices. Sections 6.5, 6.6, and 6.7 describe the procedures to correctly perform experiments using neutron, heavy ions, and laser beams, respectively. Finally, Section 6.8 concludes the chapter.

## 6.2 Effects and physical sources

The first important decision to make when preparing a physical fault injection campaign regards the source of faults at which the device should be exposed. The selection of the source strongly depends on the device operative application and on the effect that needs to be evaluated or observed.

Radiation effects can be broadly separated in transient (or soft) errors and permanent effects. Single-Event Effects (SEE) are noncumulative and caused by single events that trigger transient upsets. Total Ionizing Dose (TID) and Displacement Damage are cumulative, which means that their effects get worse over time as the system is exposed to radiation. In this chapter, we will focus on SEE, as TID experiments are well defined and just need to expose the device to a controlled source (x-rays or gamma-rays).

Soft errors do not permanently damage the circuit. They are typically radiation-induced current spikes that reverse the stored bit or change the output of a logic operation. The struck memory cell will correctly maintain the value of a new write operation, and subsequent stimuli at the struck logic cell are likely to correctly propagate. When considering a computing device, a soft error (in memory or logic) leads to (1) no effect on the program output (i.e., the fault is masked, the corrupted data is not used, or the circuit functionality is not affected), (2) a Silent Data Corruption (SDC) (i.e., an incorrect program output), or (3) a Detected Unrecoverable Error (DUE) (i.e., an detected but not correctable memory event, a crash, or device reboot). Permanent effects are caused by the long-term charge deposition in silicon (called TID) or by the activation of inherit paths from the power lines and ground (called latchup). TID is caused by the continuous exposure of the device to ionizing particles, and it typically reduces the transistor efficiency, making it gradually slower to switch (reducing the device performances and increasing the power consumption) until the transistor could not be turned on or off. A latchup is caused by a highly energetic ionizing particle that induces a short circuit between power rails, quickly increasing the current passing through the device. If proper (and prompt) actions are not taken (like cutting the device power supply or performing a power cycle of the system), the device will be literally burned with dramatic effects on the mission.

While, potentially, each particle or energy deposition could induce both soft and permanent faults, the experiences so far with current and past technology tell that neutrons-induced permanent fault probability is negligible. For terrestrial applications, then, the main concerns are soft errors. Protons and heavy ions are known to cause both soft and permanent faults, while x-rays and gamma rays are mainly used to inject charge and measure the TID. A device to be used in terrestrial application should then be characterized using neutron beams. A device to be applied in space or deep-space needs to undergo both protons and heavy-ion experiments to measure the probability of having soft errors and latchup. Then, x-rays and gamma rays campaign should be used to measure the dose at which the device starts to degrade. TID is particularly interesting for long-term applications like satellites engineered to be operative for years.

It is worth noting that modern parallel computing architectures such as Graphics Processing Units (GPUs) or Xeon Phi have some inherent reliability weaknesses [3–6]. A single particle-induced fault in the scheduler or shared memory is likely to affect the correctness of several parallel threads or kernels, leading to the corruption of multiple output values [7]. Additionally, a single corrupted thread could feed thousands of future parallel threads with erroneous data, again leading to multiple errors [8].

SRAM-based FPGAs have a peculiar fault-model. On FPGAs, in fact, neutrons can corrupt the configuration memory, changing the implemented circuit. Such an event is still *soft* but *persistent* as the configuration memory will remain corrupted until the load of a new bitstream into the device.

Previous studies have already evaluated the reliability of FPGA, GPUs, and Xeon Phi running various codes through radiation experiments [9–11] or fault injection [12–16].

Some preliminary works tried also to combine beam experimental data with fault injection. Cher *et al.* [17] use both proton irradiation and SFI to study the soft error resilience of BlueGene/Q. Proton bombardment shows that BG/Q has a mean time between correctable errors of 1.5 days validating the need for detection mechanisms. The SFI study was performed to analyze the behavior of applications without the mitigation mechanism employed on BG/Q hardware. Single and multiple bitflips are the two fault models used in the SFI study. They propose to use a more robust Error Correcting Code (ECC) on the bits or registers that show to be more sensitive. They also show that some executions of LINPACK pass the residual test providing satisfactory results even when the output does not exactly match. Fratin *et al.* [18] compare the SFI and beam experiment for a wide set of devices and codes, showing that, for commercial devices, only for some specific behaviors fault injection and beam experiments provide similar trends. Oliveira *et al.* [19] show in detail how fault injection and beam experiments are both necessary to characterize the reliability of complex computing systems as Intel Xeon Phis. Chatzidimitriou *et al.* [20] demonstrate that microarchitectural fault injection can be used to predict the SDC rate of embedded ARM processors (differences are, on average, in the order of $2\times$) while fault injection significantly underestimates the occurrence of crashes (at least one order of magnitude of difference with beam data).

## 6.3 Reliability metrics

By radiation testing, one can observe the vulnerability of a certain device under a set of particles. There are two main types of testing: static and dynamic. Static testing is when the device is powered on, but it is not executing any code or application and upsets accumulation in the memory elements of the device. Dynamic testing is when the device is executing a code and the output of the code is constantly monitored to observe errors. Both static and dynamic testings can be performed to measure the sensible area of a device, named cross section (static cross section and dynamic cross section of a device). The Cross Section is expressed with the unit of an area ($cm^2$), and it defines the vulnerable area of the device, intended as the area that, once corrupted by radiation, impacts the correct functionality of the device. The Cross Section is calculated dividing the number of observed errors (SDC, DUE, etc.) by the fluence (the amount of particles passed per unit area) received by the device. As such, the Cross Section is the probability for an impinging particle to generate an error.

If the kind of particles (including their energies spectrum) used for the experiment is similar to the kind of particles that will hit the device in its normal application on the environment, the cross section can be used to evaluate the device error rate. In fact, multiplying the Cross Section ($cm^2$) with the flux of particles in a natural operation of the device (particles/$cm^2$/h), we obtain the error rate (errors/h) of the device. From the error rate, we can calculate the device Failure In Time (FIT), i.e., errors per $10^9$ h of operation.

The Cross Section and, thus, the FIT depend only on the kind and amount of resources required for computation, without considering the code execution time [11]. As most computing devices configurations are likely to impact both the error rate and performance, it is possible to consider performance-reliability trade-offs by calculating the Mean Executions Between Failures (MEBF). MEBF is obtained by dividing the error rate by the execution time and is the number of correct executions completed before experiencing a failure [21,22]. Additionally, the amount of data produced in one execution depends on the kind of code that is being executed. It is then useful to measure the Mean Work Between Failures, which is the amount of data correctly produced before experiencing a failure [21]. The metric that better describes the device reliability depends on the application criticality.

To have a more detailed evaluation of the behavior of a device or code when exposed to radiation, it is also useful to evaluate the impact of faults on the codes' output correctness by measuring how the FIT rate is reduced when we increase the Tolerated Relative Error (TRE) [18]. A TRE of 0% considers any mismatch between the corrupted output and the expected value as a critical error. We then relax the correctness constraint accepting (corrupted) values in a given range as tolerable (corrected). As an example, if we consider a TRE of 10%, any output value between 90% and 100% of the expected value will be considered as correct or, at least, tolerable. TRE analysis is useful to compare the impact of transient faults in the execution of operations. Oliveira *et al.* [10] study the radiation effect on NVIDIA and Intel accelerators using beam tests. The study defines metrics to quantify and qualify the radiation effects observing the magnitude and how the error spread in the final output. They state that simple mismatch detection cannot accurately evaluate the radiation sensitivity of modern devices and algorithms. They also show that the reliability of some, not all, application classes are connected to the architecture while other applications are independent of the underlying architecture.

So in conclusion, it is interesting to point out that the majority of the SEEs will not cause permanent damage in the system and they can be detected, corrected, or masked by redundant-based fault-tolerant techniques implemented either in hardware or in software. Although in some cases, according to the device complexity, reset or power cycle will be needed to vanish completely the error. Some SEE will cause permanent damage to the system by provoking latchup or micro latchup.

## 6.4 General setup

While each kind of radiation, each facility, each device, and each test have its own peculiarities, there are some general guidelines and characteristics that apply to any beam experiments. Such a general setup is depicted in Figure 6.1.

The Device Under Test (DUT) is the device that should be irradiated and whose reliability measurement is of interest. Normally, the irradiation room is not easily accessible by the user. Definitely, when the beam is turned on, the user should not be in the irradiation room. This means that the setup should be remotely controlled. For neutron test, the experiment is performed in air and cables can be run between the irradiation room and the user space. Depending on the facility, these cables might be

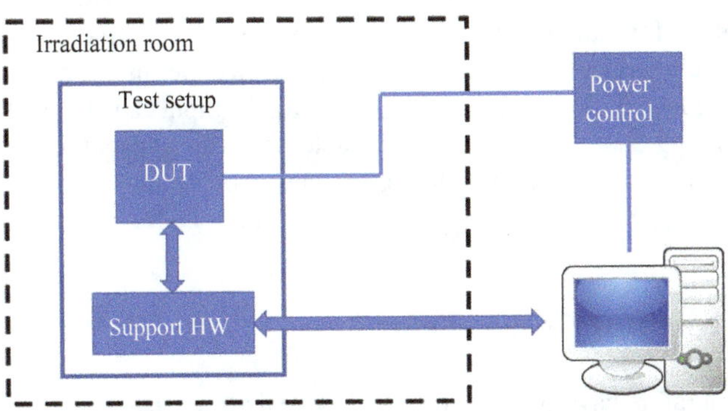


*Figure 6.1 General setup for a radiation experiment*

very long (in the order of 10 m). For heavy-ion test, the experiment must be performed in vacuum, forcing the use of few interconnections between the inside and outside of the test chamber. Testing in vacuum also limits the power dissipation of the device and should be considered.

Normally, the DUT is controlled by a support hardware that is in charge of sending input signal and controlling the output. The support hardware is placed closed to the DUT or even in the same board of the DUT. The support hardware is also in charge of communicating with a host PC and, then, with the user.

It is fundamental to include a power control system in the setup. This is necessary to perform automatic power cycle of the device in the case of unexpected current increase (latchup, i.e., a short circuit between power and ground) or to resolve hang situations that might occur during an experiment (a motherboard gets stuck, for instance).

Additionally, the gathering of data, the detection of errors or those situations that require a system reboot or a device power cycle should be automatized as much as possible. The error detection should be synchronized with the facility particle counts and should be as fast as possible to have a precise estimation of the number of particles required to produce the error. Moreover, experiments can be very long and run overnight. If the device is stuck or needs to be rebooted while the user is not in the facility would mean losing a lot of time.

When performing a radiation experiment, it is important to decide which information to log. Normally, the number of events observed (SDC, DUE, crashes, hangs, etc.) is the minimum information necessary (together with the particles counts provided by the facility) to measure the cross section and, then, the FIT rate. It is also interesting to log the corrupted value and the expected value to understand the effect of radiation in the output correctness and measure the TRE.

It is worth noting that the number of observed events should be sufficiently large not to have the calculated sensitivities dominated by statistical error. Normally, experiments should be designed to observe at least 100 events per kind per configuration. A rough estimation of the expected error rate can be made looking at previous data available on line for similar devices or technology.

One of the most difficult challenges when performing a device characterization is deciding which codes to run. Each code will produce different error rate, and even the same code can provide different error rates in different devices. While a benchmark suite for radiation experiments has not yet been defined, a previous work lists some benchmarks of interest to conduct radiation experiments on processors and programmable logic [23]. On top of this, changing the input can change the error rate of a code running on a device. Among variables that can impact the measured error rate, there are also temperature, voltage, frequency, etc. It is fundamental to design properly the setup, eventually using operative or realistic codes, input, and conditions.

## 6.5 Neutron beam experiments

Particle accelerators have been used for many years to measure and study radiation sensitivities for components and applications. As described in the other chapters of this book, SFI has also been widely used to estimate and study the resilience of hardware and applications.

There are only few facilities in the world that can provide a neutrons spectrum of energy that is similar to the terrestrial one. The most known facilities are the Los Alamos Neutron Science Center (LANSCE) at the Los Alamos National Laboratory in NM, USA, the ChipIR facility at the Rutherford Appleton Laboratory (RAL) in the Oxfordshire, UK, and TRIUMF in Vancouver, Canada. All these facilities have a neutron spallation source. Neutrons are produced by the interaction of accelerated protons beams with a target of heavy materials, typically tungsten.

As shown in Figure 6.2 (taken from [24]), the spectrum of energy distribution of neutrons produced at LANSCE or ChipIR is very similar to the one measured at sea level. This attests that the kind of neutrons produced in the accelerated facility is very similar to the kind of neutrons that reach ground. The fault model (i.e., the way the neutron perturbs computation) and the error rate (i.e., the probability for a neutron to produce an error) observed in the accelerated facilities is very similar to the ones expected at sea level. The interesting aspect of using an accelerated beam is that the amount of neutrons produced is much higher than the one at sea level (which is 13 neutrons/cm$^2$/h). The average flux of neutrons produced at LANSCE or ChipIR is in the order of $10^8$ or $10^9$ neutrons/cm$^2$/h. As a result, 1 h of beam test will provide the amount of neutrons received by a device operating in normal condition for about $10^7$ h, which is equal to more than 1,000 years.

The similarity of the neutron spectrum of LANSCE and ChipIR with the terrestrial one is helpful as it allows one to derive, from the cross section experimentally measured, the expected FIT rate of the DUT in a realistic application. In fact, by dividing the number of observed events by the received neutron fluence during the experiment, we can calculate the cross section, in cm$^2$. The cross section is basically

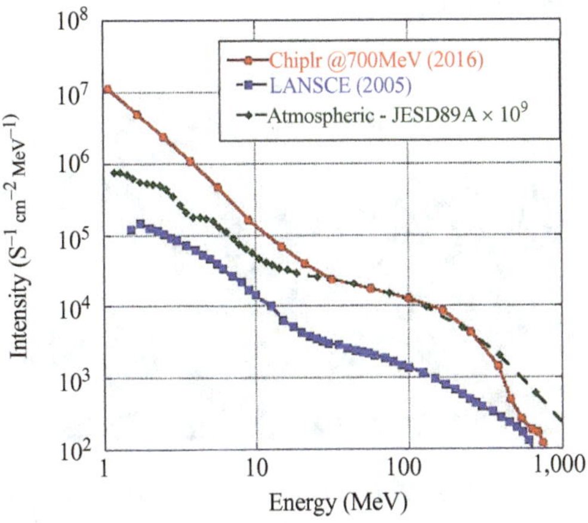


*Figure 6.2 Spectra of neutron energies produced at LANSCE and at ChipIR as compared to the atmospheric neutron one [24]*

the probability for one impinging neutron to generate an observable error. As the neutrons at LANSCE and ChipIR resemble the terrestrial ones, we can assume that this cross section would be the same in a normal application of the device. The only difference, as mentioned, is that during beam experiments the flux and, then, the error rate is higher. If we multiply the cross section (again, the probability for one neutron to generate a failure) by the expected neutron flux in the device realistic application (about 13 neutrons/cm$^2$/h at sea level), we can predict the device error rate. Multiplying this error rate by $10^9$, we calculate the device FIT rate.

The high flux allows one to gather a statistically significant amount of data in a short time. Normally, 1–2 h of experiments are sufficient to get about 100 errors. However, the high flux poses some issues. First, it is fundamental to avoid the possibility of having multiple neutrons generating errors in one code execution. This would be unrealistic, as in normal condition the flux is so low that it is basically impossible to have more than one neutron-induced corruption per code execution. Pragmatically, this is achieved tuning the experiment (distance from the source, execution time, memory utilization) to ensure to have at least 100 or 1,000 correct executions between two failures. This is a good sign that one error is observed, and it is likely to be generated by one and not more than one neutron. The second issue regards the setup stability. A high flux can impact the DUT and the supporting hardware. It is fundamental, when preparing a radiation test, to contemplate redundancy in the hardware.

One of the most interesting aspects of neutron beam experiments is that, as neutrons can penetrate several feet of concrete or lead, the user can stack multiple boards in the beam. This allows one to perform multiple experiments at the same time. Figure 6.3, taken from [23], shows a setup mounted at LANSCE. Figure 6.4 shows a particular of that setup, taken from [25]. Neutrons come from the right side of the picture and interact with all the devices. Obviously, there is an attenuation factor to be considered when stacking a lot of devices, which includes a contribution from the distance (the flux is reduced with the square of the distance) and from the upstream

*Figure 6.3 Setup at LANSCE [23]. Various devices were aligned with the neutron beam, including microcontrollers (1–4), FPGAs (5 and 9), Systems on Chip (6 and 10), Graphics Processing Units (8), and space mission controllers (11)*

*Figure 6.4* *Particular of the Setup at LANSCE for the Systems on Chip. Neutrons interact with multiple boards (the yellow beam is a cartoon)*

devices. It has been shown that thick devices as GPUs can reduce the flux of about 15% for the downstream devices.

## 6.6 Heavy ions and proton experiments

Proton facilities reach very high energies (usually between 50 and 200 MeV), and consequently devices tested under protons do not need to have their package thinned. However, if the energy of ions is not sufficient to penetrate the device and reach its active area (i.e., the transistors level), it is necessary to instrument the device removing the package and thinning the passivation layer. Heavy-ion facilities are capable of generating protons with sufficient energy to simulate solar flares and Earth's proton belt conditions. One of the main proton facilities is located at the Paul Scherrer Institut in Switzerland.

Heavy-ion facilities are generally used for testing parts destined for space orbit, where primary cosmic rays can cause significant damage to electronic devices. The energy measurement unit of heavy ion experiments is the Linear Energy Transfer (LET), which describes the amount of energy lost per unit length of track. In other words, it describes the action of radiation upon matter, or how much energy an ionizing particle transfers to the material traversed per unit distance. Thus, the LET depends on the nature of the radiation as well as on the material traversed. For the heavy-ion experiments, the package of a device must be thinned to allow that irradiated particles penetrate the active region of the silicon. It is important to have a good estimation of thickness of the materials that must be passed by the ion until this ion reaches the active layers of the transistors. This information is used in the SRIM software [26] to estimate the energy loss of the heavy ions and to calculate the effective LET values in the border of the active layers.

The most important difference among heavy-ion experiments and both neutron and proton experiments is related to the dosimetry. Devices exposed to proton testing have also accumulated doses and consequently they also have been tested for TID. Devices exposed to heavy ions usually accumulate an irrelevant amount of dose.

Static and dynamic testings can be performed under protons and heavy ions. The static and dynamic cross sections will be calculated during the tests for each specific particle used associated with its energy and the LET produced.

There is also the case in which electronic components are exposed to a very high flux of different particle species. This is the reality of the CERN's accelerators chain, where electronic components can be exposed to high-energy hadrons (protons, neutrons, pions), heavy ions, and other particles, at the same time. Aiming to simulate such complex environment, in 2015 CERN, started operating a new and unique mixed-field radiation test facility, the CERN High Energy Accelerator Mixed-field (CHARM), located at CERN, Switzerland. At CHARM, it is possible to have particles with energies near 10 GeV.

Brazil has a heavy-ion facility, which is located at the Universidade de São Paulo (USP), the Pelletron 8UD Accelerator at the Nuclear Physics Open Laboratory of University of São Paulo (LAFN-USP). At that 8-MV Pelletron accelerator a cesium-sputtering ion source provides heavy-ion beams from $^{1}H$ to $^{107}Ag$ with energies from 10 to 125 MeV with one of the beamlines designed mainly for irradiation of electronic devices [27].

Figure 6.5 shows the general experimental setup in the Pelletron Accelerator at USP, Brazil. While based on the general setup presented on Figure 6.1, this setup, however, usually has the DUT test board installed inside a vacuum chamber since the room atmosphere would interact with heavy-ions particles. The setup also includes, outside the experimental chamber, a DUT control computer and respective campaign scripts that can generate stimuli to DUT, collect diagnosis data, and record relevant events for further analysis. The setup typically uses some power switching hardware allowing for a full reset of the DUT by power cycling.

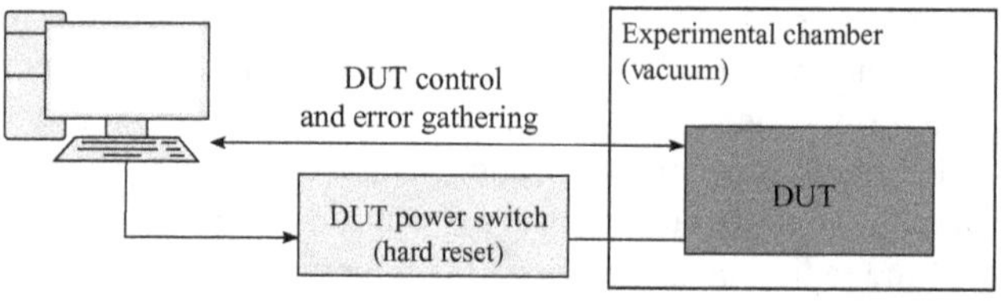


*Figure 6.5 Heavy-ion experiments setup*

Not only room atmosphere but also the integrated circuit package, usually made of epoxy and some forms of quartz, can interact with heavy-ion particles preventing it from reaching the active region of the integrated circuit of changing the experiment conditions. To avoid this, the integrated circuit package can be decapped (or deluded), on the Region of Interest (ROI).

Besides different heavy-ion species and energies, geometric parameters of the setup such as the beam angle of incidence and the DUT test board rotation can be adjusted to obtain different beam paths inside the integrated circuit. The beam incidence angle can increase the traversed length inside the integrated circuit before reaching the active region, leading to different LETs, and can also create conditions more favorable to observe multiple Memory Bits Upsets caused by a single-radiation event.

Figure 6.6 depicts the integrated circuit decapped package and the angle of incidence and rotation. The DUT test board with a decapped integrated circuit can also be seen inside the experimental chamber in Figure 6.7 that depicts the DUT control computer outside the experimental chamber and the Pelletron 0° (SAFIIRA) beamline.

Several experiments have already been conducted with Xilinx 28 nm Zynq-7000 and Artix-7 FPGA in the Pelletron Facility. Table 6.1 shows some examples of the different cocktail of ions to produce different LETs in the Zynq-7000 and Artix-7 FPGA device exposed to heavy ions.

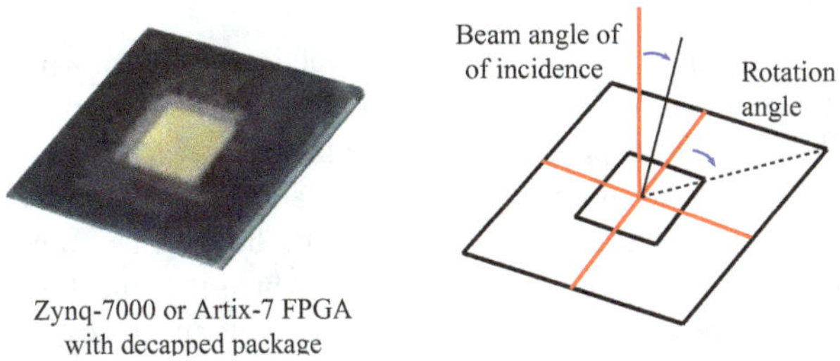


*Figure 6.6 Decapped integrated circuit under test and incidence and rotation angles*

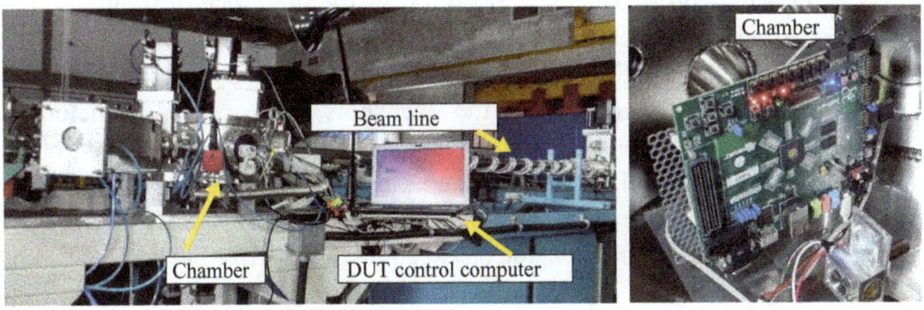


*Figure 6.7 Experimental chamber and 0° (SAFIIRA) beamline*

*Table 6.1 Heavy ions experiments on Zynq FPGA at Pelletron Accelerator*

| Ion | Energy (MeV) | Beamline | Incidence angle (°) | Rotation angle (°) | LET (MeV/mg/cm$^2$) | Penetration in Si ($\mu$m) | Works |
|---|---|---|---|---|---|---|---|
| $^{7}$Li | 20 | 30B | 60 | 90 | 3.26 | – | [28] |
| $^{11}$B | 42 | 30B | 45 | 90 | 3.74 | – | [28] |
| $^{11}$B | 42 | 30B | 60 | 90 | 6.74 | – | [28] |
| $^{12}$C | 50 | 30B | 0 | 0 | 3 | – | [28] |
| $^{12}$C | 50 | 30B | 45 | 0 | 5.21 | – | [28] |
| $^{12}$C | 50 | 30B | 60 | 0 | 9.13 | – | [28] |
| $^{12}$C | 28 | 30B | 0 | 90 | 5.25 | – | [28] |
| $^{12}$C | 35 | 30B | 0 | 0 | 4.35 | 31 | [29] |
| $^{16}$O | 60 | 30B | 45 | 90 | 9.57 | – | [28] |
| $^{16}$O | 39 | 0° | 0 | 0 | 5.5 | 24 | [30,31] |
| $^{16}$O | 51 | 30B | 0 | 0 | 5 | 29 | [32] |
| $^{16}$O | 55 | 30B | 0 | 0 | 5 | 28.5 | [33] |
| $^{16}$O | 56 | 30B | 0 | 0 | 5 | 28.5 | [34] |
| $^{16}$O | 59 | 30B | 0 | 0 | 4.5 | 28 | [35] |
| $^{19}$F | 52 | 30B | 0 | 90 | 8.26 | – | [28] |

## 6.7 Laser test

The laser beams generate a photon–material interaction instead of a particle–material interaction. Laser tests provide good accessibility of the circuit elements where faults are injected. The small laser spot and precise beam localization allow sensitive device nodes to be identified with submicron accuracy. Taking into account the complexity of modern devices, laser testing is especially useful because it can provoke charges with spatial localization and temporal precision, which is mandatory for analyzing faults that can sometimes be masked in particle accelerator beam tests. It is commonly used as an in-lab tool for injecting transient localized perturbations into a DUT by photoelectric stimulation, especially for SEE's investigations [36], security evaluation [37], and more generally to evaluate the fault tolerance of an application [38].

The most common and used laser experiment method is Two-Photon Absorption (TPA). The TPA method is usually preferred over the more classical single-photon approach because its fault occurrences are better reproducible. For laser testing, it is essential to observe at least these three aspects: to have access to the backside of the device, to observe the frequency rate of the laser shots, and know the diameter of the laser spot in correlation of the layout and technology of the target device in order to correlate to single and multiple upsets. Examples of laser experiment facilities are the one in the National Research Laboratory in the USA, the IES laser facility at the University of Montpellier in France, the laser center of the MEPhI in Russia, followed by many others.

As a practical example of a laser experiment, we can point out the work published by [39], where the embedded hardcore ARM processors from Zynq FPGA from Xilinx

(28 nm technology) were tested in the IES facility. For this experiment, the laser wavelength was 1,064 nm, with a pulse duration of 30 ps, and the beam was focused through the backside of the DUT by a 100× lens. The DUT was scanned under the laser beam using motorized stages. The laser pulse energy was set at 250 pJ. This value was found in the previous work [40] to induce between 0 and 3 bit flips per pulse in the ROI of the DUT, depending on the laser position. This energy is slightly above the energy threshold for a single bit flip. Laser pulses were triggered with a frequency of 10 or 20 Hz (depending on the benchmark under execution), without any synchronization with the DUT clock. Making use of this asynchronous approach, the methodology assures randomness concerning the arrival time of the laser pulse and the current application execution cycle. This allows the statistical coverage of any vulnerability time window without the need for unreasonable experiment time. As Figure 6.8 shows, the laser fault injections were performed in two different ROIs: the On-Chip Memory (OCM) and the L1 Data Cache memory (shared between the two processor cores of the dual-core ARM-A9 embedded at the Zynq board). This was possible due to the aforementioned precision attainable by the laser fault injection method, capable of targeting small areas of the DUT with precision.

Figure 6.9 presents a schematic of the laser fault injection experimental setup performed in the IES laser facility. Notice that it is indeed similar to the setup proposed in Figure 6.1. It is composed of a computer that controls the laser equipment and collects data coming from the Zynq board (DUT). The DUT continually communicates with the control computer to maintain a log of the errors encountered during the execution. When a fault induces an execution flow error, the DUT can get stuck in an indefinite never-ending execution (e.g., an infinite loop), which freezes and

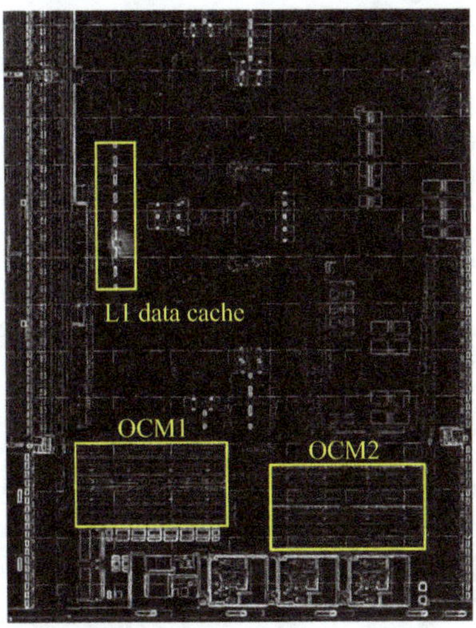


*Figure 6.8 Infrared microphotograph of the DUT showing the scanned areas (L1 Data Cache and OCM)*

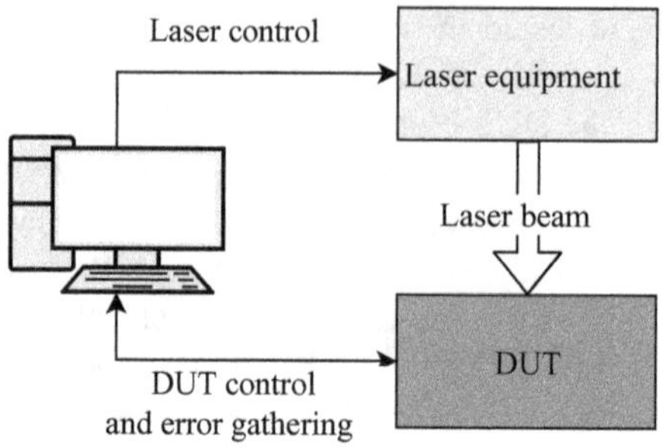


*Figure 6.9 Laser experiments setup [39]*

compromises the entire experiment process. In those cases, the DUT has to be reset, and an execution flow error is registered on the experiment log file. During the reset of the DUT, the computer pauses the laser injections to make sure that the reset is well performed (unless the researchers are interested in initialization error analysis, which was not the case). Eventually, the laser fault injection can induce a permanent fault (such as a latchup) that only vanishes after a hard reset of the DUT, i.e., needing a power cycle.

Laser fault injection experiments arise as a methodology ideal for the evaluation of individual, specific areas of the DUT. Indeed, performing an overall evaluation using the laser, with the intention to emulate faults affecting random locations of the DUT hardware, would take an extensive amount of time. Parameters such as the acceleration and deceleration phases at the extremities of each scan line of the ROI have to be taken into account to guarantee the proper pulse intensity on the $x$- and $y$-axis. Moving the laser head to different positions after each pulse takes time that depends on the distance between the two points. For practical reasons, the laser pulses are more efficient when covering consecutive points, close to each other, that define an ROI, and much less efficient when having to move to a random new position inside an enormous ROI after each pulse.

## 6.8 Conclusions

In this chapter, we have covered some of the general challenges in the physical stress of electronic devices. The benefits of using particles beams to evaluate the reliability of devices are various, such as a realistic error rate and a realistic error model. Additionally, thanks to the accelerated beam, a statistically significant amount of data can be gathered in a relatively short time (hours). However, the preparation of the setup is very challenging and, while there are some general guidelines, each facility has its own constraints. When performing beam experiments, it is fundamental to carefully design the setup. As beam experiments do not allow visibility as fault injection, it is necessary to ensure the test of all components and to test proper benchmarks. Finally, the correlation between beam experiments data and fault injection is still an open question.

## References


[1] A. Lesea, S. Drimer, J. J. Fabula, C. Carmichael, and P. Alfke, "The rosetta experiment: atmospheric soft error rate testing in differing technology FPGAs," *IEEE Transactions on Device and Materials Reliability*, vol. 5, pp. 317–328, 2005.

[2] T. Siddiqua, V. Sridharan, S. E. Raasch, *et al.*, "Lifetime memory reliability data from the field," in *2017 IEEE International Symposium on Defect and Fault Tolerance in VLSI and Nanotechnology Systems (DFT)*, pp. 1–6, Oct 2017.

[3] N. DeBardeleben, S. Blanchard, L. Monroe, *et al.*, "GPU behavior on a large HPC cluster," in *6th Workshop on Resiliency in High Performance Computing (Resilience) in Clusters, Clouds, and Grids in Conjunction With the 19th International European Conference on Parallel and Distributed Computing (Euro-Par 2013)*, (Aachen, Germany), Aug 26–30, 2013.

[4] H. J. Wunderlich, C. Braun, and S. Halder, "Efficacy and efficiency of algorithm-based fault-tolerance on GPUs," in *On-Line Testing Symposium (IOLTS), 2013 IEEE 19th International*, pp. 240–243, Jul 2013.

[5] L. A. B. Gomez, F. Cappello, L. Carro, *et al.*, "GPGPUs: How to combine high computational power with high reliability," in *2014 Design Automation and Test in Europe Conference and Exhibition* (Dresden, Germany), 2014.

[6] D. A. G. de Oliveira, L. L. Pilla, T. Santini, and P. Rech, "Evaluation and mitigation of radiation-induced soft errors in graphics processing units," *IEEE Transactions on Computers*, vol. 65, pp. 791–804, 2016.

[7] P. Rech, T. Fairbanks, H. Quinn, and L. Carro, "Threads distribution effects on graphics processing units neutron sensitivity," *IEEE Transactions on Nuclear Science*, vol. 60, pp. 4220–4225, 2013.

[8] G. Li, K. Pattabiraman, C. Y. Cher, and P. Bose, "Understanding error propagation in GPGPU applications," in *SC16: International Conference for High Performance Computing, Networking, Storage and Analysis*, pp. 240–251, Nov 2016.

[9] H. Quinn, P. Graham, J. Krone, M. Caffrey, and S. Rezgui, "Radiation-induced multi-bit upsets in SRAM-based FPGAs," *IEEE Transactions on Nuclear Science*, vol. 52, no. 6, pp. 2455–2461, 2005.

[10] D. Oliveira, L. Pilla, M. Hanzich, *et al.*, "Radiation-induced error criticality in modern HPC parallel accelerators," in *2017 IEEE International Symposium on High Performance Computer Architecture (HPCA)*, pp. 577–588, 2017. Available from: http://doi=10.1109/HPCA.2017.41.

[11] R. Baumann, "Radiation-induced soft errors in advanced semiconductor technologies," *IEEE Transactions on Device and Materials Reliability*, vol. 5, pp. 305–316, 2005.

[12] B. Fang, K. Pattabiraman, M. Ripeanu, and S. Gurumurthi, "GPU-Qin: A methodology for evaluating the error resilience of GPGPU applications," in *Performance Analysis of Systems and Software (ISPASS), 2014 IEEE International Symposium on*, pp. 221–230, Mar 2014.

[13] S. K. S. Hari, T. Tsai, M. Stephenson, S. W. Keckler, and J. Emer, "SASSIFI: An architecture-level fault injection tool for GPU application resilience evaluation," in *International Symposium on Performance Analysis of Systems and Software*, Oct 2017.

[14] D. Oliveira, V. Frattin, P. Navaux, I. Koren, and P. Rech, "CAROL-FI: An efficient fault-injection tool for vulnerability evaluation of modern HPC parallel accelerators," in *Proceedings of the Computing Frontiers Conference*, CF'17, (New York, NY, USA), pp. 295–298, ACM, 2017.

[15] M. Kaliorakis, S. Tselonis, A. Chatzidimitriou, N. Foutris, and D. Gizopoulos, "Differential fault injection on microarchitectural simulators," in *2015 IEEE International Symposium on Workload Characterization*, pp. 172–182, Oct 2015.

[16] Q. Lu, M. Farahani, J. Wei, A. Thomas, and K. Pattabiraman, "LLFI: An intermediate code-level fault injection tool for hardware faults," in *2015 IEEE International Conference on Software Quality, Reliability and Security*, pp. 11–16, Aug 2015.

[17] C. Cher, M. S. Gupta, P. Bose, and K. P. Muller, "Understanding soft error resiliency of Blue Gene/Q compute chip through hardware proton irradiation and software fault injection," in *SC '14: Proceedings of the International Conference for High Performance Computing, Networking, Storage and Analysis*, pp. 587–596, Nov 2014.

[18] V. Fratin, D. Oliveira, C. Lunardi, F. Santos, G. Rodrigues, and P. Rech, "Code-dependent and architecture-dependent reliability behaviors," in *2018 48th Annual IEEE/IFIP International Conference on Dependable Systems and Networks (DSN)*, pp. 13–26, Jun 2018.

[19] D. Oliveira, L. Pilla, N. DeBardeleben, *et al.*, "Experimental and analytical study of Xeon Phi reliability," in *Proceedings of the International Conference for High Performance Computing, Networking, Storage and Analysis*, SC '17, (New York, NY, USA), pp. 28:1–28:12, ACM, 2017.

[20] A. Chatzidimitriou, P. Bodmann, G. Papadimitriou, D. Gizopoulos, and P. Rech, "Demystifying soft error assessment strategies on ARM CPUs: Microarchitectural fault injection vs. neutron beam experiments," in *2019 49th Annual IEEE/IFIP International Conference on Dependable Systems and Networks (DSN)*, pp. 26–38, Jun 2019.

[21] G. Reis, J. Chang, N. Vachharajani, and S. Mukherjee, "Design and evaluation of hybrid fault-detection systems," in *Proceedings of the 2005 International Symposium on Computer Architecture, ISCA'05*, pp. 148–159, IEEE Press, 2005.

[22] P. Rech, L. L. Pilla, P. O. A. Navaux, and L. Carro, "Impact of GPUs parallelism management on safety-critical and HPC applications reliability," in *IEEE International Conference on Dependable Systems and Networks (DSN 2014)*, (Atlanta, GA, USA), 2014.

[23] H. Quinn, W. H. Robinson, P. Rech, *et al.*, "Using benchmarks for radiation testing of microprocessors and FPGAs," *IEEE Transactions on Nuclear Science*, vol. 62, pp. 2547–2554, 2015.

[24] C. Cazzaniga and C. D. Frost, "Progress of the scientific commissioning of a fast neutron beamline for chip irradiation," *Journal of Physics: Conference Series*, vol. 1021, p. 012037, 2018.

[25] T. Santini, P. Rech, L. Carro, and F. R. Wagner, "Exploiting cache conflicts to reduce radiation sensitivity of operating systems on embedded systems," in *Proceedings of the 2015 International Conference on Compilers, Architecture and Synthesis for Embedded Systems*, CASES '15 (Piscataway, NJ, USA), pp. 49–58, IEEE Press, 2015.

[26] J. F. Ziegler, M. Ziegler, and J. Biersack, "SRIM" – The stopping and range of ions in matter (2010)," *Nuclear Instruments and Methods in Physics Research Section B: Beam Interactions with Materials and Atoms*, vol. 268, no. 11, pp. 1818–1823, 2010. 19th International Conference on Ion Beam Analysis.

[27] N. H. Medina, V. A. P. Aguiar, N. Added, *et al.*, "Experimental setups for single event effect studies," *Journal of Nuclear Physics, Material Sciences, Radiation and Applications*, vol. 4, pp. 13–23, 2016.

[28] J. Tonfat, F. L. Kastensmidt, L. Artola, *et al.*, "Analyzing the influence of the angles of incidence and rotation on MBU events induced by low LET heavy ions in a 28-nm SRAM-based FPGA," *IEEE Transactions on Nuclear Science*, vol. 64, pp. 2161–2168, 2017.

[29] F. M. Lins, L. A. Tambara, F. L. Kastensmidt, and P. Rech, "Register file criticality and compiler optimization effects on embedded microprocessor reliability," *IEEE Transactions on Nuclear Science*, vol. 64, pp. 2179–2187, 2017.

[30] L. A. C. Benites, F. Benevenuti, A. B. de Oliveira, *et al.*, "Reliability calculation with respect to functional failures induced by radiation in TMR Arm Cortex-M0 soft-core embedded into SRAM-based FPGA," *IEEE Transactions on Nuclear Science*, vol. 66, pp. 1433–1440, 2019.

[31] A. B. de Oliveira, G. S. Rodrigues, F. L. Kastensmidt, *et al.*, "Lockstep dual-core ARM A9: Implementation and resilience analysis under heavy ion-induced soft errors," *IEEE Transactions on Nuclear Science*, vol. 65, pp. 1783–1790, 2018.

[32] E. Chielle, F. Rosa, G. S. Rodrigues, *et al.*, "Reliability on arm processors against soft errors through SIHFT techniques," *IEEE Transactions on Nuclear Science*, vol. 63, pp. 2208–2216, 2016.

[33] L. A. Tambara, J. Tonfat, A. Santos, *et al.*, "Analyzing reliability and performance trade-offs of HLS-based designs in SRAM-based FPGAs under soft errors," *IEEE Transactions on Nuclear Science*, vol. 64, pp. 874–881, 2017.

[34] F. Libano, P. Rech, L. Tambara, J. Tonfat, and F. Kastensmidt, "On the reliability of linear regression and pattern recognition feedforward artificial neural networks in FPGAs," *IEEE Transactions on Nuclear Science*, vol. 65, pp. 288–295, 2018.

[35] L. A. Tambara, P. Rech, E. Chielle, J. Tonfat, and F. L. Kastensmidt, "Analyzing the impact of radiation-induced failures in programmable SoCs," *IEEE Transactions on Nuclear Science*, vol. 63, pp. 2217–2224, 2016.

[36] S. P. Buchner, F. Miller, V. Pouget, and D. P. McMorrow, "Pulsed-laser testing for single-event effects investigations," *IEEE Transactions on Nuclear Science*, vol. 60, pp. 1852–1875, 2013.

[37] E. Trichina and R. Korkikyan, "Multi fault laser attacks on protected CRT-RSA," in *2010 Workshop on Fault Diagnosis and Tolerance in Cryptography*, pp. 75–86, Aug 2010.

[38] V. Pouget, A. Douin, G. Foucard, *et al.*, "Dynamic testing of an SRAM-based FPGA by time-resolved laser fault injection," in *2008 14th IEEE International On-Line Testing Symposium*, pp. 295–301, Jul 2008.

[39] G. Rodrigues, J. Fonseca, F. Kastensmidt, V. Pouget, A. Bosio, and S. Hamdioui, "Approximate TMR based on successive approximation and loop perforation in microprocessors," *Microelectronics Reliability*, vol. 100–101, p. 113385, 2019.

[40] V. Pouget, S. Jonathas, R. Job, J. Vaillé, F. Wrobel, and F. Saigné, "Structural pattern extraction from asynchronous two-photon laser fault injection using spectral analysis," *Microelectronics Reliability*, vol. 76–77, no. Supplement C, pp. 650–654, 2017.

# Chapter 7
# Soft error modeling and simulation

*Mojtaba Ebrahimi*[1] *and Mehdi Tahoori*[1]

Although the sources of soft errors are device-level interactions, the generated errors could propagate and cause system-level failures. As a result, it is very important to analyze the impact of soft errors using a device to system-level approach. Therefore, an efficient soft error vulnerability estimation technique has to be able to accurately model the error generation at device-level as well as the masking behavior at higher abstraction levels. The proposed cross-layer Soft Error Rate (SER) analysis platform employs a combination of empirical models at the device level, error site analysis at chip layout, analytical Error Propagation (EP) at logic level, and fault simulation/emulation at the architecture/application level to provide the detailed contribution of each component (flip-flops, combinational gates, and memory arrays) to the overall SER. At each stage in the modeling hierarchy, an appropriate level of abstraction is used to propagate the effect of errors to the next higher level.

## 7.1 Introduction

While soft errors are generated at device level, they could propagate and impair the output of the running workload. As a result, it is important to evaluate the impact of this kind of errors using a combined knowledge of various layers in a cross-layer approach. Development of a fast and accurate soft error analysis requires a detailed information about underlying technology, functionality of the Register Transfer Level (RTL) code, and also the running workload characteristics. Among these, the underlying technology is only important in error generation phase as well as the first cycle of EP. In addition, the error generation phase is independent of the functionality of the design. Considering such dependencies, as shown in Figure 7.1, the error generation and propagation phases could be divided into three phases: error generation at device-level, EP in first cycle, and EP after first cycle.

The proposed cross-layer platform is shown in Figure 7.2. For error generation, all the cells were analyzed using an accurate SER analysis tool to obtain the raw Fault Injection Technique (FIT) rates. Also, Multiple Event Transient (MET) error sites are extracted according to the layout information. Then, by employing an

[1]Department of Computer Science, Karlsruhe Institute of Technology, Karlsruhe, Germany

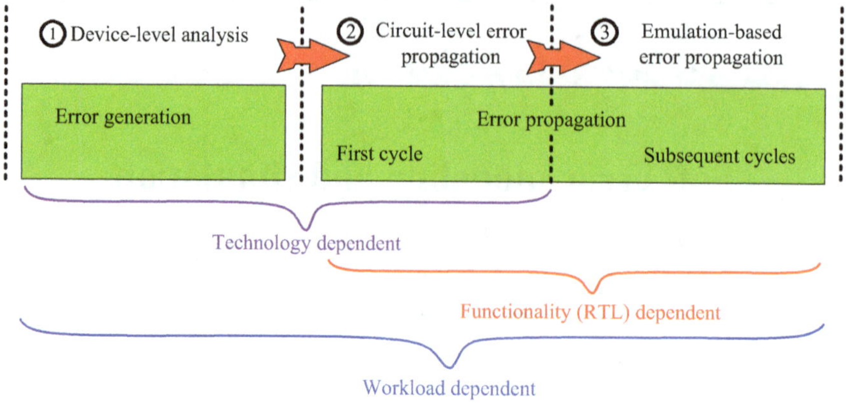


*Figure 7.1 Reasoning behind proposed cross-layer soft error analysis platform*

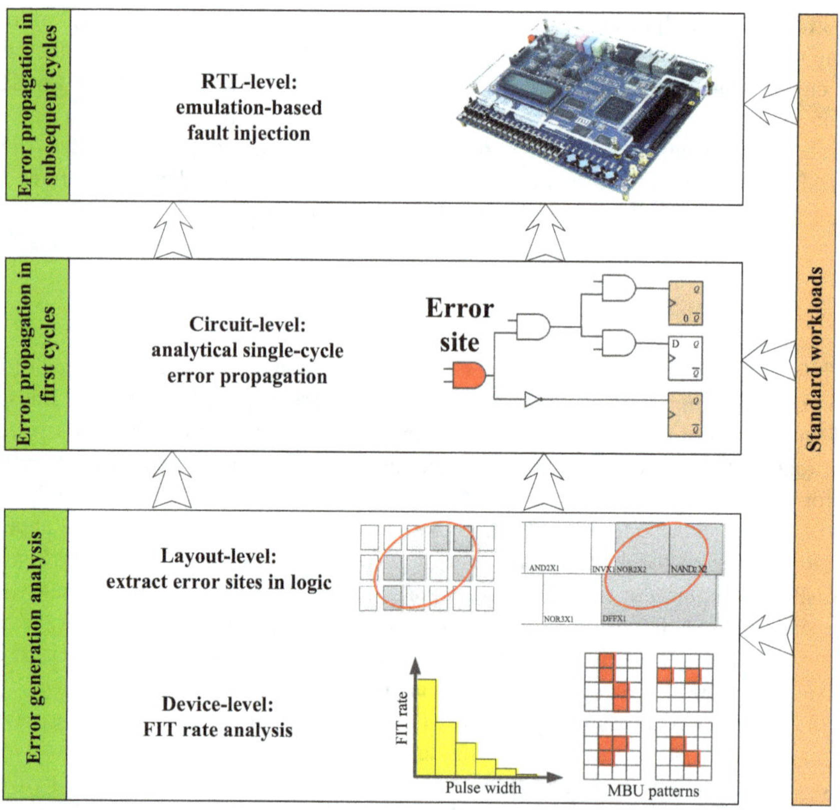


*Figure 7.2 Rationality behind proposed cross-layer soft error analysis platform*

accurate analytical model considering electrical, logical, and latching-window masking, errors are propagated from combinational logic toward flip-flops and memory units. In addition, temporal masking of each flip-flop is determined using a novel event-driven EP approach according to the post-layout timing information. Finally, an emulation-based tool is employed to inject errors in flip-flops and Static Random Access Memory (SRAM) arrays. This tool is able to propagate errors and determine their effect (masked, latent, and failure) with respect to an actual workload. Then, the SERs of all components are computed based on the results obtained during these steps.

The rest of this chapter is organized as follows: Section 7.2 describes the device-level FIT rate analysis tool. Section 7.3 introduces our proposed layout-level Multiple Transient (MT) error site identification method for random logic. The single-cycle circuit-level analysis for flip-flops and combinational gates are presented in Section 7.4 and Section 7.5, respectively. The proposed emulation-based platform is explained in Section 7.6 followed by a novel importance sampling technique for accelerating fault injection experiments in Section 7.7.

## 7.2 FIT rate analysis at device level

The first step of SER estimation is to characterize the intrinsic FIT rate of each library cell under multiple load conditions. This is done using an accurate FIT rate analysis tool [1]. This tool uses a process response model that is generated based on a set of device-level simulations. Using this response model, the tool is able to characterize the sensitivity of standard cells by simulating a series of current pulses in a Simulation Program with Integrated Circuit Emphasis (SPICE) netlist of the cell, taking into account the physical layout. The layout information is used by a FIT rate analysis tool to identify which transistors in the standard cell are physically adjacent. The adjacency information is exploited to consider charge sharing effects that can result in multiple current sources to be added to the netlist to model the fault. The tool also accesses a nuclear database that enables it to compute the interactions between neutrons and the atoms in the device, the secondary particles that are produced, and the distribution of generated current pulses. The accuracy of this tool for raw FIT rate analysis has been validated on several commercial processes.

For combinational gates, this tool provides the distribution of pulses with various widths. This distribution is dependent on the output load seen by the cell; thus, each cell is characterized for multiple loads. During the full-circuit analysis, for each gate, based on its load, an interpolation is performed between the two closest load conditions that were simulated for the given cell type. In order to accelerate the analysis flow in the next phases, the pulse width distribution curve is discretized into a few pulse widths with a distance of $d$. Smaller values of $d$ result in higher accuracy but larger runtime. Therefore, an appropriate value for this variable must be selected considering the trade-off between runtime and accuracy.

Figure 7.3 shows the pulse width distribution of few gates in Nangate 45 nm standard cell library for a typical output load of 3 fF and at a supply voltage of 1 V. From this figure, it becomes clear that alpha particles mostly generate very

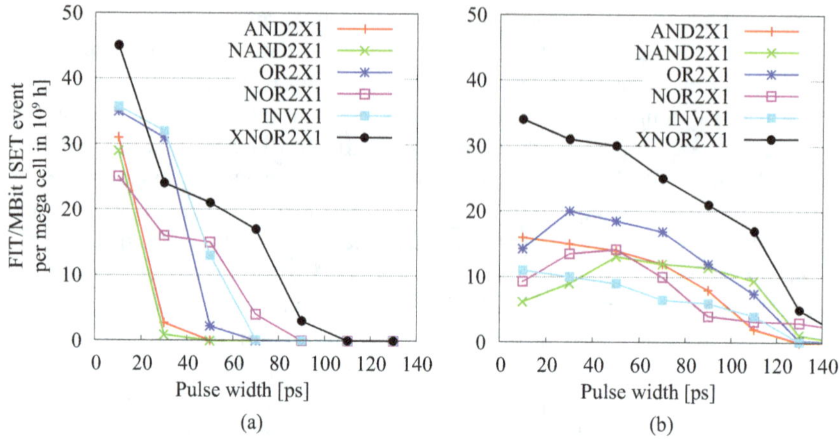


*Figure 7.3 Pulse width distribution for few standard cells from Nangate 45 nm standard cell library for output load of 3 fF and supply voltage of 1 V: (a) alpha and (b) neutron*

*Table 7.1 Neutron and alpha-induced FIT rates of different components in a 45-nm technology in terms of FIT/Mbit per billion hours of operation*

| | | **Neutron** | **Alpha** | **Overall** |
|---|---|---|---|---|
| Single-port SRAM cell | SEU | 150 | 351 | 501–693 |
| | MBU | 140 | 52 | 192 |
| Dual-port SRAM cell | SEU | 211 | 448 | 659–867 |
| | MBU | 160 | 48 | 208 |
| Flip-flop | | 119–217 | 181–216 | 300–435 |

short pulse widths that are very likely to be electrically masked. On the other hand, neutron-induced transient pulses have considerably longer widths which thus have a higher chance to be latched in the downstream flip-flops.

In the case of flip-flops and latches, this tool reports the Single Event Upset (SEU) rate. For single- and multi-port SRAMs, the tool provides SEU and Multiple Bit Upset (MBU) rates as well as MBU patterns and their occurrence probability. The neutron and alpha-induced FIT rates of different components in a 45-nm technology are reported in Table 7.1. As shown, for a single-port SRAM cell that is employed for caches and their tags, almost half of the neutron-induced soft errors are in the form of MBU, while the ratio is considerably smaller for alpha particles. A similar trend could be seen for dual-port SRAM cell that is used to implement the register-file. For the case of flip-flops, we reported the range of FIT rate as there are different types of flip-flops in Nangate 45 nm standard cell library. Moreover, both SRAMs and flip-flops are more sensitive to alpha particles in this technology node.

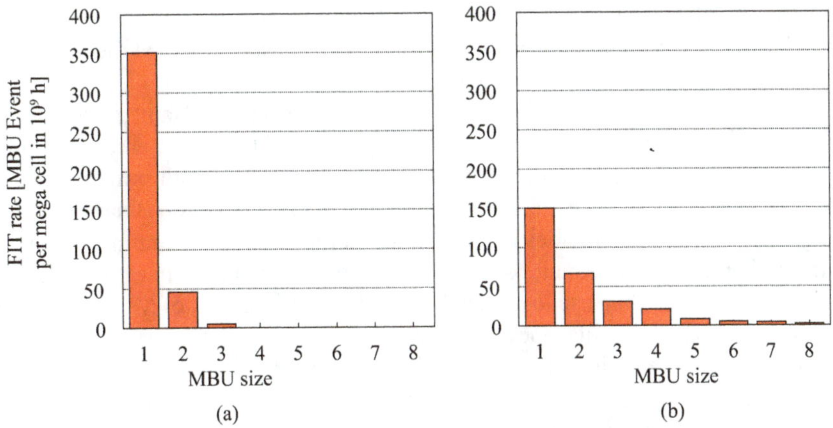


*Figure 7.4 MBU size distribution for alpha and neutron particles in a typical 45-nm single-port memory: (a) alpha and (b) neutron*

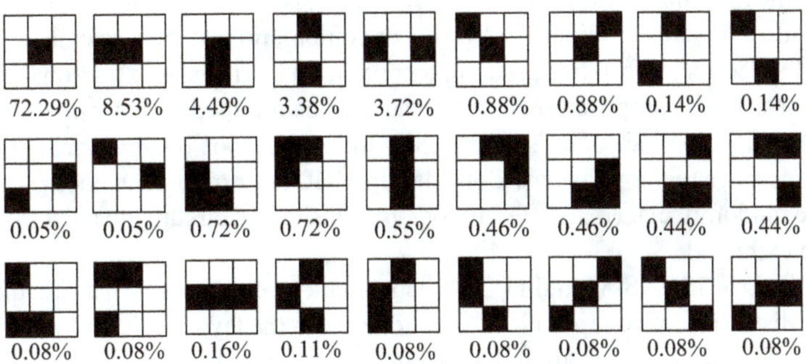


*Figure 7.5 Most common MBU patterns for alpha and neutron particles*

The FIT rates of different MBU sizes for alpha- and neutron-induced errors are demonstrated in Figure 7.4. In addition, some of the most common MBU patterns and their contribution to the overall FIT rate are reported in Figure 7.5.

## 7.3 Multiple transient error site identification using layout information

With miniaturization of device geometries in nanoscale technologies, it is very likely that a high-energy particle strike affects several adjacent cells in a circuit resulting in MBU in sequential elements or MET in combinational gates. Although the FIT rate analysis tool is able to accurately model MBUs, due to the irregular structure of random logic, it cannot analyze METs as it requires a detailed information about the location of cells in the circuit layout. Given the fact that a considerable fraction of particle strikes in random logic results in METs [2,3], their impacts at the layout and logic levels must be accurately modeled.

The analytical techniques presented in [4,5] address the MET fault model in logic circuits. The Error Propagation Probability (EPP) based technique presented in [4] propagates the error probabilities from the error sites toward both primary outputs and sequential elements. The technique presented in [5] is based on Binary Decision Diagrams (BDDs) and provides more accuracy at the expense of runtime, compared to the earlier method. The major shortcoming of these techniques in estimating SER due to METs is that they use logic-level netlist for identification of MET error sites, neglecting layout-level adjacency of error sites. Such assumption can significantly underestimate the circuit SER as we will experimentally demonstrate later in this section. Additionally, the distribution of affected error sites and the number of affected cells completely depend on the layout-level details and cannot be extracted from the logic-level netlist. For example, our experiments show that a considerable fraction of particles simultaneously affects both combinational gates and flip-flops, which is completely ignored in the previous techniques. Ignoring such cases can further increase the SER inaccuracy. In the rest of this thesis, the occurrence of MT errors at sequential elements, combinational gates, or combination of them is called MTs.

In this section, a fast and accurate technique MT error site extraction technique is proposed. In the proposed technique, the surface affected by a particle in combinational and sequential logic is estimated with oval shapes obtained from available MBU patterns in memory arrays. Considering MBU patterns' occurrence probability, ovals are randomly placed at different locations inside the circuit and a list of affected cells are extracted.

Analysis of ISCAS'89 and ITC'99 benchmark circuits reveal that less than 10% of the netlist adjacent cells are physically adjacent in the layout. Also, more than 60% of physically adjacent cells are not adjacent in the netlist. Experimental results show that neglecting layout adjacency can cause inaccuracy up to 36.04% in the circuit overall SER.

### *7.3.1 Motivation for layout-based MT analysis and mitigation*

An important step in MT analysis is identifying physically adjacent cells as error sites. Although the layout of the circuit is necessary for the accurate identification of adjacent cells, the previous MT analysis techniques presented in [4,5] employ some heuristic approaches to extract the list of adjacent cells from the netlist. In these techniques, four categories, including a gate and its fan-in (GFI), a gate and its fan-out (GFO), common fan-ins of a gate (CFI), and common fan-out of a gate (CFO), are considered as adjacent nodes for MT error sites. Figure 7.6 shows several examples of netlist-based adjacencies. In these techniques, a gate is first selected as primary error site and then its MT pair is randomly selected among its netlist adjacent cells.

In order to check the accuracy and layout-relevance of this model, i.e., extraction of adjacent cells from the logic netlist, the layouts of several circuits selected from ISCAS'89 and ITC'99 benchmarks synthesized with respect to Nangate 45 nm have

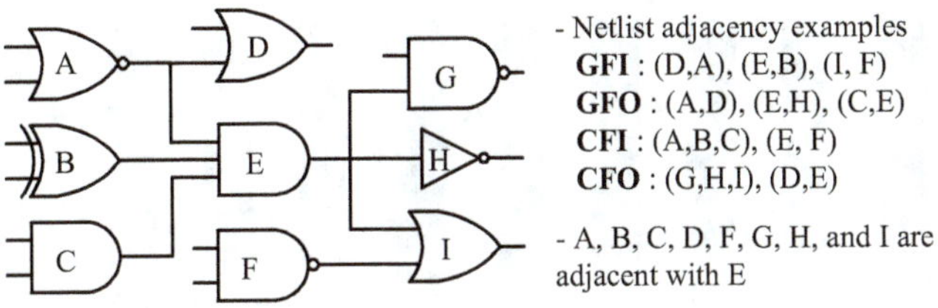


*Figure 7.6 Examples of different adjacency scenarios considered in the netlist*

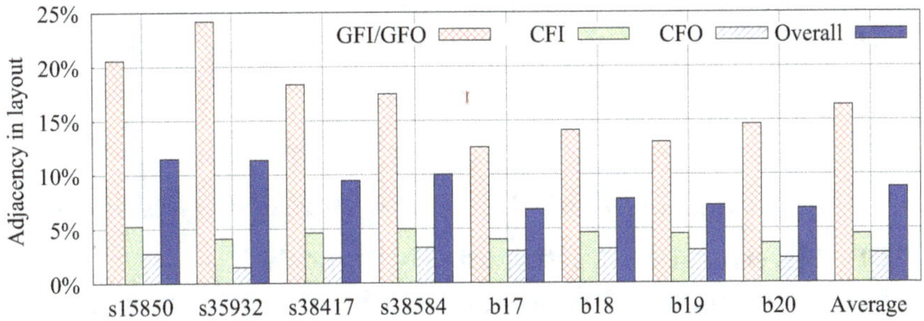


*Figure 7.7 Relevance of different adjacency scenarios considered in the netlist to layout adjacency*

been comprehensively analyzed (the details of this analysis framework are provided in Section 7.3.3). In this regard, all possible adjacent pairs for different netlist adjacency categories are first extracted from the netlist, and then the physical adjacency of each pair in the circuit layout is investigated. Since the order of adjacent pairs is not important in this investigation, both GFI and GFO categories are equal and are assumed as a single category called GFI/GFO. In the example given in Figure 7.6, E is the fan-out of gate B and gate B is a fan-in of gate E. As a result, the pair (B,E) belongs to both GFI and GFO categories. Figure 7.7 shows the results of this experiment. As can be seen, on average, only less than 10% of netlist adjacent pairs are also adjacent in the circuit layout. Also, the probability of physical adjacency of GFI/GFO category is much higher than that of CFI and CFO categories.

There is another experiment conducted in which all layout adjacencies on the circuit layout are first extracted, and then for each physical adjacency, its netlist adjacency category is investigated (Figure 7.8). It can be inferred from Figure 7.8 that more than 60% of the physical adjacencies do not belong to the previously defined netlist adjacency categories.

From these two experiments, it is clear that there is no statistically meaningful correlation between netlist and layout adjacencies. This means that for an accurate MT analysis, netlist-level analysis for MT error site abstraction is not sufficient, and layout-level analysis must be performed. To address this issue, we propose a fast and accurate layout-based MT modeling technique.

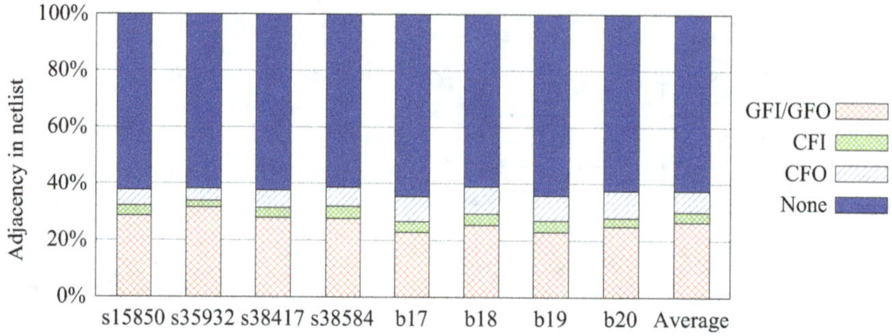


*Figure 7.8 Relevance of extracted adjacencies from layout to netlist-level MT error models*

## *7.3.2 Proposed layout-based MT error site extraction technique*

### 7.3.2.1 MT error site extraction using MBU analysis

The first step for accurate MT modeling is extracting physically adjacent error sites from the circuit layout. This requires to have MT patterns projected in the circuit layout and their occurrence probabilities. To the best of our knowledge, no field results about MT patterns on combinational and sequential logic have been reported in the literature. This could be mainly due to the fact that the logic gates and sequential elements (such as flip-flops) have significantly less observability as compared to regular structures such as SRAM-based memory arrays. Since memory arrays are more regular and dense than logic structures and also have a full observability, the affected area can be accurately estimated.

In the proposed technique, we try to use available MBU patterns in memory arrays for identification of MT error sites in logic circuits. In this method, it is assumed that the surface affected by a particle strike is almost the same for the memory arrays and logic cells in different technology nodes. In order to support this assumption, the average affected areas of SRAM for neutron-induced particle that strikes across different technology nodes are reported in Figure 7.9. These results are extracted according to the cell area and MBU size information reported in [6]. Basically SRAM cells act as a guideline for determining the affected area, and smaller cell size could provide more accurate results due to high resolution in estimating the area. As can be seen from this figure, the estimated affected area reduces from 250 to 90 nm, which is mainly due to a larger SRAM cell size, and hence, less resolution in the area estimation. However, the estimated affected area is relatively constant from 90 to 22 nm, independent of the cell size. This clearly shows that the affected area is independent of the underlaying SRAM technology. Hence, it could be concluded that the affected area would also be the same for the irregular logic circuits.

The main idea behind our work is to extract the affected area in a memory array and then consider all logic cells covered by a similar surface to be MT error sites. In this regard, the affected area for each MBU pattern is first extracted. Predominant

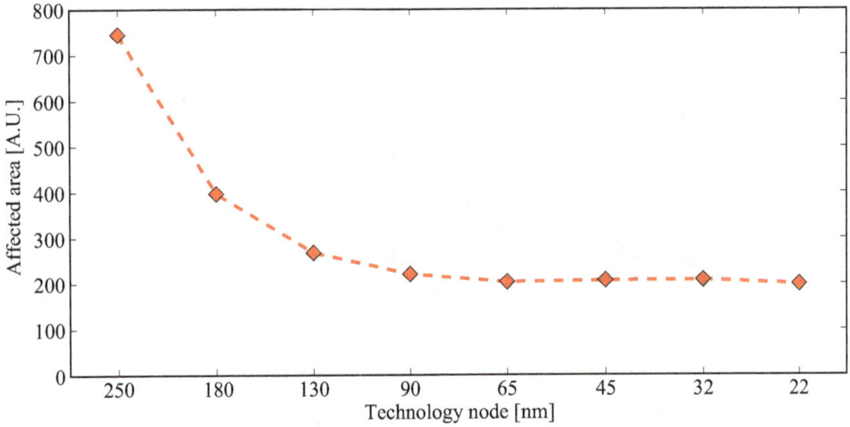


*Figure 7.9 Average affected area for neutron particles across different technology nodes based on the cell and MBU size information reported in [6]*

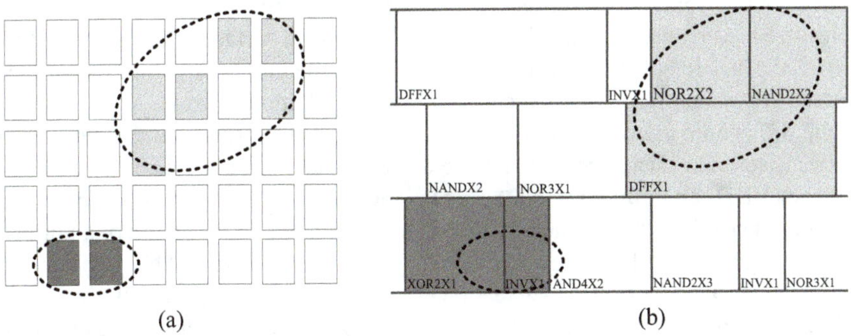


*Figure 7.10 Extraction of MT error sites from existing MBU patterns: (a) two MBU patterns in SRAM and (b) equivalent MT error sites in circuit layout*

MBU patterns in memory arrays have been comprehensively studied using neutron beam-based accelerated SER estimation [7–9]. The number of affected bits by a single strike depends on several parameters, including particle type and its energy, strike angle, cell type, cell size, and output load [10]. Given memory cell dimension as well as vertical and horizontal distance between adjacent cells, for each MBU pattern, the surface affected by the strike can be calculated. Most of the MBU patterns, especially MBUs with more than 8 bits, can be effectively covered by an oval surface. As a result, in our approach as shown in Figure 7.10, all cells affected by an MBU are first surrounded by an oval. Then, during SER estimation, the same oval is transferred to random locations inside the logic layout, and all cells affected by the MT are listed as error sites.

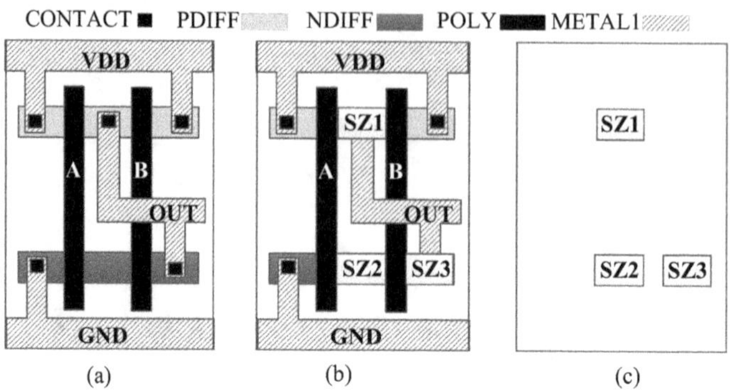


*Figure 7.11 Sensitive Zone (SZ) extraction for a NAND gate: (a) cell layout, (b) identification of diffusion parts that are not connected to supply voltage, and (c) sensitive zones (adapted from [11])*

#### 7.3.2.2 Library characterization

In the circuit layout, only a subset of cells that has an overlap with the oval surface is considered as error sites. A cell has overlap with an oval surface if at least one of the *sensitive zones* to soft errors (N-diffusion and P-diffusion) falls within the surface. In fact, when a particle strikes a cell, it causes an additional charge to be collected in the diffusion parts of the transistor that in turn disturbs the normal operation of the transistor. The charge collected to the diffusion parts that are connected to supply voltage (VDD) and ground (GND) pins is evacuated and does not affect the circuit behavior. The other parts of the diffusion can be disturbed by collection of additional charge. Figure 7.11 shows how sensitive zones for a not and (NAND) gate layout are extracted based on this explanation. Using this approach, all cells inside the technology library are characterized and sensitive zones' coordinations are extracted.

#### 7.3.2.3 Overall flow

The main steps of the proposed layout-based SER estimation are summarized in Algorithm 1. Initially, the list of ovals and their occurrence probability are extracted from existing MBU patterns (line 1) and then technology library is characterized for identification of sensitive zones (line 2). These two steps are performed once in advance, and their results are used for all circuits to be analyzed in the same technology and library settings.

Due to the large number of cells and sensitive zones inside industrial-size circuits, a hierarchical approach is employed to minimize the time needed for identification of error sites affected by an MT (line 3). In this approach, the entire layout area is divided into smaller grids and the list of cells inside each grid is extracted. During SER estimation, instead of searching among a large number of cells, in the first step, for each grid it is checked whether it has an overlap with the oval surface and the list of layout grids overlapped by the oval surface is extracted. Then, the list of overlapped cells is extracted by investigation of cells inside overlapped layout grids. At the end,

**Algorithm 1:** Layout-based SER estimation

```
1  Extract a surrounding oval for each MBU pattern;
2  Extract sensitive zones of each cell by library characterization;
3  Divide layout into smaller grids and extract the list of cells in each grid;
4  while sampling error < predefined value AND number of cells covered by at least one MT < 99.9%
     do
5  |  Randomly select an oval based on their occurrence probability;
6  |  Place the oval in a random location on the layout;
7  |  Find the list of grids which have overlap with the oval;
8  |  Search overlapped grids cell lists and construct overlapped cells list;
9  |  Remove cells without overlapped sensitive zones from cell list;
10 |  Mark all cells in overlapped cell list as covered by an MT;
11 |  Propagate MT at logic-level and calculate failure probability;
12 end
13 Report average failure probability
```

those cells that have no overlapping sensitive zone with the oval surface are eliminated from the target cell list. The remaining cells will be used as candidate MT fault sites (lines 7–9).

## 7.3.3 *Experimental results of MT modeling*

Using the proposed layout-based MT error site extraction, we have performed an extensive analysis on the impact of particle energy on the MT error sites. By performing experiments on both Nangate 45 nm (Planner) and 15 nm (FinFET) technology nodes, we were able to show the impact of technology scaling on MT size. Additionally, the impact of netlist adjacency assumption on the overall SER of the circuit is investigated.

### 7.3.3.1 Experimental setup

In order to show the scalability of the proposed approach, we have evaluated the largest available benchmark circuits in ISCAS'89 and ITC'99 benchmarks suites. For each benchmark, the Hardware Description Language (HDL) description of the circuit is first synthesized using a Synopsys Design Compiler for the maximum possible operational clock frequency.* Then, the layout of the netlist is extracted using Cadence SoC Encounter by considering a utilization ratio of 75% (typically about 65%–85% [12]). The minimum clock period and the chip area for each circuit are reported in Table 7.2. According to these results, the area and frequency are decreased by around 4× for the circuit synthesized for 15 nm compared to 45 nm.

*This requires two rounds of synthesis: first round is for obtaining the maximum operational frequency. In this round, the clock period is set to zero to force the synthesis tool to implement all paths as short as possible. The maximum negative slack obtained shows the maximum operational frequency. Then, another synthesis operation is performed in the second round to minimize the area and power for the maximum operation frequency.

*Table 7.2 Minimum clock period and area of benchmark circuits synthesized into 45- and 15 nm Nangate standard cell libraries*

| Benchmark | Nangate 45 nm | | Nangate 15 nm | |
|---|---|---|---|---|
| | Clock period (ps) | Area ($\mu m^2$) | Clock period (ps) | Area ($\mu m^2$) |
| s15850 | 825 | 9,134 | 184 | 2,723 |
| s35932 | 487 | 32,735 | 105 | 9,669 |
| s38417 | 924 | 27,760 | 139 | 9,320 |
| s38584 | 610 | 28,709 | 121 | 8,754 |
| b17 | 1,357 | 32,175 | 331 | 8,451 |
| b18 | 2,133 | 101,200 | 549 | 27,200 |
| b19 | 2,046 | 194,360 | 530 | 52,232 |
| b20 | 1,734 | 29,433 | 452 | 8,259 |

The MBU patterns for particles with 22, 37, 95, and 144 MeV provided by [7] are used during the layout-based MT extraction. This information is given to our layout-based SER estimation to calculate the overall SER of the circuit according to Algorithm 1. In our framework, SER estimation analysis terminates when the maximum inaccuracy of the Monte Carlo is less than 0.5%. In this experiment, the layout is divided into $30 \times 30\ \mu m^2$ grids, where each grid includes around 800 cells.

#### 7.3.3.2 MBU patterns and MT error sites

As mentioned earlier, in order to extract MT error sites, the areas affected by MBU patterns are first extracted and then a surrendering oval for each MBU pattern is constructed. These ovals are used for identification of MT error sites. For this purpose, detailed information about different MBU patterns in a memory array is necessary for the identification of MT error sites.

Radaelli *et al.* [7] have reported a detailed information about predominant MBU patterns in a 150-nm technology SRAM device and their occurrence probability for particles with 22, 47, 95, and 144 MeV energies. Considering the SRAM cell dimensions, the area affected by each MBU surrounding oval can be accurately estimated. For these cases, the oval shapes and their occurrence probability (the same as the occurrence probability of the corresponding MBU pattern) are computed. Table 7.3 shows the average area affected by each particle energy obtained by weighted averaging of oval surfaces based on their occurrence probability.

The area affected by a particle strike is mostly a function of particle energy, while the strength of the transient pulse mostly depends on other parameters such as diffusion volume (width, length, and depth) and load capacitance [6]. As a result, the affected area information acquired for a 150-nm SRAM technology can also be used for the logic area affected by a particle strike with the same energy in the 45- and 15-nm technology nodes. Although the affected area remains constant, however,

*Table 7.3 Average area affected with different particle energies*

| **Particle energy (MeV)** | **Average affected area ($\mu m^2$)** |
|---|---|
| 22 | 1.178 |
| 47 | 1.902 |
| 95 | 2.903 |
| 144 | 4.613 |

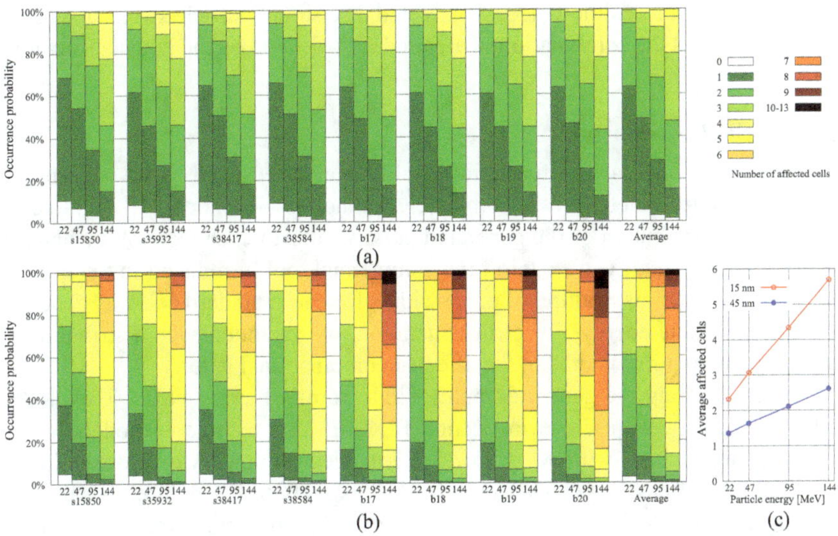


*Figure 7.12 Distribution of a number of cells (flip-flops or combinational gates) affected by particle strikes with 22, 47, 95, and 144 MeV energies for benchmark circuits synthesized with respect to (a) 45 nm and (b) 15 nm Nangate standard cell library, and (c) an average number of affected cells per technology node*

due to the technology downscaling, the number of affected cells increases in smaller technologies.

By randomly locating these ovals on the circuit layout according to their occurrence probability, different combinations of affected combinational gates and flip-flops are extracted and identified as MT error sites. Figure 7.12 shows the occurrence probability of a different number of affected cells for particle strikes with 22, 47, 95, and 144 MeV energies. As is expected, by increasing the particle strike energy, the occurrence probabilities of SET and SEU decrease significantly and MT becomes predominant. In addition, the MT ratio in 15 nm is significantly larger than that in 45 nm.

In this experiment, the maximum number of affected cells at 45 and 15 nm is 8 and 13 cells, respectively. In order to clearly show the impact of technology scaling on MT rate, an average number of affected cells for the aforementioned particle energies are reported in Figure 7.12(c). This figure reveals that the average number of affected cells is almost doubled for all particle energies from 45 to 15 nm. Hence, accurate modeling and mitigation of MTs are more challenging in smaller technology nodes with a larger number of affected cells.

In the previous netlist-based techniques [4,5], it is assumed that a particle strike leads to either MET on combinational logic or MBU on sequential cells. Also, the number and type of affected cells was a function of the particle energy rather than a function of the layout. However, our results show that (1) adjacent combinational and sequential cells can be simultaneously affected by a single-particle strike, and (2) the number, type, and the combination of affected cells depend on the layout structure as well.

#### 7.3.3.3 SER estimation

Although the affected area information could be acquired from MBU patterns, the pulse width of generated SETs is completely technology dependent [13,14]. In order to obtain the distribution of the generated pulse widths in each technology node for a certain particle energy, we exploit the commercial tool introduced in Section 7.2. Using this tool, the distribution of pulse widths for each standard cell was generated for a given particle energy. For error generation, we exploited such distributions to generate random pulse widths in the affected combinational gates.

For propagating the generated pulse width from MT error sites, we exploit an analytical technique presented in Section FIXME. This technique models logical, electrical, and latching-window masking in a unified framework by propagating the errors for several clock cycles (here, ten). In this section, the failure probability is defined according to [15,16] as the probability of propagation from error sites to primary outputs during first few cycles after error occurrence.

#### 7.3.3.4 Impact of SET/SEU vs. MT model on overall SER

In order to explore the importance of MTs and technology scaling impact on the failure probabilities, overall SERs extracted for particles with 22, 47, 95, and 144 MeV energies are compared with the case that the simple SET/SEU model is considered for both 45- and 15-nm technology nodes (Figure 7.13). In the case of SET/SEU, a single error is injected in each gate/flip-flop and the average of the failure probabilities of all cells are reported as the circuit failure probability.

The results shown in Figure 7.13 reveal that the average failure probability significantly increases from 45 to 15 nm (by more than 2× in some cases). One reason for such an increase is that the number of affected cells is almost twice in the smaller technology node (as shown in Figure 7.12(c)). Another interesting reason is higher operational frequency and smaller latching window of the circuits designed with Nangate 15 nm (see Table 7.2), which significantly affects the error latching probability. The average latching windows of flip-flops in 45 and 15 nm are 42.4 and 12.7 ps, respectively. As shown in [17], the combinational logic SER linearly grows

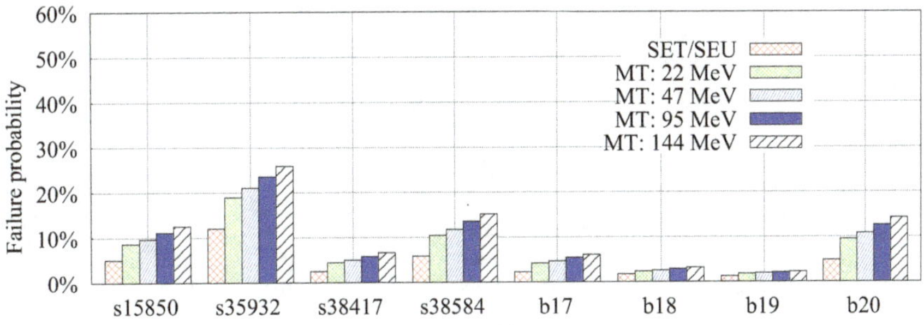


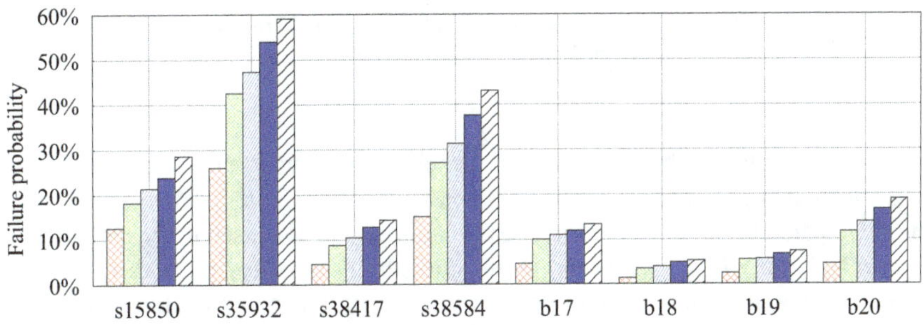


*Figure 7.13 Average failure probabilities of different particle energies for circuits synthesized with respect to Nangate (a) 45 nm and (b) 15 nm standard cell*

by increasing the frequency due to an increased ratio of latching window to the entire clock period.

The results also reveal that the circuit failure probability does not linearly increase with the particle energy. As an example, on average, the SER in the presence of 47 MeV particles is only 1.15× greater than when considering 22 MeV particles. This can be explained by the number of affected cells for different particle energies as reported in Figure 7.12.

One reasonable solution to show the SER trend from 45 to 15 nm is to investigate the trade-off between failure probability increase and the decrease in the particle strike rate across different technology nodes. As shown in Table 7.2, the area and clock period of circuits in 15 nm are reduced by 3.5× and 4.6× compared to 45 nm. This means 15.8× less particles strikes for a certain task running on a circuit fabricated in 15 nm due to the smaller area and shorter runtime. On the other hand, the failure probability grows on average by 2.2× from 45 to 15 nm (see Figure 7.13). Hence, the overall trend would be 7.2× reduction in the overall SER. These results are in-line with that of the existing studies [14,18,19].

#### 7.3.3.5 Impact of netlist adjacency assumption on SER

In order to investigate the effect of netlist adjacency on the overall SER, we have implemented a netlist-based approach. The EP method of the netlist-based approach

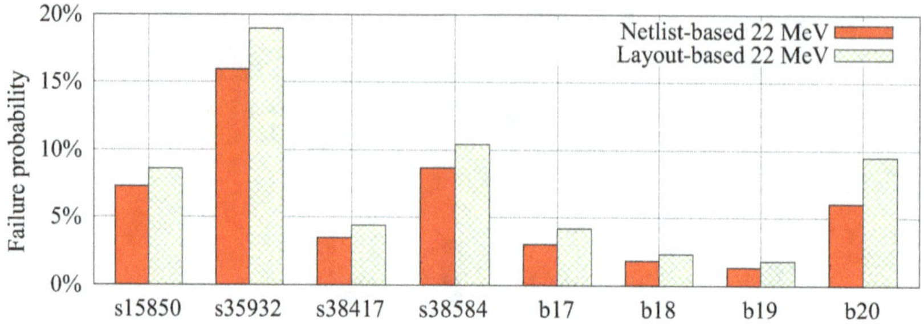


*Figure 7.14 Comparison of overall SER obtained by netlist- and layout-based approaches for circuits synthesized for Nangate 45 nm standard cell library and particle with 22 MeV energy*

is similar to the layout-based approach. Although different combinations of affected cells and their occurrence probability are unknown at the netlist-level, in order to have a fair comparison, the same occurrence probabilities are also used in the netlist-based approach. Figure 7.14 reports the failure probability obtained by both netlist- and layout-based approaches. As can be seen, the netlist-based approach always underestimates the overall failure probability. Our analysis reveals that there are two main reasons for this underestimation. First, when there are simultaneous errors at the outputs of the CFI pairs, these transient pulses reach at the same time to the inputs of the fan-out gate. In this case, the propagated transients are either completely masked, attenuated, or at least converged to one transient pulse. However, as shown in Figure 7.6, most of CFI pairs are not physically adjacent in the layout. Second, the forward cones of netlist adjacent pairs are highly overlapped and share similar paths from error sites to the circuit outputs. This can increase the chance that several errors are masked due to one kind of masking (e.g., logical masking in a common gate in the forward cone of both error sites). When MT occurs in error sites that have nonoverlapping forward cones, they are independent, and the probability of masking is much lower.

On average, netlist-based MT analysis has an inaccuracy of 22.34% which is as high as 36.04% for b20 benchmark. Since the information regarding the occurrence probability of different affected gate/flip-flops does not exist at the netlist level, the inaccuracy of those techniques could be even higher.

## 7.4 Propagating flip-flop errors at circuit level

Since the soft error vulnerability of flip-flops is quite nonuniform [20], a fast and accurate SER estimation technique is required to identify the most vulnerable flip-flops and replace them with robust alternatives. The combined effect of interdependent and correlated and logical and temporal masking factors determines the vulnerability of flip-flops to soft errors and has to be considered in a unified manner to reach a reasonable accuracy. These factors are interdependent and highly correlated as they are related to the logic state and delay of the components (logic gates and flip-flops) on the EP paths. As a result, an accurate soft error estimation technique should be able to capture a combined effect of these factors.

While both logical and temporal masking factors can prevent EP in the error generation cycle, once the error is latched in the downstream flip-flops at the end of that cycle, the errors can only be logically masked in the subsequent cycles. The logical and temporal masking factors are dependent on the logical value and the delay of the EP paths. When there is no activity on a certain part of the circuit, EP behavior on that part is the same as the previous cycle. Therefore, computing EP possibility for all paths in every clock cycle is unnecessary, and adjusting EP paths according to the logic events serves the same purpose. In this regard, in order to compute the vulnerabilities of flip-flops to soft errors, we propose to embed transient EP in an event-driven logic simulator.

Our objective is not only to consider the logic masking in the propagation of the errors, but also to accurately compute the vulnerable time interval within each cycle according to the temporal masking factor. For this purpose, in the cycle of error occurrence, the delay required for an error to reach each downstream flip-flop has to be computed. We perform such a timing analysis in the proposed event-driven EP approach. Thus, the proposed approach not only identifies the vulnerable cycles but also determines the vulnerable time interval within the vulnerable cycles.

In the proposed approach, errors originating from different flip-flops are concurrently propagated across multiple clock cycles, so there would be no need to repeat EP for different flip-flops and for different cycles of the error occurrence. This results in multifold speedup over Statistical Fault Injection (SFI). While conventional Concurrent Fault Simulation (CFS) techniques reduce the simulation time by only evaluating multiple flip-flops simultaneously, the proposed approach reduces the number of EPs by decreasing the number of error injection cycles as well. Even when there is a logic event, the EP behavior changes for the subset of EP paths. These result in significant speedup (two orders of magnitude as shown in experimental results section) over conventional CFS techniques.

## *7.4.1 Event-driven logic simulation*

We exploit a very accurate logic simulation method that implements the three-valued logic system ("0," "1," and "X") and considers different rise and fall delay (i.e., $(t_r, t_f)$) for each input–output pair of a component (flip-flop or a combinational gate). The detailed timing information are obtained from a Standard Delay Format (SDF) file. Similar to other event-driven logic simulators, we exploit the *time wheel* data structure to store the list of events to be simulated and their corresponding simulation time. When an event is simulated, which changes the logical value of a component output, for each fan-out node of that output, an event is inserted in the time wheel with respect to the delay of that component. This event-driven simulator accepts a synthesized gate-level netlist and an SDF file as inputs and performs the timing-aware post-synthesis logic simulation.

### 7.4.2 Error propagation from single flip-flop

We first explain the proposed EP approach for a single flip-flop as the error site and in the next subsection, we show how multiple error sites (flip-flops) are evaluated concurrently in one pass without sacrificing the accuracy.

In the proposed approach, we do not explicitly inject errors during workload simulation. Instead, we compute the list of time intervals in which an SEU error on the target flip-flop is propagated to each node in the circuit. Each node in the circuit has an associated error list, and each element of this list is represented as a tuple of $[FF_i, c, (t_s, t_e)]$ that means that if an SEU occurs in flip-flop $FF_i$ during time interval of $(t_s, t_e)$ at $c$ cycles ago, it will propagate to the node that is assigned. $(t_s, t_e)$ is a subinterval of one clock period, i.e. $(0, clock\_period)$, and determines the temporal masking factor.

Figure 7.15(a) demonstrates an example of error lists associated with internal nodes. Each list has two elements, both originating from the same flip-flop $FF_S$. In all examples used in this section, the clock period is assumed to be 1,000 ps. The tuple $[FF_S, 0, (0, 500)]$ at node $a$ shows that an SEU occurring in the first half of the current cycle propagates to this node. An error occurred in the last 40 ps of $(0, 500)$ interval is temporally masked due to the delay of the inverter and cannot reach node $b$. Thus, errors occurred in $(0, 460)$ can only reach this node. The effect of temporal masking factor during EP is shown in Figure 7.15(b). While SEU errors in one to seven cycles ago are masked and cannot reach these nodes, error in the time interval of $(0, 800)$ of eight cycle ago is also propagated to node $a$. As mentioned earlier, the

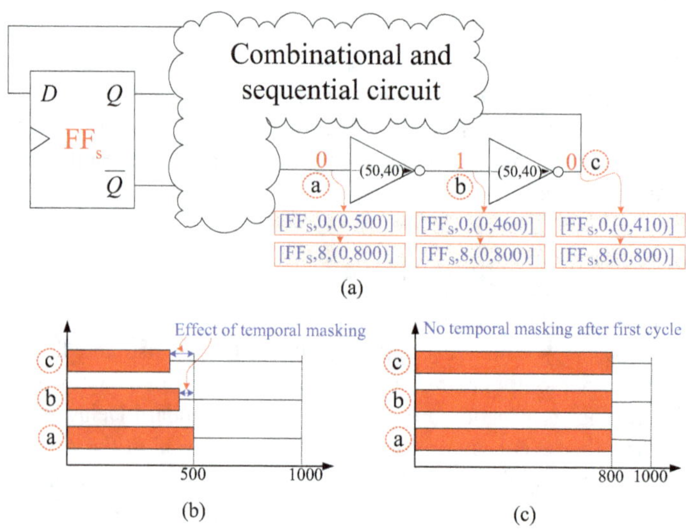


*Figure 7.15 A conceptual example of error propagation in proposed approach: (a) an example of error lists, (b) vulnerable interval of $FF_S$ in current cycle, and (c) vulnerable interval of $FF_S$ in eight cycle before*

temporal masking factor is only important in the first cycle; hence, it has no effect on the propagation of this error and it is propagated to nodes *b* and *c* without any change (see Figure 7.15(c)).

The error lists of the nodes are updated according to data events during the event-driven logic simulation. As we will show later, the data events and the update of error lists are highly correlated. This means that error lists remain unchanged when there is no data event to be simulated. On the other hand, only the error lists associated with the nodes with data event will change.

During the workload simulation, the vulnerability of the target flip-flop is computed according to the errors propagated to Primary Outputs (POs). For example, if $[FF_S, 5, (0, 500)]$ is added to a PO node at cycle $k$ and removed at cycle $k + 100$, flip-flop $FF_S$ is vulnerable at time intervals of $(0, 500)$ of cycles $k - 5$ to $k + 95$ as error requires five cycles to propagate to this node. By combining of the errors propagated to the POs, the overall vulnerability of the target flip-flop is computed.

#### 7.4.2.1 Error generation at target flip-flop

In case the value stored in a flip-flop has the state of "X," its value is *do not care* for the correct execution of the workload. Therefore, errors are only generated when the flip-flop stores either "0" or "1." An error occurring inside a flip-flop requires some time to propagate to its output(s). We assume that this time is equal to $Clk \rightarrow Q$ and $Clk \rightarrow \overline{Q}$ delays for $Q$ and $\overline{Q}$ nodes, respectively.

As an example, consider the flip-flop shown in Figure 7.16. The rise and fall pair, i.e., $(t_r, t_f)$, for $Clk \rightarrow Q$ and $Clk \rightarrow \overline{Q}$, are (117,97) and (96,89) ps, respectively. When the flip-flop output is "X," there is no error at the output nodes. When the flip-flop stores "1," an SEU would change its $Q$ ($\overline{Q}$) output from "1" ("0") to "0" ("1") that requires $t_f^{Clk \rightarrow Q}$ =97 ($t_r^{Clk \rightarrow \overline{Q}}$= 96) ps to reach this node. This means that if an SEU occurs within 97 ps (96 ps) before the clock edge, it does not have enough time to propagate to $Q$ ($\overline{Q}$) node. Therefore, for the clock period of 1,000 ps, an SEU in the time interval of [0,903] ([0,904]) could propagate to $Q$ ($\overline{Q}$) node. The case in which the flip-flop stores "0" can be analyzed similarly.

When the value stored in the target flip-flop changes at positive clock edge, an event is inserted at $Q$ and $\overline{Q}$ outputs in the time wheel with respect to $Clk \rightarrow Q$ and $Clk \rightarrow \overline{Q}$ delays. When such an event is simulated, the error list of the corresponding target flip-flop output is updated. The update could be the elimination of an existing element (state: "1"→"X" or "0"→"X"), update of an existing element (state: "1"→"0" or "0"→"1"), or adding a new element (state: "X"→"0" or "X"→"1").

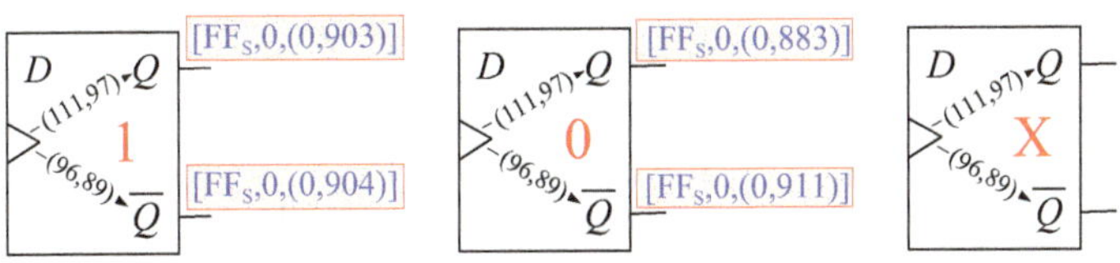


*Figure 7.16 Error propagation to target flip-flop output*

These error lists are only updated when there is a data event at flip-flop output; otherwise, the EP behavior is the same as that of the previous cycle and recomputation is unnecessary.

#### 7.4.2.2 Error propagation through combinational gates in first cycle

In case an error reaches an input of a combinational gate, depending on the logic value of other inputs, it might propagate to its output. However, when there is no event, except a special case that is detailed later, the EP behavior remains the same. As a result, when simulating an event on a gate, we should also update the error list of the output node with respect to those of the all inputs.

Let us assume that $[FF_i, 0, (t_s, t_e)]$ is an element in the input error list of a gate and the other inputs of the gate are not in the controlling state. This error propagates to the output node of the gate with respect to rise or fall delay ($d$) of the gate dependent on the error polarity and gate functionality. Although errors occurred at time interval of $(t_s, t_e)$ propagate to the gate input, due to the gate delay, errors in $(t_e - d, t_e)$ are temporally masked and cannot reach the gate output. Hence, $[FF_i, 0, (t_s, t_e - d)]$ is added to the output error list.

An example of propagation behavior in combinational gates, when error occurs in flip-flop $FF_S$, is shown in Figure 7.17. In step (a), the target flip-flop has a value of "X" and, as discussed earlier, there is no error at its output error list and hence there is nothing to be propagated in combinational gates. In step (b), the state of this flip-flop is changed to "1" and the flip-flop output error list contains $[FF_S, 0, (0, 903)]$. The other inputs of the NAND and NOR gates are not in the controlling state and the error is propagated through these gates when the corresponding logic events are simulated from the time-wheel. In step (b), the output of the NAND gate is "0." In

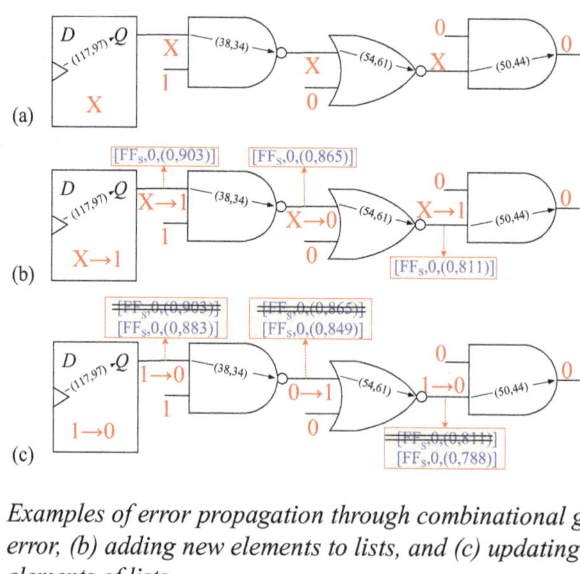


*Figure 7.17 Examples of error propagation through combinational gates: (a) no error, (b) adding new elements to lists, and (c) updating existing elements of lists*

the case of an SEU error in $FF_S$, the output of this gate would change to "1" that requires a delay equal to $t_r = 38$ ps. As a result, the output event for this gate would be $[FF_S, 0, (0, 903 - 38 = 865)]$. The similar concept applies to the NOR gate output while for the AND gate, since the other input is in controlling state, the error is masked and does not propagate to its output. In step (c), the value of flip-flop changes to "0." In this case, all the error lists are updated by replacing the new timing intervals with the old ones.

Although in the example shown in Figure 7.17, the update of error lists is always done during the simulation of an event in the corresponding node, there are some exceptional cases where an error list has to be updated while there is no event at that node. Figure 7.18 demonstrates an example of such cases. In step (a), there is an error in the first input of the NOR gate that is logically masked as the second input is in the controlling state. In step (b), the second input of the NOR gate changes to "0." This input is not in the controlling state anymore and the error could propagate through the gate. However, this change does not affect the gate output, and there is no event on this node. In order to propagate this kind of errors, a *fake event* is inserted in the logic-simulator time wheel. The fake events assign the previous value of a node to it and, hence, have no effect on the functionality of the event-driven simulator. After simulating the fake event, the error list of the node is updated. In step (b), there are three fake events demonstrated by either "0"→"0" or "1"→"1" to add an error to the error list. In step (c), an event changes the first input of AND gate to "0." Consequently, the error in the second input cannot propagate to the output of this gate. In this case, similar to step (b), fake events are employed to update the error list when there is no event; however, it results in the elimination of some errors due to a logical masking factor.

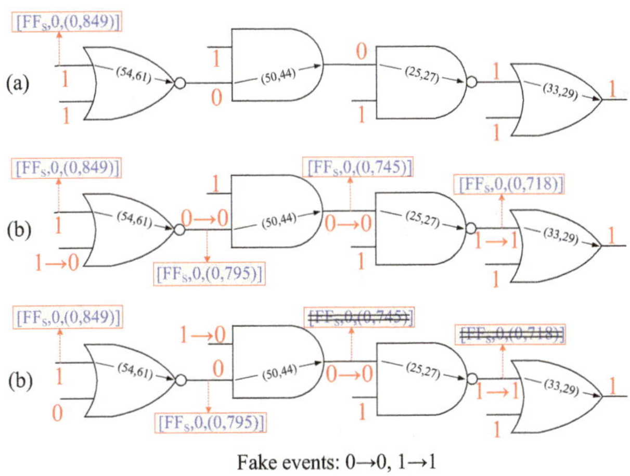


*Figure 7.18 Error propagation with fake events: (a) error is logically masked, (b) error is propagated with fake events, and (c) error is eliminated with fake events*

#### 7.4.2.3 Error propagation after first cycle

Temporal masking only occurs at the cycle of error occurrence and once the error is latched in the flip-flops, only logical masking can prevent the EP. All errors in a flip-flop input error list are propagated to its output. While propagating the errors to flip-flop output, the number of cycles (i.e., $c$) for that error is incremented to show that it belongs to previous cycles while the time interval remains the same as that of the corresponding error in the input. Indeed, in the cycle error occurs, the timing intervals are adjusted with respect to the gate delays during EP to estimate the temporal masking factor. Once error reached to an input of a downstream flip-flop, its time interval shows the amount of temporal masking factor. For instance, the error $[FF_i, 0, (0, t_e)]$ in the input of a flip-flop shows that an SEU occurring in the time interval of $(t_e, clock\_period)$ is going to be temporally masked. Once an error is latched in the flip-flops, the propagation in the combinational logic does not change its vulnerable time interval, and it could only be masked by means of the logical masking factor.

Figure 7.19 shows EP through flip-flops. As shown, all errors in the input error list propagate to the output list by incrementing the number of cycles. Besides the errors propagated from the input, the target flip-flop has also new elements in its error list for the errors generated in the current cycle. Figure 7.20 shows the error list of

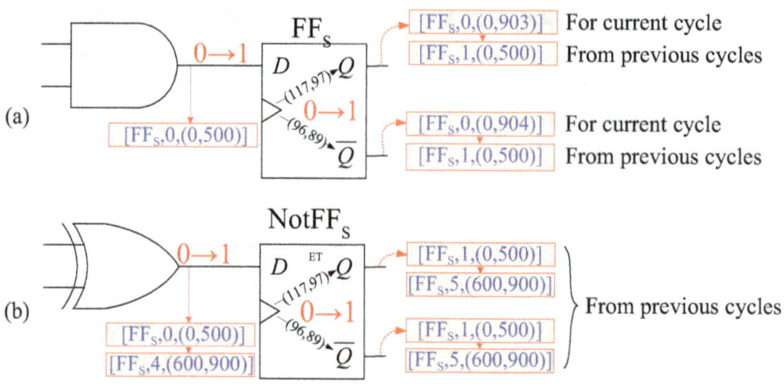


*Figure 7.19 Error propagation through (a) target flip-flop and (b) nontarget flip-flop*

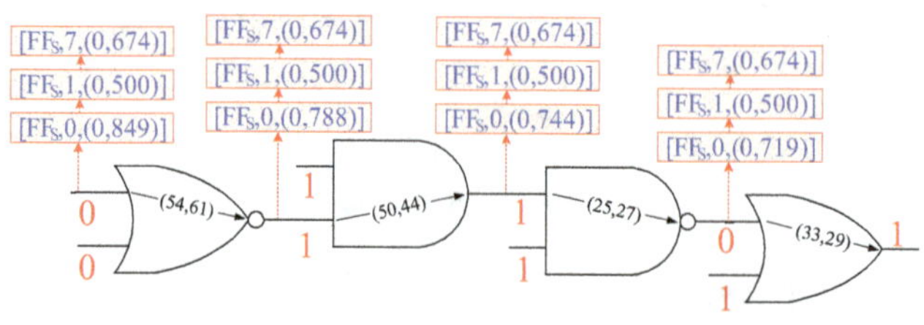


*Figure 7.20 Error propagation through combinational gates after first cycle*

different nodes of a combinational logic for errors occurred in the current cycle and those propagated from previous cycles. As can be seen, for the error occurred in this cycle (i.e., $[FF_i, 0, (0, 849)]$), the timing interval is adjusted during the propagation step to consider the effect of a temporal masking factor, while for those propagated from one and seven cycles ago, the timing interval is unchanged.

#### 7.4.2.4 Vulnerability time computation

The vulnerability of target flip-flop is estimated with respect to the error list of POs. Let us assume that $[FF_S, c, (t_1, t_2)]$ is an error appeared at a PO at cycle $k_1$ and removed from this node at cycle $k_2$. Considering the fact that the propagation of the error takes $c$ cycle, it would be clear that errors occurred at $(t_1, t_2)$ time interval of $(k_1 - c, k_2 - c)$ cycles of flip-flop $FF_S$ are propagated to this PO. For each error in the PO, the vulnerable time intervals and their corresponding clock cycles are computed. The union of all these time intervals is the overall vulnerable intervals of flip-flop $FF_S$. The vulnerability of this flip-flop is the sum of all vulnerable intervals over the overall workload execution time.

### *7.4.3 Concurrent transient error propagation from multiple flip-flops*

In Section 7.4.2, only vulnerability of one flip-flop is computed during the workload execution. Therefore, the event-driven EP has to be repeated for each flip-flop in the design as the error site. Considering the fact that an error only affects a small subset of the nodes, error lists of most of the nodes are empty and major proportion of the runtime is spent for logic simulation rather than EP. By concurrent transient EP from multiple flip-flops, the time spent for recomputation of logic events for each flip-flop could be significantly reduced. In this case, errors from different flip-flops are propagated simultaneously; however, they do not affect each other. This means that in case one gate has errors from multiple flip-flops in its inputs, their effects at the output are independently computed.

In order to expedite the operation of finding events that could affect each other (i.e., from the same flip-flop and the same cycle), the error lists of the nodes are stored in a sorted order with respect to the flip-flop index and cycles. During the EP, these lists are processed in the sorted order and for each pair of flip-flop that have the same flip-flop and cycle, the combined effect of these errors is computed. Figure 7.21 demonstrates an example of concurrent EP on a combinational gate. As can be seen, two errors are propagated from $FF_1$ in the current cycle and these create one output.

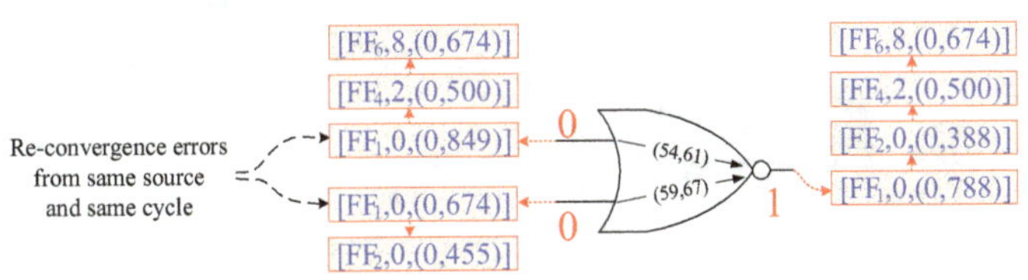


*Figure 7.21 An example of concurrent error propagation*

All the remaining errors have unique flip-flop index and propagation cycle, and hence, are processed independently.

### *7.4.4 Experimental results*

In order to show the effectiveness of the proposed approach, we conducted an experiment to evaluate the vulnerability of all flip-flops of the OR1200 processor with respect to four MiBench workloads.

#### 7.4.4.1 Experimental setup

The OR1200 core is synthesized with Nangate45 nm standard cell library using Synopsys Design Compiler. The resulted netlist has 2693 flip-flops and 30,986 combinational gates. The minimum clock period for the synthesized netlist is 1.39 ns. Four workloads from the MiBench [21] benchmark suite, namely *Bitcount*, *CRC32*, *QSort*, and *Stringsearch*, are simulated on the OR1200 processor during our experiments. All experiments are performed on a workstation with Intel Xeon E5540 2.53 GHz processor and 16 GB RAM.

#### 7.4.4.2 Validation of error propagation results

The EP capability of the proposed approach is compared with fault injection results using ModelSim to validate both masking factors for 2,000 randomly injected errors. Each error is traced for 100 cycles, and propagation behavior is compared with our simulator, which showed 100% accuracy.

#### 7.4.4.3 Runtime of transient error simulator

Figure 7.22 shows runtime of event-driven EP for concurrently evaluating various numbers of flip-flops and propagation cycles during the execution of Stringsearch workload. As shown, the runtime of the approach grows linearly with the number of evaluated flip-flop until 2,000 and then becomes exponential. Also, it has linear relation with the limit for the number of propagation cycles.

Since the runtime of the proposed approach is highly dependent on the number of evaluated flip-flops, evaluating all flip-flops in one pass might not be the best option

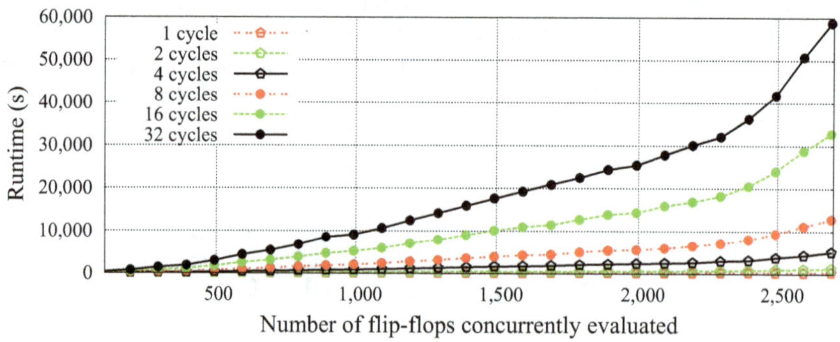


*Figure 7.22 Runtime vs. number of flip-flops concurrently evaluated*

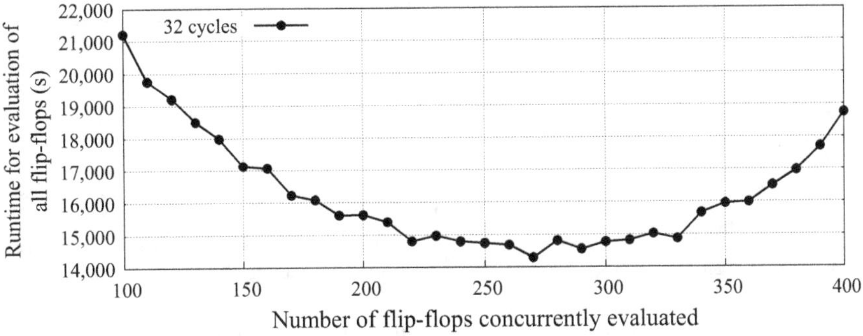


*Figure 7.23 Runtime for evaluation of all the designs for different numbers of concurrently evaluated flip-flops*

from the runtime point of view. Indeed, simultaneous evaluation of a large number of flip-flops increases the computation complexity as error events in the input of a gate have to be sorted to identify the reconvergent events. By having more events in the inputs of a gate, the sorting algorithm requires more time as it has a complexity of $O(n^2)$ where $n$ is the number of error events to be sorted. In order to show this effect, we estimated the overall runtime required for evaluating all OR1200 flip-flops with respect to Stringsearch workload by concurrently evaluating different number of flip-flops. The results of this assessment are presented in Figure 7.23. As shown, evaluation of flip-flops in 10 sets with 270, flip-flops gives the minimum runtime.

#### 7.4.4.4 Independent analysis of masking factors

An important feature of the proposed approach is the combined analysis of the logical and temporal masking factors. In order to show the benefits of this approach, we compared the soft error vulnerability obtained by this approach by that of the independent analysis of these two factors. For the independent analysis, the logical masking factor is computed by our tool as the ratio of the number of vulnerable cycles over all cycles (i.e., by ignoring computed vulnerable time intervals inside vulnerable cycles which represent the temporal masking factor). The temporal masking factor is extracted with respect to shortest path starting from the target flip-flop according to the method detailed in [22]. As shown in Figure 7.24, the independent analysis imposes inaccuracy by up to 67.8% in vulnerability estimation as error does not necessarily propagate through short paths and typically there is a considerable difference in the delay of paths in a circuit.

#### 7.4.4.5 Comparison with other techniques

In order to show the effectiveness of the proposed approach in terms of accuracy and runtime, we experimentally compared two versions of our technique with two other techniques:

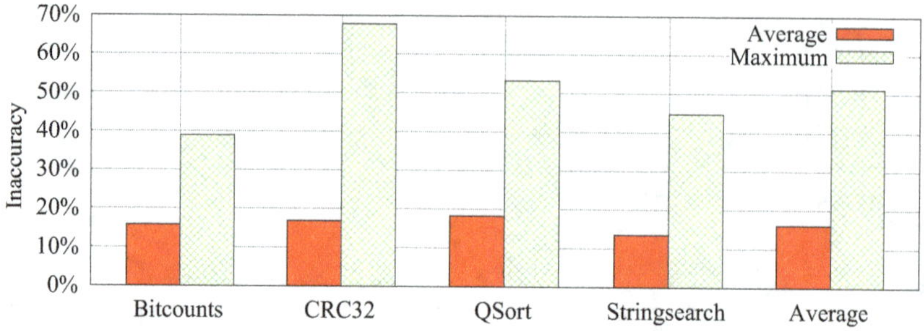


*Figure 7.24 Average and maximum inaccuracy imposed due to independent analysis of logic and temporal masking factors*

*Table 7.4 Average and maximum inaccuracy of different techniques*

| Workload | Bitcounts | | CRC32 | | QSort | | Stringsearch | | Average | |
|---|---|---|---|---|---|---|---|---|---|---|
| | Avg. (%) | Max. (%) | Avg. (%) | Max. (%) | Avg. (%) | Max. (%) | Avg. (%) | Max. (%) | Avg. (%) | Max. (%) |
| Proposed-16 | 0.2 | 0.7 | 0.6 | 1.7 | 1.7 | 3.6 | 0.5 | 1.1 | **0.7** | **1.8** |
| Proposed-32 | 0.1 | 0.5 | 0.1 | 0.4 | 0.8 | 2.1 | 0.3 | 0.9 | **0.4** | **0.9** |
| CFS-5% | Maximum inaccuracy of 5.00% with confidence-level of 95% | | | | | | | | | |
| SFI-5% | that requires 1,034,112 fault injection according to [24] | | | | | | | | | |

- Proposed-16 and Proposed-32 propagate the error for 16 and 32 cycles, respectively, and in each iteration of the EP 270, flip-flops are evaluated concurrently.
- SFI is a statistical fault injection built on top of our logic simulator.
- CFS [23] is employed to expedite the SFI by concurrently propagating errors from different flip-flops. However, this technique needs to repeat the vulnerability assessment for each cycle and even for random times within each cycle in order to compute the temporal masking factor.

The average and the maximum inaccuracies of different techniques for evaluation of OR1200 flip-flops are reported in Table 7.4. Additionally, the runtime of each technique for evaluation of different workloads is reported in Table 7.5. As can be seen, CFS provides 133× speedup over SFI for simulating the same number of errors. Among these, our proposed approach provides four orders of magnitude speedup with respect to SFI while it is highly accurate. The only source of inaccuracy in our proposed approach is the errors that require a large number of simulated cycles to cause a failure. This approach is much faster than CSF as it skips unnecessary computations when there is no logic event. These features make our technique appropriate solution for soft error vulnerability of flip-flops.

*Table 7.5 Comparison of runtime of different techniques (minutes) (speedups are w.r.t. SFI)*

| Workload | | Bitcounts | CRC32 | QSort | Stringsearch | Average |
|---|---|---|---|---|---|---|
| Proposed-16 | Absolute | 178 | 233 | 669 | 76 | |
| | Speedup | 91,356 | 88,370 | 101,308 | 90,767 | **92,950** |
| Proposed-32 | Absolute | 560 | 698 | 2,164 | 239 | |
| | Speedup | 29,037 | 29,599 | 31,346 | 29,091 | **29,768** |
| CFS-5% | Absolute | 141,628 | 144,389 | 517,976 | 48,114 | |
| | Speedup | 115 | 143 | 131 | 144 | **133** |
| SFI-5% | Absolute | 16,287,264 | 20,647,769 | 67,854,982 | 6,928,550 | |

## 7.5 Propagating combinational gates errors at circuit level

A transient pulse in combinational gates is only latched in flip-flops if it is not masked by any of the logical (where another input of a gate is in controlling state), electrical (pulse electrically attenuated during propagation), or latching-window masking (pulse is not captured in the flip-flops as it is not arrived in the latching-window of receiving elements) factors [25].

Since the electrical and latching-window masking behavior of combinational logic is completely different in a conventional emulation environment [26], SPICE-based fault injection is the most accurate approach for analyzing SER of these elements. However, as the number of combinational gates is much more than flip-flops and also the EP should be repeated for different pulse widths, reaching a reasonable level of accuracy with such fault injections is intractable even for a small circuit with a few thousand gates.

In this regards, we employed a combination of previously proposed analytical techniques to propagate the errors in the combinational logic in the first cycle. The techniques presented in [27–29] have been used for modeling logical, electrical, and latching-window masking factors, respectively.

The single-cycle analysis was repeated for different pulse widths. At the end of single-cycle analysis for pulse width $w$, we obtained $SCPP^g(w, s)$ that is the probability of propagation of the transient pulse from gate $g$ to all flip-flops in set $s$.

A running example of EP using the employed single-cycle analytical technique is shown in Figure 7.25. In this example, an SET with a pulse width of 50 ps is propagated from gate A toward flip-flop inputs. For each gate in the forward cone of the error site, logical and electrical masking and the vulnerability window are computed.

**Logical masking**: The probability of EP to the gate output is computed with respect to the signal and error probabilities of its input nodes according to the rules provided in [27]. For instance, in the running example, one input of the AND gate

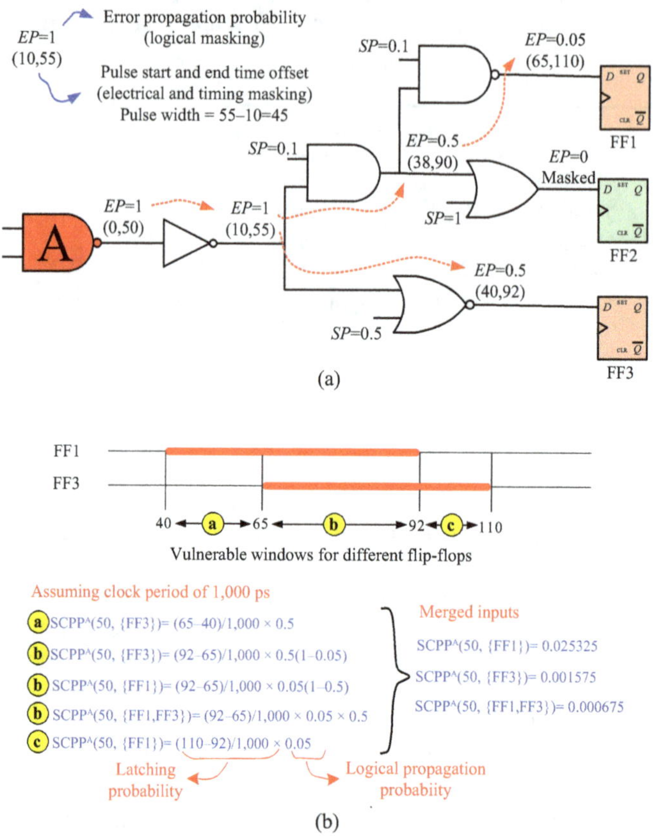


*Figure 7.25 Error propagation using single-cycle analysis: (a) error propagation from error site to flip-flops' inputs and (b) computation of Single-Cycle Propagation Probabilities*

has $EP = 1$ while the other input has signal probability of 0.5. Hence, only 50% of errors are propagated to the output of this gate and EP of the output node is 0.5.

**Electrical masking**: The pulse width of the transient error at the output of a gate is computed according to the equation-based model described in [28]. This technique approximates a transient pulse with a trapezoidal waveform by computing width, amplitude, rise and fall time of the transient pulse. Basically, for large pulse widths, as described in [28], the transient pulse at the output of a combinational gate could be accurately modeled by $pw + t_f - t_r$ where $pw$ is the input pulse width, and $t_r$ and $t_f$ are the rise and fall delays of the combinational gate, respectively. The resulting transient pulse can be either attenuated or broadened depending on the ratio of rise and fall delays of the gate.

**Vulnerable window**: An error might reach the inputs of several flip-flops at the same time or at different time intervals. These cases result in significantly different failure probabilities and, hence, have to be accurately identified. Therefore, the vulnerable window for each node has to be computed, and based on the intersection of vulnerable windows of different flip-flop inputs, various latching possibilities could be investigated. Since the error occurrence time is random, the relative start and end of vulnerable window of nodes are described with respect to the error occurrence time. The relative start time of the transient pulse is set to zero for the error origin (gate A in Figure 7.25), and it is computed for other gates in the forward cone with respect to the time required for error to propagate to those nodes. Considering the pulse width ($pw$) computed by electrical masking model as well as the relative start time ($t_s$), a pair of relative start and end time of vulnerable window denoted by ($t_s, t_e = t_s + pw$) could be specified. In the running example, for the error origin, this is $(0, 50)$, as we want to evaluate the EP for pulse width of 50 ps. The average propagation delay of inverter gate is 10 ps, and the width of transient pulse at the output node is 45 ps according to the electrical masking model; hence, the vulnerable window for the output node would be $(10, 10 + 45) = (10, 55)$.

Once the error is propagated to the inputs of flip-flops, the latching probabilities for different sets of flip-flops are computed. The latching probability could be computed as (*pulse width*/*clock period*) $\times$ *EP* as detailed in [29]. In the running example, the error propagates to FF3 with probability of 0.5 during $(40, 65)$ interval, hence, by assuming clock period of 1,000 ps, $SCPP^A(50, FF1) = (65 - 40)/1,000 \times 0.5$. In the interval $(65, 92)$, the error is propagated to both FF1 and FF3 flip-flops. In such cases, we have to compute the probability of error logically propagating to each subset of flip-flops. As shown in Figure 7.25, error could be propagated to either FF1 or FF3 or both of them. The error computations for all three cases are shown in this figure. As shown, compared to the case with only one vulnerable flip-flop, the only difference is in the computation of logical masking (*EP*). For example, the probability that error is logically propagated to both flip-flops is $EP = 0.5 \times 0.05$ which is computed according to input EP of two flip-flops. Once all Single-Cycle Propagation Probabilities (SCPPs) are computed, all similar cases are merged by a summation of similar cases to obtain one unique number for each subset of flip-flops (see Figure 7.25).

In order to show the efficiency of the employed single-cycle analysis, it is compared with an SPICE-based FIT [30]. In this technique, the SPICE netlist of the forward-cone of the target error site is constructed by combining the existing SPICE description of the standard cells. Then the fault injection is performed on the resulting SPICE netlist of the forward-cone. Comparison of the employed analytical techniques with SPICE-based fault injection for 100 randomly selected combinational gates from the OR1200 processor shows an average inaccuracy of 2.1% while offering five orders of magnitude speedup compared to SPICE simulations.

## 7.6 Emulation-based fault injection platform

Once we have analyzed how the errors propagate to flip-flops and SRAM cells, we perform an emulation-based fault injection using our novel platform on those cells to extract the effect of architecture and application masking. Emulation-based FITs are either reconfiguration-based [31–34] or instrumentation-based [35–37]. Reconfiguration-based techniques rely on an internal mechanism of Xilinx Field Programmable Gate Arrays (FPGAs). These techniques change the configuration bits of the FPGA device using full or partial reconfiguration in such a way that a target fault model is applied on the desired fault site. In these techniques, the reconfiguration process is the speed bottleneck [38].

In instrumentation-based techniques, fault injector circuits called saboteurs are added to each fault site. By activation of each saboteur, a fault is injected in its corresponding fault site. One disadvantage of instrumentation-based techniques is that, unlike reconfiguration-based techniques which are completely dependent on commercial tools, a special tool should be developed for the placement of saboteurs in the design, because none of saboteurs presented so far are supported by commercial tools. The development of a noncommercial tool is a very time-consuming task. Additionally, since noncommercial tools do not pass the exhaustive software test process, the possibility of the existence of bugs in such tools is much more than commercial tools. To address this issue, we present an instrumentation-based technique that completely relies on commercial tools for the placement of saboteurs.

In this section, we briefly introduce our instrumentation-based fault injection platform that completely relies on commercial tools for the placement of saboteurs. This technique uses two Altera debugging facilities, including In-system Memory Content Editor (MCE) and In-system Sources and Probes (SAP) for fault injection purpose. Since these debugging facilities have no impact on the behavior of the design, we call the proposed technique, Shadow Components-based FIT (SCFIT). Unlike existing instrumentation-based techniques, SCFIT completely relies on commercial tools to implement the saboteurs. Reliance on commercial tools makes the development of the proposed technique much easier and faster than that of other instrumentation-based techniques. SCFIT has the capability of fault injection in both flip-flops and memory units. For fault injection in flip-flops, a scan chain is inserted in all Device Under Test (DUE) flip-flops and fault is injected in the target flip-flop value during scan chain circulation. For fault injection in memory units, single-port memories are converted to dual-port memories using MCE facility. The additional port is used to write the faulty data at the fault injection time and to read memory contents whenever is needed. In SCFIT, the SAP facility has the responsibility of establishing communication between Quartus II software in host computer and fault injection board inside FPGA board.

Experimental results of OpenRISC 1200 processor evaluation show that the SCIT technique is four orders of magnitude faster than a simulation-based technique. This enables the designers to inject more faults and get more accurate results in a limited time period.

### *7.6.1 Shadow components*

Altera Corporation provides a complete set of design debugging facilities that can be easily adapted to any design. All these debugging facilities are integrated in the

Quartus software to ease the design verification process. These facilities utilize combination of available memory, logic, and routing resources to help the designers in the verification process by increasing the controllability and observability of the design. We have exploited two of these debugging facilities, namely, SAP and MCE to propose a fast, flexible, and easy-to-develop FPGA-based FIT. These facilities are used to control and observe the value of the control registers and single-port memories in the proposed technique.

A simplified block diagram of a design connected to these facilities is shown in Figure 7.26. As shown, both SAP and MCE are instrumented by a JTAG controller. The JTAG controller acts as an interface between debugging facilities inside FPGA board and the Quartus II software on the host computer.

The MCE facility allows the designer to easily observe and modify memory contents. This tool converts a single-port memory to a dual-port memory. The Block diagram of an ordinary single-port memory and an MCE dual-port memory are depicted in Figure 7.27. The additional port is connected to the JTAG controller in order to observe and modify the memory contents.

The SAP facility is a perfect choice for toggling control signals. Each SAP unit contains several source ports for driving values into and several probe ports for sampling values from the signals that are connected to the ports. Figure 7.28 shows a simplified SAP unit connected to an internal node in the design. As can be seen, SAP provides single-cycle sample and single-cycle write to selected logic nodes. The SAP

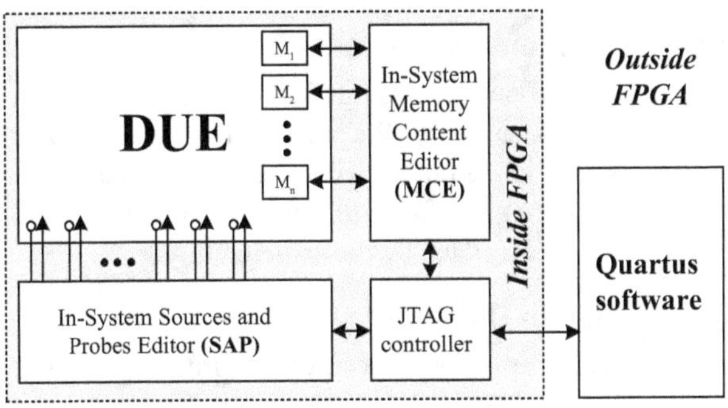


*Figure 7.26 MCE and SAP facilities connected to a design in FPGA (adapted from [39])*

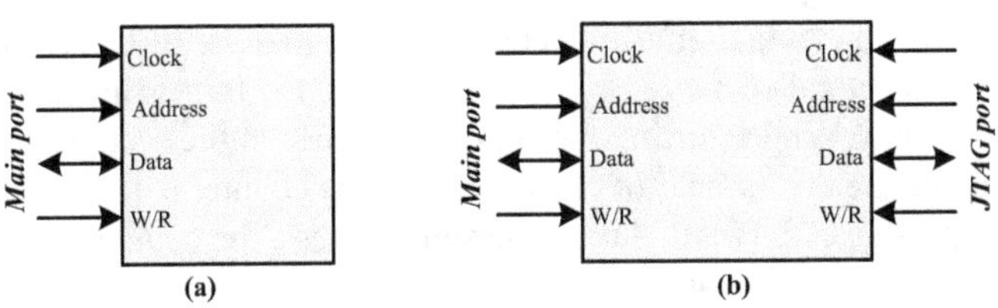


*Figure 7.27 (a) A single-port memory and (b) an MCE dual-port memory*

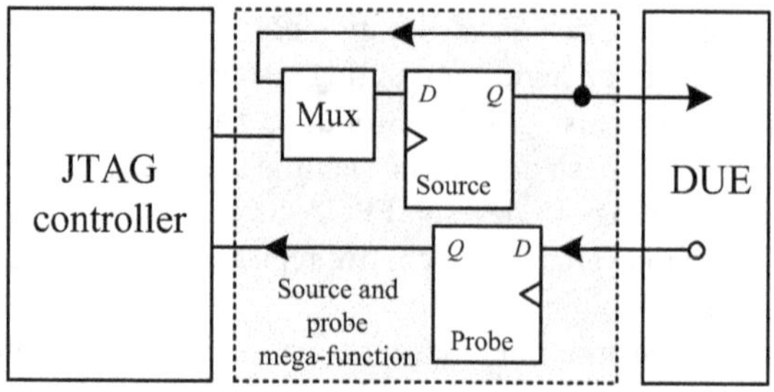


*Figure 7.28 An SAP unit connected to DUE*

allows designers to easily implement control signals in the design as virtual stimuli. By means of this facility, we can observe and change the value of control registers employed in our fault injection board.

### 7.6.2 Shadow components-based fault injection technique

SCFIT uses different approaches for fault injection in memory units and flip-flops. Fault injection in memory units can be easily done using the MCE feature of the Altera FPGAs, while a scan chain is employed for fault injection in flip-flops. A simple and general purpose fault injection board is designed to ease the communication between the host computer and the FPGA board.

#### 7.6.2.1 Fault injection in memory units

As mentioned in the previous section, using MCE the contents of the memory units can be observed or modified at any time. We have used this feature for fault injection in memory units. In this regard, in the fault injection time, the value of target address(es) is read using the MCE feature and after fault injection in the target bit(s), the faulty value is written back to its prior place by the same facility.

#### 7.6.2.2 Fault injection in flip-flops

For fault injection in DUE flip-flops, a scan chain is inserted in DUE. Scan chain is one of the well-known and widely used techniques in design for testability. In this technique, in test mode, all flip-flops in the circuit form a long shift-register, while in the normal mode, the functionality of the circuit is preserved. By scan chain insertion in DUE, faults can be injected during complete circulation of the scan chain by inversion of the target flip-flop value.

Figure 7.29 shows a fault injection controller designed for SEU injection in DUE flip-flops. A register called Target_FF is used to store the target flip-flip number for injecting SEU fault. While shifting the flip-flops values in the scan chain, a counter is used to enumerate the number of clock cycles. This counter is first loaded by total number of flip-flops. Afterward, during the shift process, the counter is decremented by one at each clock cycle. Whenever the value of Counter and Target_FF are matched, the fault-free value of the fault site can be accessed at the scan chain output port. By inversion of this value for one clock cycle, the value of target flip-flop is inverted. After

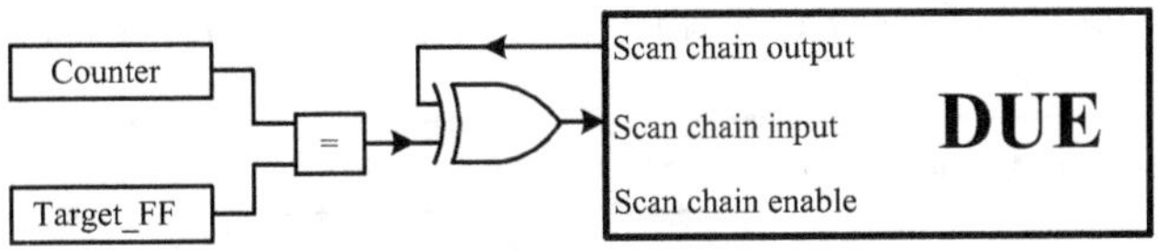


*Figure 7.29 Fault injection controller for fault injection in flip-flops*

*Table 7.6 FPGA resources used by OR1200 with and without SCFIT*

| Resource | Without SCFIT | With SCFIT | Overhead (%) |
|---|---|---|---|
| Memory bits | 712,384 | 716,274 | 0.5 |
| Logic elements | 26,357 | 28,002 | 6.2 |

fault injection, circulation continues until the circulation process is completed. Once the Counter reaches zero, the fault injection process is completed and all flip-flops except the target flip-flop have their previous values.

For injecting an MBU fault in several flip-flops, one approach is to repeat the operation of SEU injection for each target flip-flop. Another approach is to use multiple Target_FF registers and whenever one of these matches with the counter value, a bit-flip is injected at scan chain input. We prefer the first approach as time needed for multiple circulations is negligible in emulation environment, and it has simpler circuitry than the second one.

### *7.6.3 Experimental results*

As a case study, we have evaluated the soft error vulnerability of the OR1200 processor using SCFIT. Five workloads from MiBench embedded benchmark suite [21], including Bitcount, CRC32, FFT, QSort, and Stringsearch, run on OR1200 during the fault injection campaigns.

An Altera Cyclone IV FPGA board is employed as our platform in SCFIT. Table 7.6 shows FPGA resources used by OR1200 in two cases: first, it is normally programed on the FPGA without any debugging facility; second, it is modified according to SCFIT and then programed on the FPGA. As can be seen in this table, applying SCFIT to OR1200 does not significantly affect the allocated resources.

The FPGA board is connected to a host computer with an Intel Xeon E5540 2.53 GHz processor and 16 GB RAM. In order to verify the functionality of SCFIT, it is compared with a simulation-based FIT. The simulation-based fault injections are done using the ModelSim simulator on the same computer. Comparison of 1,000 randomly injected faults using both simulation-based technique and SCFIT reveals that the effects of faults in two cases are exactly the same, while fault injection in simulation-based platform is much slower than the proposed technique. Table 7.7 shows the average time required for each fault injection for different benchmarks.

*Table 7.7 Runtime of SCFIT vs. simulation-based fault injection (seconds)*

| Workload | Cycles | Simulation-based | SCFIT | Speedup |
|---|---|---|---|---|
| Stringsearch | 209,935 | 1,476.4 | 1.79 | 825× |
| Bitcount | 532,327 | 5,481.5 | 1.80 | 3,045× |
| QSort | 2,065,603 | 16,836.8 | 1.99 | 8,461× |
| CRC32 | 11,391,957 | 66,627.0 | 2.64 | 25,237× |
| FFT | 26,137,266 | 176,883.3 | 3.73 | 47,421× |
| Average | | | | 15,998× |

As can be seen from this table, each FPGA-based fault injection takes about 1.79–3.73 s to be done, which is on average four orders of magnitude faster than the simulation-based fault injection. Our experiments in [40,41] on the nonprotected and protected implementations of a Leon2 processor showed the same amount of speedup.

## 7.7 Fault injection acceleration

Fault injection experiments are conventionally based on Monte Carlo simulation. Typically fault injection results show a very small vulnerability to soft errors [42–44]. This means that most of the fault injection experiments have no system-level impact. For a desired level of accuracy, the Monte Carlo simulations could be accelerated to a large extent by exploiting an importance sampling technique that skips most of unnecessary fault injections.

The Architecturally Correct Execution (ACE) analysis is able to classify the time intervals to those that might result in failure (i.e., ACE) and those that definitely have no chance to cause a failure (i.e., un-ACE). In our proposed approach, the ACE analysis is leveraged as an importance sampling technique to identify the majority of non-vulnerable intervals. Afterward, fault injection experiment is only conducted on the ACE intervals to obtain the failure probability. The fault injection sampling spaces for both conventional fault injection and proposed approach are presented in Figure 7.30. As shown, the proposed approach skips unnecessary fault injection iterations in un-ACE intervals. This technique significantly reduces the number of fault injection iterations for providing the same level of accuracy.

### *7.7.1 Workflow*

The workflow of the proposed approach is illustrated in Figure 7.31. In this approach, a list of fault injection samples is randomly selected first. Afterward, by performing an ACE analysis, the ratio of ACE intervals to the entire workload execution time (denoted as $\alpha$) is determined. In addition, the list of fault injection samples is pruned by eliminating samples belonging to un-ACE intervals. Then, fault injection is performed on the pruned fault injection list.

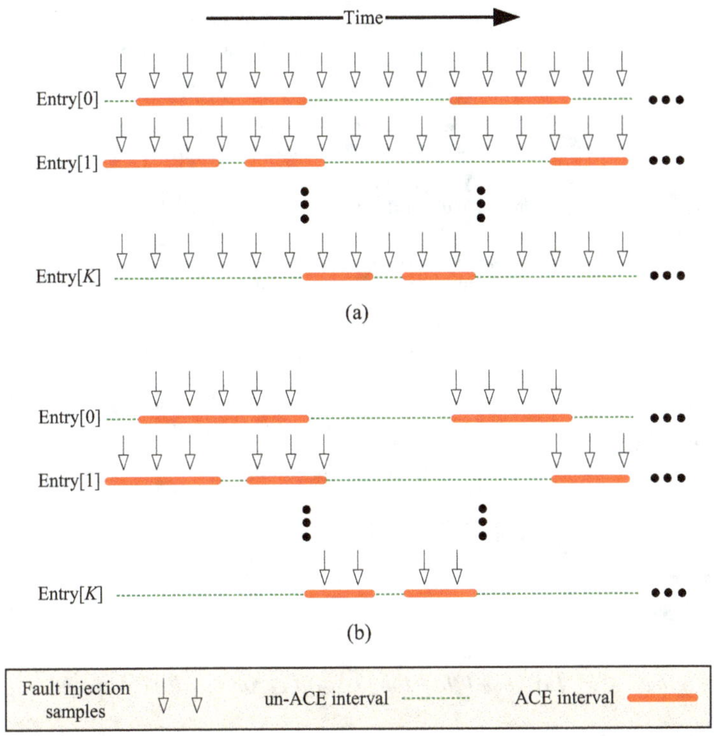


*Figure 7.30 Conventional fault injection vs. proposed approach: (a) typical fault injection approach injects faults in random locations and time and (b) our proposed approach injects faults only in ACE locations and time*

## 7.7.2 Analytical modeling

In order to obtain the speedup of our proposed technique, we present an analytical model that accepts the ratio of ACE intervals (i.e., $\alpha$) as input and provides the number of required error injection to reach a desired level of accuracy. The number of samples to reach the maximum sampling inaccuracy of $e$ for failure probability of $p$ and cutoff point of $t$ (depends on the confidence level) in the conventional fault injection approach (i.e., without importance sampling) could be computed as [24]

$$n = \frac{t^2 \times p \times (1-p)}{e^2}. \tag{7.1}$$

Let us assume that the ACE analysis results show that the memory array is in ACE state with probability of $\alpha$. In this case, the fault injection has to be performed to evaluate what ratio of the ACE time intervals leads to a failure. However, the failure probability ($p'$) and desired sample inaccuracy ($e'$) are different from that of a typical

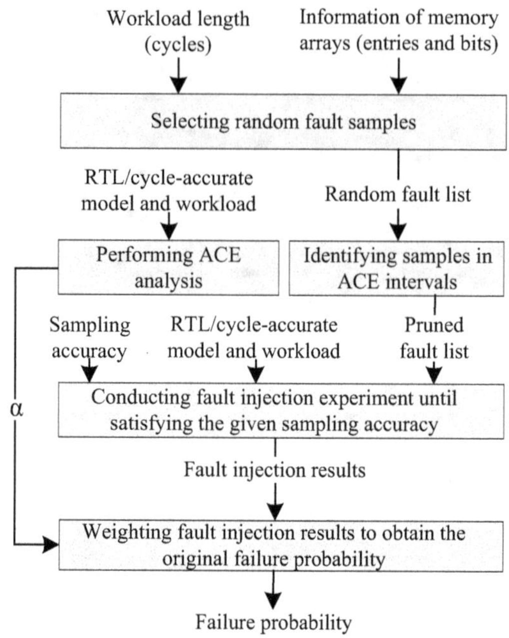


*Figure 7.31 Proposed flow for fast soft error sensitivity evaluation*

fault injection. In this case, $p'$ is the failure probability when faults are only injected in ACE time intervals (i.e., $p' = P(fail|ACE)$). According to *Bayes' law* [45]:

$$p' = P(fail|ACE) = \frac{P(ACE|fail) \times P(fail)}{P(ACE)} \tag{7.2}$$

where $P(fail) = p$ and $P(ACE) = \alpha$. In addition, $P(ACE|fail)$ is theoretically equal to one as all failures are related to ACE intervals. Therefore, $p' = p/\alpha$.

In addition, the desired sampling inaccuracy for the fault injection after ACE analysis is not the same. Fault injection on ACE intervals provides the sampling inaccuracy for the results in the range of [0,1], while the share of this results in the final results in the range of [0,$\alpha$]. Due to the fact that the results of non-ACE intervals have theoretically no inaccuracy, whereas that of ACE intervals could have inaccuracy of $e$ in the range of [0,$\alpha$]; hence $e' = e/\alpha$ in the range of [0,1], where $\alpha \leq 1$, $e \leq e'$. This means that the fault injection only on ACE intervals (proposed approach) requires less accurate results to provide the same level of accuracy on the overall results.

Considering the aforementioned points, the number of fault injection iterations required by our proposed approach is

$$n_p = \frac{t^2 \times p' \times (1-p')}{e'^2} = \frac{t^2 \times \frac{p}{\alpha} \times (1-\frac{p}{\alpha})}{(e/\alpha)^2} = \frac{t^2 \times p \times (\alpha - p)}{e^2}. \tag{7.3}$$

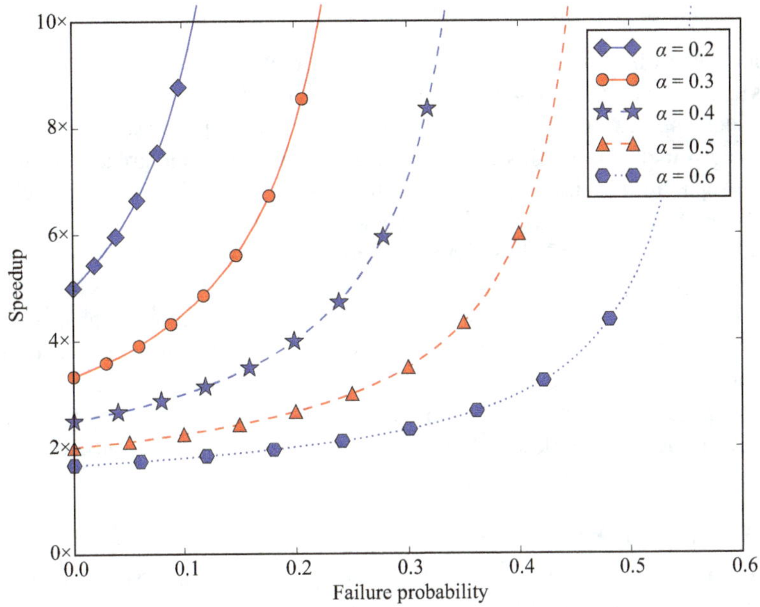


*Figure 7.32 Speedup of the proposed approach for different values of p and α*

The speedup of our approach over conventional fault injection approach could be estimated with respect to the reduction in the number of fault injection iterations according to (7.1) and (7.3). Besides the iterations required for the fault injection, the proposed approach also needs one additional iteration to compute the ratio of ACE intervals. However, fault injection campaigns typically have to provide the results with small inaccuracy that mandates a larger number of fault injection iterations. Hence, this additional iteration could be safely ignored in computing speedup (i.e., $1 \ll n_{proposed}$). The speedup of our proposed approach over conventional fault injection approach is computed according to

$$\text{speedup} = \frac{n}{n_p + 1} \approx \frac{n}{n_p} = \frac{1 - p}{\alpha - p}. \tag{7.4}$$

Figure 7.32 depicts the speedup obtained for the typical value of $\alpha$ (i.e., 0.2–0.6). As shown, the speedup for small values of $p$ is very close to $1/\alpha$ and as $p \rightarrow \alpha$, speedup $\rightarrow \infty$. However, as shown in the previous studies [46,47] and also confirmed with our experimental results (demonstrated in Section 7.7.3), the ACE analysis significantly overestimates the failure probability. This means that the value of $p$ is typically much smaller than $\alpha$ and speedup $\approx 1/\alpha$.

### 7.7.3 Case study: fault injection on memory arrays of Leon3

In order to show the effectiveness of the proposed approach, we evaluated the speedup of this technique for assessing the soft error sensitivity of cache arrays and register-file of the Leon3 processor. Leon3 is an open-source embedded processor with an in-order seven stage pipeline unit. This 32-bit processor implements Harvard architecture, i.e. it has a separate instruction and data caches. Both data and instruction caches are write-through with direct-mapped policy and have a size of 1,024×32 bits, and their tags are stored in a 128×28 bits memory array. The register-file of this processor has 128 words.

In order to compute ACE intervals, the RTL implementation of the processor is modified to include an ACE model for each on-chip memory array. For register-file, we implemented the generic ACE model, while for instruction and data cache, the cache hit/miss status is also taken into account according to the description in [48]. After executing each workload, the value of total ACE intervals for each memory entry is read and the value of $\alpha$ is computed.

#### 7.7.3.1 Speedup analysis

In order to obtain the speedup of our proposed approach for assessing soft error vulnerability of on-chip memory arrays of a Leon3 processor, the runtime of a pure fault injection (i.e., conventional) is compared with that of our approach for the maximum tolerable sampling inaccuracy of 1% ($e = 0.01$) and confidence interval of 95%.

The speedups obtained for evaluating different workloads are illustrated in Figure 7.33. The results reveal that there is a significant variation among different workloads and also between different components. By employing the proposed

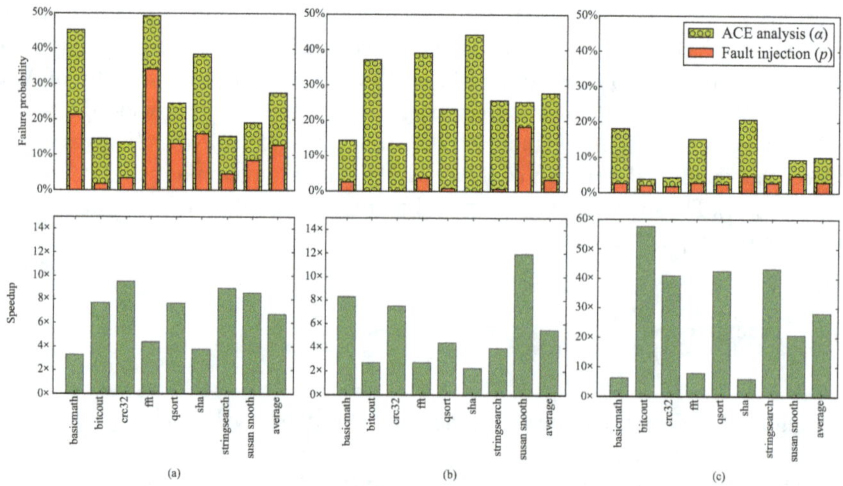


*Figure 7.33 Failure probability obtained by ACE and fault injection as well as the speedup of ACE before FI technique for (a) instruction cache, (b) data cache, and (c) register-file of a Leon3 processor*

approach, the soft error sensitivity assessment is expedited for instruction cache, data cache, and register-file by 6.7×, 5.5×, 28.2×, respectively (13.5×, on average).

The speedup obtained for register-file is significantly higher than the two other components as generic ACE model is very accurate (i.e., less pessimistic) for this component. This is mainly because all data read from register-file are operands required by the execution unit and have a very high chance to affect the workload output. In contrast, an error in data/instruction cache has to be propagated through several units to reach the execution unit of the processor. Each of these intermediate units masks a subset of the errors that cannot be captured by our employed ACE model and results in a larger gap between $\alpha$ and $p$ in these two components.

Our experimental results are in-line with the analytical study provided in Section 7.7.2. For the cases in which there is considerable gap between $\alpha$ and $p$, the speedup is mostly dependent on $\alpha$, i.e., a smaller value of $\alpha$ results in a higher speedup and vice versa. For instance, in data cache, all workloads, excluding *susan_smooth* have speedup close to $1/\alpha$. In contrast, the speedup for the cases with a small gap between $\alpha$ and $p$ is strongly dependent on both parameters and is considerably higher than the other cases. This significant difference in speedup could be clearly seen for *susan_smooth* and *stringsearch* workloads in data caches that have almost the same value of $\alpha$, but considerable difference in the value of $p$. As shown, the speedup for *susan_smooth* workload is considerably higher than *stringsearch*. These results could be interpreted similar to Figure 7.32.

## 7.8 Conclusions

In this chapter, a comprehensive SER analysis platform was presented, which models both error generation and propagation behaviors. This platform consists of a combination of empirical models at the device level, error site analysis at chip layout, analytical EP at logic level, and fault simulation/emulation at the architecture/application level. By having an appropriate model of abstraction at each step, we are able to accurately propagate the errors and observe their application-level effects.

The proposed platform could evaluate a relatively complex design using its hierarchical modeling methodology. This platform enabled us to accurately model the intrinsic SER as well as all the masking factors for all memory arrays, flip-flops, and combinational gates. The results of SER analysis of an embedded processor using this approach will be presented in the next chapter.

## References

[1] iRoC Technologies. TFIT: Cell Level Soft Error Analysis, http://iroctech.com; 2015.

[2] Harada R, Mitsuyama Y, Hashimoto M, *et al.* Neutron induced single event multiple transients with voltage scaling and body biasing. In: International Reliability Physics Symposium. IEEE; 2011. p. 3C–4.

[3] Rossi D, Omana M, Toma F, *et al.* Multiple transient faults in logic: an issue for next generation ICs? In: Proceedings of International Symposium on Defect and Fault Tolerance in VLSI Systems. IEEE; 2005. p. 352–360.

[4] Fazeli M, Ahmadian SN, Miremadi SG, *et al.* Soft error rate estimation of digital circuits in the presence of multiple event transients. In: Design, Automation and Test in Europe Conference. IEEE; 2011. p. 1–6.

[5] Miskov-Zivanov N and Marculescu D. Multiple transient faults in combinational and sequential circuits: a systematic approach. IEEE Transactions on Computer-Aided Design of Integrated Circuits and Systems. 2010;29(10):1614–1627.

[6] Ibe E, Taniguchi H, Yahagi Y, *et al.* Impact of scaling on neutron-induced soft error in SRAMs from a 250 nm to a 22 nm design rule. IEEE Transactions on Electron Devices. 2010;57(7):1527–1538.

[7] Radaelli D, Puchner H, Wong S, *et al.* Investigation of multi-bit upsets in a 150 nm technology SRAM device. IEEE Transactions on Nuclear Science. 2005;52(6):2433–2437.

[8] Giot D, Roche P, Gasiot G, *et al.* Heavy ion testing and 3-D simulations of multiple cell upset in 65 nm standard SRAMs. IEEE Transactions on Nuclear Science. 2008;55(4):2048–2054.

[9] Hubert G, Artola L, and Regis D. Impact of scaling on the soft error sensitivity of bulk, FDSOI and FinFET technologies due to atmospheric radiation. Integration, the VLSI Journal. 2015.

[10] Maestro JA and Reviriego P. Study of the effects of MBUs on the reliability of a 150 nm SRAM device. In: IEEE Design Automation Conference; 2008. p. 930–935.

[11] Rusu C, Bougerol A, Anghel L, *et al.* Multiple event transient induced by nuclear reactions in CMOS logic cells. In: IEEE International On-Line Testing Symposium; 2007. p. 137–145.

[12] Clarke M, Hammerschlag D, Rardon M, and Sood A. Eliminating routing congestion issues with logic synthesis. In: Cadence White Paper; 2014.

[13] Puchner H, Radaelli D, and Chatila A. Alpha-particle SEU performance of SRAM with triple well. IEEE Transactions on Nuclear Science. 2004;51(6): 3525–3528.

[14] Fang YP and Oates AS. Neutron-induced charge collection simulation of bulk FinFET SRAMs compared with conventional planar SRAMs. IEEE Transactions on Device and Materials Reliability. 2011;11(4):551–554.

[15] Fazeli M, Miremadi SG, Asadi H, *et al.* A fast and accurate multi-cycle soft error rate estimation approach to resilient embedded systems design. In: Proceedings of the International Conference on Dependable Systems and Networks. IEEE/IFIP; 2010. p. 131–140.

[16] Asadi H, Tahoori MB, Fazeli M, *et al.* Efficient algorithms to accurately compute derating factors of digital circuits. Microelectronics Reliability. 2012:1215–1226.

[17] Mahatme N, Jagannathan S, Loveless T, *et al.* Comparison of combinational and sequential error rates for a deep submicron process. IEEE Transactions on Nuclear Science. 2011;58(6):2719–2725.

[18] Mahatme N, Gaspard N, Assis T, *et al.* Impact of technology scaling on the combinational logic soft error rate. In: International Reliability Physics Symposium. IEEE; 2014. p. 5F-2.

[19] Seifert N, Gill B, Jahinuzzaman S, *et al.* Soft error susceptibilities of 22 nm tri-gate devices. IEEE Transactions on Nuclear Science. 2012;59(6): 2666–2673.

[20] Ebrahimi M, Evans A, Tahoori MB, *et al.* Comprehensive analysis of alpha and neutron particle-induced soft errors in an embedded processor at nanoscales. In: Design, Automation and Test in Europe Conference. IEEE; 2014. p. 1–6.

[21] Guthaus MR, Ringenberg JS, Ernst D, *et al.* MiBench: A free, commercially representative embedded benchmark suite. In: IEEE International Workshop on Workload Characterization; 2001. p. 3–14.

[22] Silburt AL, Evans A, Perryman I, *et al.* Design for soft error resiliency in internet core routers. IEEE Transactions on Nuclear Science. 2009;56(6):3551–3555.

[23] Cha H, Rudnick EM, Patel JH, *et al.* A gate-level simulation environment for alpha-particle-induced transient faults. IEEE Transactions on Computers. 1996;45(11):1248–1256.

[24] Leveugle R, Calvez A, Maistri P, *et al.* Statistical fault injection: quantified error and confidence. In: Design, Automation & Test in Europe Conference. IEEE; 2009. p. 502–506.

[25] Ferlet-Cavrois V, Massengill LW, and Gouker P. Single event transients in digital CMOS: a review. IEEE Transactions on Nuclear Science. 2013;60(3):1767.

[26] Entrena L, Garcia-Valderas M, Fernandez-Cardenal R, *et al.* Soft error sensitivity evaluation of microprocessors by multilevel emulation-based fault injection. IEEE Transactions on Computers. 2012;61(3):313–322.

[27] Chen L, Ebrahimi M, and Tahoori MB. CEP: correlated error propagation for hierarchical soft error analysis. Journal of Electronic Testing. 2013;29(2): 143–158.

[28] Rajaraman R, Kim JS, Vijaykrishnan N, *et al.* SEAT-LA: a soft error analysis tool for combinational logic. In: IEEE International Conference on VLSI Design; 2006.

[29] Costenaro E, Evans A, Alexandrescu D, *et al.* Towards a hierarchical and scalable approach for modeling the effects of SETs. In: IEEE Workshop on Silicon Errors in Logic-System Effects (SELSE); 2013.

[30] Kiamehr S, Ebrahimi M, Firouzi F, *et al.* Chip-level modeling and analysis of electrical masking of soft errors. In: VLSI Test Symposium. IEEE; 2013.

[31] Cheng K, Huang S, and Dai W. Fault emulation: a new methodology for fault grading. IEEE Transactions on Computer-aided Design of Integrated Circuits and Systems (TCAD). 1999;18(10):1487–1495.

[32] Antoni L, Leveugle R, and Feher B. Using run-time reconfiguration for fault injection applications. IEEE Transactions on Instrumentation and Measurement. 2003;52(5):1468–1473.

[33] Andres DD, Ruiz JC, Gil D, *et al.* Fault emulation for dependability evaluation of VLSI systems. IEEE Transactions on Very Large Scale Integration Systems (TVLSI). 2008;16(4):422–431.

[34] Sterpone L and Violante M. A new partial reconfiguration-based fault-injection system to evaluate SEU effects in SRAM-based FPGAs. IEEE Transactions on Nuclear Science (TNS). 2007;54(4):965–970.
[35] Hwang S, Hong J, and Wu C. Sequential circuit fault simulation using logic emulation. IEEE Transactions on Computer-Aided Design of Integrated Circuits and Systems (TCAD). 1998;17(8):724–736.
[36] Civera P, Macchiarulo L, Rebaudengo M, *et al.* Exploiting circuit emulation for fast hardness evaluation. IEEE Transactions on Nuclear Science (TNS). 2001;48(6):2210–2216.
[37] Ejlali A, Miremadi SG, Zarandi H, *et al.* A hybrid fault injection approach based on simulation and emulation co-operation. In: International Conference on Dependable Systems and Network; 2003. p. 479–488.
[38] Lopez-Ongil C, Entrena L, Garcia-Valderas M, *et al.* A unified environment for fault injection at any design level based on emulation. IEEE Transactions on Nuclear Science (TNS). 2007;54(4):946–950.
[39] Quartus, I. I. Handbook Version 11.0: Design and Synthesis, Altera Corporation; 2012.
[40] Mohammadi A, Ebrahimi M, Ejlali A, *et al.* SCFIT: a FPGA-based fault injection technique for SEU fault model. In: Proceedings of Design, Automation and Test in Europe Conference (DATE). IEEE; 2012. p. 586–589.
[41] Ebrahimi M, Miremadi SG, Asadi H, *et al.* Low-cost scan-chain-based technique to recover multiple errors in TMR systems. IEEE Transactions on Very Large Scale Integration (VLSI) Systems. 2013;21(8):1454–1468.
[42] Reick K, Sanda PN, Swaney S, *et al.* Fault-tolerant design of the IBM Power6 microprocessor. IEEE Micro. 2008;28(2):30–38.
[43] Cher CY, Muller KP, Haring RA, *et al.* Soft error resiliency characterization and improvement on IBM BlueGene/Q processor using accelerated proton irradiation. In: International Test Conference. IEEE; 2014. p. 1–6.
[44] Ebrahimi M, Sayed N, Rashvand M, *et al.* Fault injection acceleration by architectural importance sampling. In: International Conference on Hardware-Software Codesign and System Synthesis; 2015.
[45] Stuart A and Ord JK. Kendall's Advanced Theory of Statistics, Volume I: Distribution Theory. Arnold, London; 1994.
[46] George NJ, Elks CR, Johnson BW, *et al.* Transient fault models and AVF estimation revisited. In: International Conference on Dependable Systems and Networks (DSN); 2010. p. 477–486.
[47] Wang NJ, Mahesri A, and Patel SJ. Examining ACE analysis reliability estimates using fault-injection. In: ACM SIGARCH Computer Architecture News. vol. 35; 2007. p. 460–469.
[48] Biswas A, Racunas P, Cheveresan R, *et al.* Computing architectural vulnerability factors for address-based structures. In: International Symposium on Computer Architecture. IEEE; 2005. p. 532–543.

# Chapter 8
# Microarchitecture-level reliability assessment of multi-core processors

*Athanasios Chatzidimitriou*[1] *and Dimitris Gizopoulos*[1]

One of the primary purposes of reliability evaluation is to identify and protect vulnerable components of a system. At the hardware level, the results of the evaluation can lead to design revisions that aim to increase the fault tolerance of the system. Every potential change is subject to validation and additionally requires more iterations of reliability evaluation. This back-and-forth process is expensive, especially when considering that hardware design changes require significant amounts of time to be applied.

To address this problem, microarchitecture-level reliability assessment has been proposed. Instead of assessing the actual hardware design, the evaluation can be performed at microarchitecture-level (or performance) models that are often available very early in the design chain and are both flexible and allow high observability. The existence of reliability evaluation results before the actual design implementation enables early reliability-related design decisions and significantly decreases the cost of redesign cycles. But the absence of transistor-level detail on the evaluation inheritably results to some accuracy loss.

Only components that are accurately modeled at the microarchitecture level, which are mostly memory elements (Static Random-Access Memory (SRAM) arrays, flops, and latches), can be assessed. Combinational logic and sequential elements are (in majority) functionally modeled, and thus, they cannot be evaluated at microarchitecture level. Fortunately, literature suggests that only a small portion (<10%) of failures sources from these elements, which implies that the accuracy loss that can be attributed to the un-modeled resources at microarchitecture-level is limited. In this chapter, we present the throughput, capabilities, and accuracy of microarchitecture-level reliability assessment, and how it can be effectively used at early design stages.

## 8.1 Introduction

Reliability has been identified by the US Department of Energy as one of the ten significant challenges for exascale performance computing [1]. A lack of understanding or the underestimation of devices and applications error rate may lead to lower scientific

[1]Department of Informatics and Telecommunications, University of Athens, Athens, Greece

productivity of large-scale HPC (High Performance Computing) servers, resulting in significant monetary loss [2]. Consequently, reliability estimation has become one of the major challenges of microprocessor design.

The assessment of a system can be achieved through various techniques, varying in design maturity, level of accuracy, assessment speed, and observability. Since reliability is an essential factor that can be used to guide design decisions and potentially initiate re-design cycles, the ability to have an assessment as early as possible in the design phase is very significant. If, however, the design is at an early stage, a lot of the implementation details may be missing, and the accuracy of the assessment may be inadequate. On the other hand, the assessment can be performed at the end product with all of the implementation details in place, but if this reveals a significant issue, the cost of a redesign iteration can be disastrous.

It is quite common to use simulation of the design to perform reliability assessment, before the availability of the end product. Since the simulation models are based on the design, the level of maturity is reflected in the level of detail in the simulation. Early design phases can have models with less detail available, while later design stages can provide more detailed models. The level of detail in the simulation has a direct impact on the simulation's performance, something that raises another problem for the reliability assessment.

Highly detailed models, such as Register Transfer Level (RTL), are prolonged to simulate, and they are available late in the design chain. On the other hand, less detailed models, such as microarchitecture-level models, are faster and available early, but the level of abstraction that is used implicitly constitutes a source of error in the modeling of the system itself.

Statistical Fault Injection (SFI) [3] on models of a microprocessor at different detail levels has been extensively used to evaluate the vulnerability of hardware architectures and software codes [4–9]. By injecting faults in the hardware blocks of a microprocessor or particular parts of a software code, it is possible to measure the probability for faults to impact the application output or the system responsiveness, i.e., the Architectural Vulnerability Factor (AVF) [10]. Fault injection can then identify codes, code portions, or architectural resources that are more likely, once corrupted, to affect the system reliability. This information is extremely useful to design dedicated and efficient hardening solutions. It has been shown that fault injection at the microarchitecture level is a very effective (fast and accurate method) for early assessment of the reliability of a microprocessor [9,11,12]. However, faults can typically be injected in a limited set of resources and, due to time constraints, simplified error models (such as single bit flip) are usually adopted. In other words, if fault injection is not validated or tuned with physical experimental data, the efforts could potentially lead to imprecise results.

## 8.2 Background

Reliability metrics quantify the frequency of end-to-end failures. The most commonly used metric is Failures In Time (FIT) that is defined as the failure rate per 1 billion hours of operation. By stating "end-to-end," we consider a sequence of events, including the introduction of a threat, the activation, and finally the propagation of it to a system failure. Not all threats lead to failure. There are several steps in this sequence of events and multiple derating factors in this chain. The terms used to describe a dependability threat are defined as follows:

- **Failure:** An event that occurs when the delivered service does not meet its requirements. A service fails either because it does not comply with the functional specification, or because this specification did not adequately describe the system function. Correct service is delivered when the service implements the system function [13].
- **Error:** Part of the entire state of the system that may lead to its subsequent service failure. It is important to note that many errors do not reach the system's external state and thus do not cause failure [13].
- **Fault:** Adjudged or hypothesized cause of an error. A fault is active when it causes an error; otherwise, it is dormant [13].

The earliest form of a reliability threat is a fault. Faults can be classified according to different properties: the phase of creation or occurrence, the system boundaries, the phenomenological cause, the dimension, the objective, the intent, the capability, and the persistence. The types of faults that are being considered for expressing a microprocessor's reliability are as follows:

- Manufacturing defects: Open or short circuits, parametric failures.
- Physical deterioration: Wear-out effects like Negative Bias Temperature Instability (NBTI), electromigration, Time Dependent Dielectric Breakout (TDDB), and Hot Carrier Injection (HCI), etc.
- Physical interference: Soft-errors and Electromagnetic Interferences (EMIs).

Faults at the microarchitecture level can be classified as *transient*, *permanent*, and *intermittent* faults [14]. Transient faults (or Single-Event Upsets) are caused by temporary environmental conditions, such as cosmic rays and EMI. Permanent faults correspond to irreversible hardware defects, while intermittent faults correspond due to marginal or unstable hardware that have a finite duration. Depending on the abstraction level that is being considered, each of these fault categories can further be subcategorized to more classes.

Figure 8.1 illustrates the chain of reliability threats and the mechanisms of their development. The conditions and parameters that can affect reliability ultimately have an impact on one or more nodes of this chain. For instance, fabrication technology, aging, and environmental conditions can affect the rate of introduced faults in the system, as well as the models of these faults. On the other hand, the hardware design and workloads can have an impact on the probability of a fault to cause an error and ultimately, a failure.

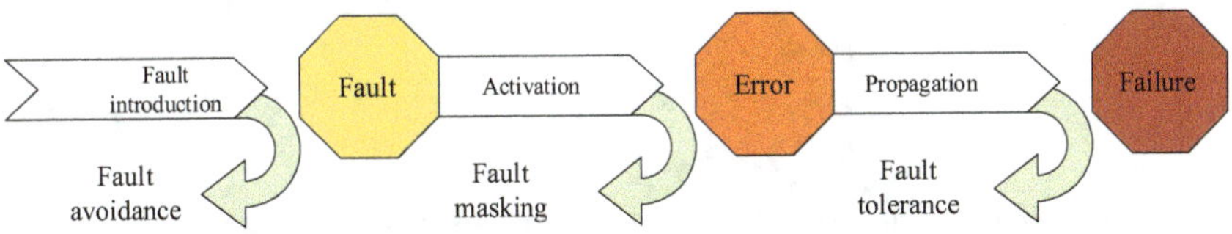


*Figure 8.1 Chain of reliability threats*

### *8.2.1 Threats and vulnerability*

We can identify two key ingredients for calculating the reliability of a system: the fault rate and the probability of these faults to cause a failure. In a complex system like a microprocessor, faults can be introduced by multiple causes. The overall reliability is a summary of all failures that are caused by all of the fault models introduced in a system. For instance, two identical systems with different age will have the same amount of failures coming from soft-errors, but a different number of failures caused by wear-out phenomena. As a result, the two systems will have different overall reliability. We will use the abstract term *raw FIT rate* to describe the rate of faults of a particular fault model that are introduced in a system or component.

The second attribute that is used to calculate the reliability, apart from the fault rate, is the probability of a fault to cause a system failure. This probability can be expressed as the vulnerability of a particular system (a combination of hardware and running software) toward a type of faults. The AVF is a reliability metric for microprocessor hardware structures that depends on the design parameters and the workload characteristics (program structure and input data). AVF expresses the probability that a fault in a hardware structure generates a program-visible error [10]. AVF can be estimated by Architecturally Correct Execution (ACE) analysis. In a hardware structure, a bit that eventually propagates to a visible output is defined as ACE bit. In contrary, un-ACE bits are those that do not propagate to an architecturally visible point. Faults in those bits cannot harm the reliability of the system. The *ACEness* (being ACE or un-ACE) of a bit incorporates the notion of time, and thus, each bit is characterized as ACE or un-ACE for a certain period of time.

Using this definition, AVF can then be calculated using the following formula:

$$\text{AVF} = \frac{\text{ACE bits of hardware structure} \times \text{clock cycles}}{\text{Total bits of hardware structure} \times \text{clock cycles}}$$

With the definition of AVF in place, we can calculate the FIT of a hardware component as:

$$\text{FIT}_{\text{component}} = \text{AVF} \times \text{raw FIT}_{\text{bit}} \times \text{Size (in bits)}$$

Figure 8.2 illustrates the relation between FIT, AVF, and raw FIT.

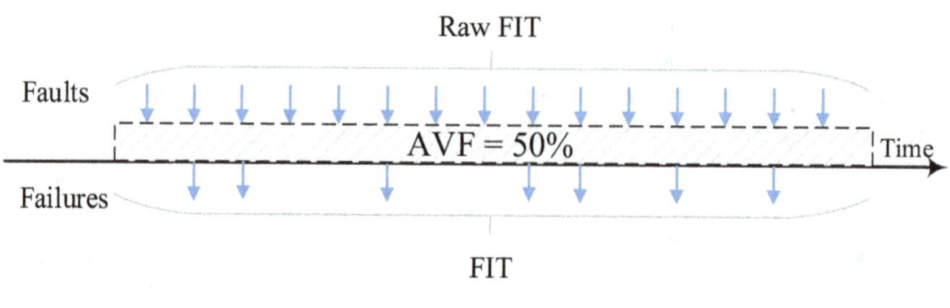


*Figure 8.2 Failures In Time calculation using raw FIT and AVF*

## 8.3 Fault-effect classes

Every system has a unique specification of what is considered a failure and what is not, and thus, vulnerability assessment should be adjusted accordingly. Depending on the requirements of the assessment, the fault effect can be determined using different metrics and observations. For instance, a fault injection may corrupt the program output without any indications, causing a Silent Data Corruption (SDC). On a simulation environment, this case can be detected with a comparison of the golden output against the fault injection one. If the output is different, we can tell that it is corrupt. In some workloads however, a data corruption may not always be severe like, for instance, on an approximate computation. Instead, only corruptions that exceed a specific threshold may be dangerous and should be considered. In such a case, it is not enough to simply compare the outputs for any differences but should also quantify how big this difference is. In addition, assume a critical real-time application that should not violate timing deadlines. A fault that does not have a functional effect on the system may still cause severe delays, which could lead to violation of these constraints and lead to system failure.

It is evident that there is no single rule for classifying the fault effects. However, there are some commonly used fault classes that cover the majority of the requirements. These classes are as follows:

- **Masked:** The simulation was executed with no deviations from the fault-free simulation. The fault did not affect the system or the application in any observable way. The result of a masked simulation is identical to the golden simulation. Masked class summarizes the non-vulnerable cases.
- **SDC:** The simulation was completed successfully, but the program output was different compared to the fault-free simulation, without any observable indications of this effect.
- **Detectable Unrecoverable Error (DUE):** An error was detected during the system operation. This class requires a protection mechanism capable of detecting faults. Instruction Set Architecture (ISA) exceptions can also be considered as DUE, wherever applicable.
- **Crash:** A simulation that did not reach the end of the workload, as it was disturbed by a catastrophic event. As a result, no program output was produced. The crash may refer to process crash (killed process) or system crash (kernel panic).
- **Assert:** The simulation was unexpectedly terminated due to a high-level condition that the simulator was unable to handle. If, for instance, a physical address that is outside the system map is requested, the simulator cannot tell how a system would behave and raise an assertion. The outcome of such cases is not explicitly defined, and each implementation could handle it differently.
- **Timeout:** The simulation did not finish within a certain amount of time, equal to four times the fault-free execution time. These simulations are externally stopped to solve possible deadlock or livelock situations.

The summary of SDC, DUE, crash, Assert, and Timeout equals to the AVF. Figure 8.3 shows the decisions that a parser goes through in order to determine the fault-effect

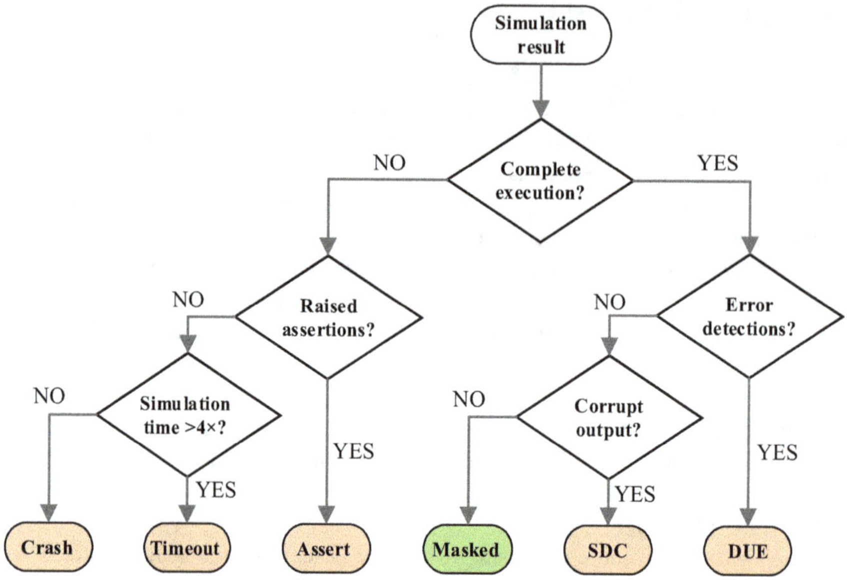


*Figure 8.3 Fault-effect classification decision flowchart*

class. As shown in the flowchart, if a fault is detected, the simulation is classified as DUE, no matter if the detected fault caused any abnormalities (true DUE) or not (false DUE). In order to add this more detailed classification, an additional check on the program output would be required. The fault-effect classes are subjects to change, depending on the requirements and system capabilities. Systems that lack the capability of detecting faults could completely discard the DUE class.

## 8.4 Statistical fault injection

A microarchitecture-level simulation consists of a hardware platform and the software that it is running. In order to capture the effect of one or more faults of any type and combination on a system, it is sufficient to artificially inject them during a simulation and observe the outcome. This process is a *fault-injection simulation*. We define as *faultmask* the set of information that is used to describe one or more faults to be injected on a single-fault-injection simulation. Each fault can be injected at any requested simulation time and target. Once the simulation completes, all of the generated outputs and system states (including simulation statistics, output files, terminal IO, Operating System (OS) verbose faults, messages, exceptions) are stored for later processing.

Each fault-injection simulation provides enough information to classify a fault effect for the injected fault and to determine the ACEness of the stricken bit(s). When considering single faults, these can hit any single bit of a structure at any given time. In order to tell what is the probability of any fault to cause a malfunction (calculate

AVF), we need to quantify how many of these bits are ACE and how many are un-ACE. Considering that modern microprocessors operate in frequencies of gigahertz magnitude and are often capable of executing more than One Instruction Per Cycle (1 IPC), even small programs require millions or even billions of clock cycles. Multiplied by the size of hardware structures, which are usually in the magnitude of kilobits, the total population can easily reach the magnitude of many trillions. Even in native hardware speeds, a program with 1-s execution time on a 1 GHz processor would require more than 30,000 years to assess a 1 kbit structure. Considering that simulation is at least three orders of magnitude slower, it is practically unfeasible to make an exhaustive ACE characterization of the entire population.

Instead, SFI aims to provide an estimation of the vulnerability using statistical sampling. Like all statistical problems, it is difficult to choose an appropriate balance between speed and accuracy. A large statistical sample suffers from the same time issues as the large population but delivers more accurate results, while a smaller sample requires less time but comes along with higher error margins. In an effort to aid designers that aim to use SFI, Leveugle *et al.* [3] have proposed using existing mathematical frameworks that are already established in other fields where sampling is necessary, adapted to the special requirements of SFI. They have proposed using a formula that calculates the sample size, based on a given population, confidence level, and error interval. With respect to:

- the initial population $N$,
- the estimated proportion $p$ of individuals in the population that have a given characteristic,
- the margin of error $e$,
- the cutoff point (or critical value) $t$ that corresponds to the confidence level, calculated as inverse $t$-distribution (with respect to the normal distribution).

The sample size $n$ can then be calculated as:

$$n = \frac{N}{1 + e^2 \times (N - 1/t^2 \times p \times (1 - p))}$$

The $p$ parameter corresponds to an estimate of the true value being searched, which in most of our cases is AVF. This is typically unknown *a priori*. A common practice is to use a fixed estimation of $p = 0.5$ as this conservatively maximizes the sample size and will ensure that the size population complies with the requested error margin and confidence. If, however, the expected proportion is significantly smaller, or larger, the calculation can be further refined to better approach the results. In practice, it is enough to choose an initial population based on $p = 0.5$, proceed with the assessment, and based on the result, revisit the calculation to refine the actual error margin. To quantify the error margin, given a sample size $n$, confidence level $t$, and estimated proportion $p$, the following formula can be used:

$$e = \sqrt{\frac{p \times (1 - p)}{n} \times \frac{N - n}{N - 1}}$$

Every fault-injection campaign consists of a series of faultmasks to be injected, which constitute the statistical sample, the simulations for every faultmask, and the output results and their parsing, to estimate the vulnerability or other metrics. To be rational, the injected faultmasks must correspond to realistic effects that are being modeled. For instance, a reliability assessment on single event upsets, caused by particle strikes, is expected to introduce transient faults, both single and spatial multi-bit. A fault-injection campaign of single-bit transient faults on a standard distribution is required to evaluate the single-bit effects, while separate spatial multi-bit fault-injection campaigns are required to evaluate the Multi-Bit Upsets (MBUs). While the generation of faultmasks for the first fault-injection campaign is trivial, the second campaign requires additional information and processing. Depending on the fabrication technology, the size (how many bits are affected), distance, and frequency of MBUs are subject to change. Alpha particles, protons, and neutrons are also expected to deliver different patterns of MBUs. All these attributes must be taken into consideration when generating the faultmasks.

On another example, assume a reliability evaluation of a system that is being aggressively under volted. Every SRAM cell is expected to function normally above a voltage threshold that is unique for every cell and depends on the fabrication process variation. At a given voltage setting, a number of cells are expected not to be functional. But the distribution of these cells, the voltage level and the number depend on the fabrication technology and level of expected variability. A set of faultmasks that corresponds to actual observations and is expected to occur on the target technology can provide useful insights on their system-level effects, while a simplistic random fault distribution does not, as it has no chances to happen in the first place, on the under-volted system.

Faultmask generation is a very important process as eventually instruments the scenario that is being evaluated. Knowing the vulnerability of a system level effect that is not realistic is not useful. The fault generators must be designed with attention to detail, to consider all the available information on the system level effects that they are modeling. Once the faultmasks are in place, the next step is the fault-injection simulations. The controller provides the appropriate inputs for simulation as well as keeping copies of the produced outputs for later processing. Finally, the parsing of the results takes place where a parser component goes through all of the produced results to classify each fault injection to one of the available fault-effect classes. Each fault-injection campaign consists of three discrete phases: preparation (or campaign initialization), simulation, and parsing, as shown in Figure 8.4. Microarchitecture-level SFI is a time-consuming process.

## 8.5 Cross-layer and single-layer evaluation

When considering the reliability threat chain, AVF quantifies the full-system vulnerability, which includes both the phases of activation and propagation to the program output. It involves the combination of hardware and software and for instance, assuming that a fault is presented inside the register file of an out-of-order processor, the

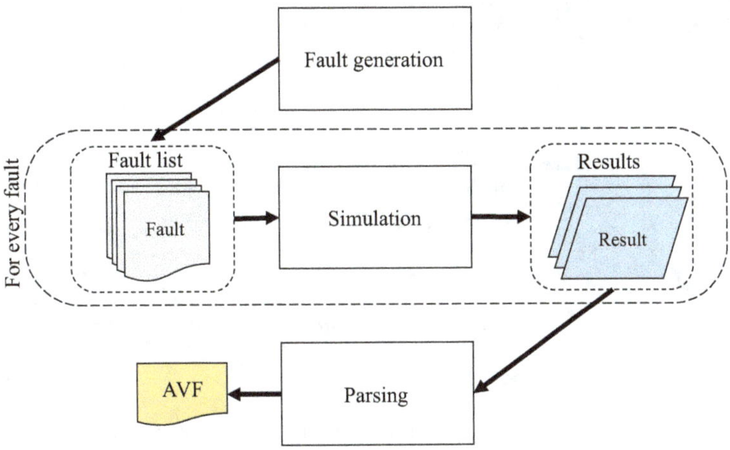


*Figure 8.4 Fault injection campaign high-level layout*

hardware may speculatively use the fault and then discard the result due to a pipeline flush. In such a case, the fault will not result in failure. But even if the fault is activated and propagated to the software level, the software algorithm may not, on its turn, use the corrupt data (discard the fault) and again deliver the same result. In this example, we can see that the fault may get masked in two separate levels: the hardware and the software.

It is often useful to separate the full system into layers and calculate AVF based on the individual vulnerability factors of each level of the system stack. This concept has been defined as system vulnerability stack [15]. Similarly, to the architectural vulnerability, a fault may be introduced at the hardware level and has a probability to reach (turn visible) the software level. A fault that is visible in the software level has a probability of reaching the program output, etc. In the stack, a bit is assigned vulnerability at every layer to which it is visible. If it is visible at every layer of the system, it is vulnerable for the system. This end-to-end probability expresses the AVF, but each level individually has a different probability of propagating the fault to the next level. If we consider that the frequency of faults introduced on each level depends on the faults that were introduced in the previous level and the vulnerability of that level, we can build a stack to describe the vulnerability of a system. Figure 8.5 shows how the system vulnerability stack consists of different layers, each having a separate vulnerability factor. Notice that in the software level, we can identify multiple sub-layers. Since software vulnerability is not the primary focus of this work, from now on, we will consider the software vulnerability a unifying factor that includes all of the subcategories (Virtual machine, OS, and application).

The Hardware Vulnerability Factor (HVF) of a hardware structure is the fraction of faults in the structure that are either activated within the hardware layer or exposed to a higher layer [15]. A hardware-visible fault is exposed to the user program once it reaches a *software resource*. According to [16], a *software resource* is defined as

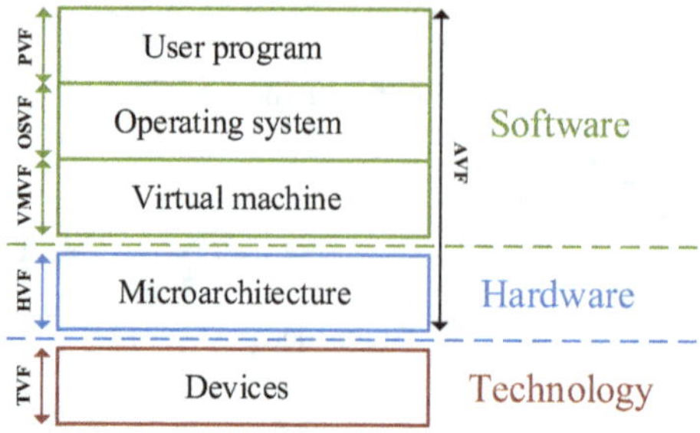


*Figure 8.5 System vulnerability stack*

any independently-addressable architectural structure. *Software masking* can then be defined as the probability of a fault involved in the program flow to be masked by the program. The Program Vulnerability Factor (PVF) [17] is the probability of a fault that is introduced in the software level to result on an output corruption and is equal to 1—software masking. The product of HVF and PVF results to AVF.

The concept of system vulnerability stack suggests that it is rational to use different tools and methodologies for the assessment of each individual layer. Software-level tools can be used to determine the PVF, while hardware models for the HVF. While this is true, it hides several risks. While the vulnerability of the software appears to be independent of the hardware, the actual faults and their models that are propagated from the hardware are not. A PVF analysis, for instance, can determine that the vulnerability against data corruptions inside architectural registers is 50%, while the vulnerability of opcode corruptions is 90%. Unless it is known *a priori* how many of the faults that reach the software layer are from the first category and how many from the second, it is impossible to determine the reliability.

On the opposite direction, the HVF is not independent of the software, as the software is the force that decides how much and in which way the hardware will be utilized. This involves most, if not all, of the hardware components, including the pipeline, branch-prediction units, cache hierarchy, and translation buffers. As a result, any effort of reducing the vulnerability of the software, given a fixed hardware design that does not consider the HVF in every iteration, has risks of providing wrong results. A change in the software may alter the hardware utilization windows in such a way that the balance of fault types that reach the software level also changes. If, for example, a program of the previous example (with 50% vulnerability on data corruptions and 90% on opcode corruptions) is modified to deliver increased reliability against the latter case, from 90% to 70%, but this alteration changes the balance of faults introduced from the hardware, from 50% data-50% opcode, to 20% data and 80% opcode, then the final AVF will have increased, instead of the opposite.

While the system vulnerability stack allows assessment of each layer individually, this should not be confused that each assessment can be performed independently to the other layers, and any change in any of the system's components can affect multiple layers.

In microarchitecture-level fault injection, the faults are injected directly to the system's hardware components and belong to the hardware layer. As the simulation includes both hardware and software, it is possible to determine whether this fault corrupted the final program output, which equals to an AVF estimation. But in addition to that the high level of observability allows placement of observation points anywhere in the system, which also enables identification on when a fault crosses a layer, giving the ability of an HVF estimation as well. In fact, by proper placement of observation points, microarchitecture-level fault injection can be used to deliver any intermediate vulnerability estimation inside the system vulnerability stack.

## 8.6 Assessment throughput

Microarchitecture-level SFI is a time-consuming process. The assessment time is relative to the size of the statistical sample and the length of the workload. In the simplest approach, this time will equal to:

$$\text{Assessment time} = \text{simulation time} \times \text{number of faults}$$

Table 8.1 shows an approximation of the simulation throughput (in clock cycles) on different abstraction layers. Architecture level does not include hardware modeling, and it is assumed that it executes 1 IPC. We can see how the throughput difference between two abstraction layers is two to three orders of magnitude. In practice, a simulation of a workload with 50 million instructions and an average IPC of 1 would take 25 ms to execute on a native system, 2.5 s on an architecture-level simulator, about 4 min on a microarchitecture-level simulator, and around 20 h on an RTL simulator. These times can significantly vary depending on the workload, especially for the hardware models (microarchitecture and RTL). If the model could achieve an average of 2 IPC for the benchmark, the simulation times would drop to half. The table shows a rough estimation of how fast or how slow each model is.

In order to achieve statistical significance, a large number of simulations are required. We have described how to calculate the number of faults required. Assuming that we want to achieve a result with a 3% error margin and 99% confidence. For the workload of the previous example (with 50 million cycles execution time) and for a component, e.g., register file of 128 registers ($\times 32 = 4{,}096$-bit size), the required

*Table 8.1 Simulation throughput of different abstraction layers*

| **Abstraction level** | **Model** | **Performance (cycles/s)** |
|---|---|---|
| Software (native) | Modern processors | $2 \times 10^9$ |
| Architecture | Gem5 atomic model | $2 \times 10^7$ |
| Microarchitecture | Gem5 detailed out-of-order model | $2 \times 10^5$ |
| RTL | NCSIM simulator | $6 \times 10^2$ |

*Table 8.2 Time required in different abstraction levels for the reliability assessment a 4-kbit hardware component, for a 1-IPC benchmark of 50 million instructions*

| Abstraction level | Simulation time | Assessment time | | |
|---|---|---|---|---|
| | | 5% error (663 samples) | 3% error (1,843 samples) | 1% error (16,587 samples) |
| Software (native) | 25 ms | 16.5 s | 46 s | 7 min |
| Architecture | 2.5 s | 28 min | 76 min | 11.5 h |
| Microarchitecture | 4 min | 44 h | 5 days | 46 days |
| RTL | 20 h | 18 months | 4 years | 38 years |

Calculated samples assuming 99% confidence level 5%, 3%, and 1% error margin.

number of faults is 1,843. Table 8.2 shows the required time for different error margins of this example setup. The fully detailed RTL modeling requires more than a year to provide an assessment with a 5% error margin, for a single hardware structure and a single workload. Also, notice that the workload used in this example is very short and would execute in only 25 ms on a real system.

Considering that a microprocessor assessment should include characterization of multiple components and workloads (that could easily be in the order of billion instructions), the time required for the assessment is practically infeasible at this level of detail. But even the more abstract microarchitecture level, which is the fastest model that includes a hardware model, still significantly suffers from the very long simulation times. There are two strategies that can be used to reduce the reliability evaluation time:

- reduction of the simulation time
- reduction of the fault sample size

Fortunately, it is possible to achieve both without sacrificing the accuracy of the estimation. Most of the simulations have identical portions and it is not necessary to rerun parts that can be identified as identical. The same principle can be applied on faults, as it is possible to identify faults that are expected to have the same effect and group them, instead of rerunning them.

### 8.6.1 Simulation acceleration

There are three possible significant events that, along with the beginning and end of the simulation, have a special role on the simulation timeline. These are as follows:

- Fault-injection event: This corresponds to the time of the bit-flip occurrence.
- First access event: The first access of the faulty entry, which can be a read or write. This event may not occur at all during a simulation.

- Visible fault-effect event: The first visible effect of the fault on the program flow. This event can only occur optionally after a read access of the fault.

These events mark the bounds of five simulation phases. The fault-injection periods defined by the events are as follows:

**Pre-fault period:** The period starts with the simulation and ends at fault-injection time. The pre-fault period is out of interest for reliability estimation since it has no residing faults. The simulation is identical to the fault-free (golden) simulation.

**Idle period:** This period starts at fault-injection time and ends either on the first access of the faulty entry or at the end of simulation; whichever comes first. A fault has been injected into the system and resides silently unused during idle period.

**Manifestation period:** The first access of the faulty entry marks the beginning of the manifestation period. During this period, a fault has been accessed and has chances to affect the software layer, unless it gets software masked. In the first case, the period ends at the time a corruption is being committed by the software and for the last, the period ends along with the simulation, if no corruptions occur.

**Corruption period:** By the time a fault reaches the program flow, corruption period starts. From this point on, the fault resides in the software layer and may cause corruption of the program output, system crash or can be masked.

**Inactive period:** If the fault gets overwritten and no longer exists on the simulation, the rest of the simulation is marked as inactive. Fault simulations that reach this period are certainly masked.

Idle and manifestation are the most important periods, since they include the fault injection, fault propagation, and fault-effect events. Although manifestation period describes the complete lifetime of a fault inside the microarchitecture, idle period remains an integral part of the simulation because it includes all of the hardware access patterns and results to if, when and where the fault will reach the microarchitecture. These periods include all of the required information to characterize a fault effect for the hardware layer. However, it must be pointed out that these periods also include parts without interest. For instance, cases where a fault is propagated to the microarchitecture but gets hardware masked occupy simulation time with inactive faults and no effect.

Notice that the described periods only refer to the first occurrence of these events, which however may not depend on each other. This means that the first visible effect of a fault may be due to the second or third access of the fault and not necessarily the first. Part (a) of Figure 8.6 illustrates the description of these periods and part (b) presents the portion that each period occupies on the simulation timeline. Pre-fault period is estimated to consume 49% of total simulation time and corruption and inactive periods to follow with average of 38% [12]. The most critical periods to determine the fault propagation (idle and manifestation) only hold an average of 13% of simulation time, which shows that there is a significant acceleration potential.

Each period has some different attributes to which the simulator can exploit and either reduce some of the monitoring overhead, or even apply more aggressive approaches, such as fast forwarding or outcome prediction.

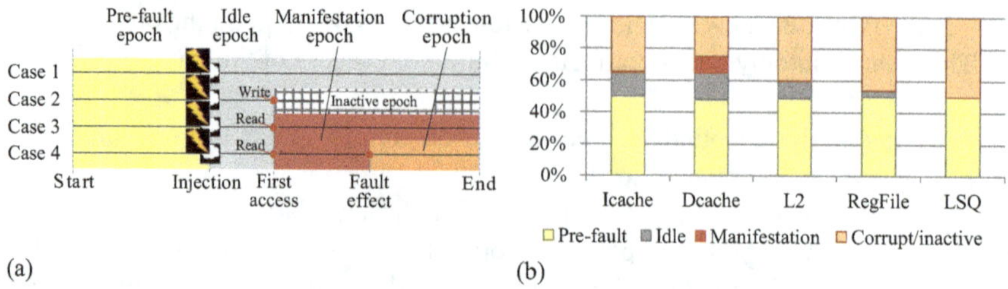


*Figure 8.6 (a) Fault injection timeline and (b) breakdown of the simulation period duration*

The features can be categorized as direct features, implying pure acceleration without accuracy penalty and indirect features, which can lead to some inaccuracy on the results but significantly reduce the simulation times. Notice that all of these features can be enabled or disabled, giving the flexibility of different combinations, according to the estimation requirements.

#### 8.6.1.1 Direct features

**Checkpoint fast-forwarding:** The pre-fault period does not contain any faults and thus, it is out of interest for the reliability assessment. The simulator knows in prior the time of the fault injection and therefore can delimit the pre-fault period. Checkpointing is widely used for skipping insignificant parts of the simulation and fast-forwarding the execution.

**Early stop on overwrite/invalid:** Injected faults that get overwritten cannot harm the system state and they can be considered as masked prior to the completion. The same applies to inactive structure entries, which will get overwritten upon allocation. The period that follows the overwriting is the inactive period that can be safely discarded. Skipping the inactive period can save up to 50% of campaign time compared to the traditional fault injection without any impact on the final result.

#### 8.6.1.2 Indirect features

**Early stop on program corruption:** By properly placing an observation point on the instruction commit flow, it is possible to identify and skip or accelerate the corruption period. Any kind of mismatch that is observed to the program flow marks the fault-effect event. Stopping at that time changes the estimation to HVF, and the software masking portion of AVF is ignored.

**Early switch to functional emulation:** Unlike the previous features that target to skip, early switch intends to fast-forward a period, and more precisely, the corruption period. Some microarchitecture-level simulators, such as Gem5, support simple functional CPU models that have no microarchitectural details and have a throughput up to 20× faster than the detailed models.

Following the early stop on program corruption feature, the missing software-masking part can be covered by the same tool. By switching to emulation mode, the AVF is estimated using microarchitecture-level detail for the HVF and

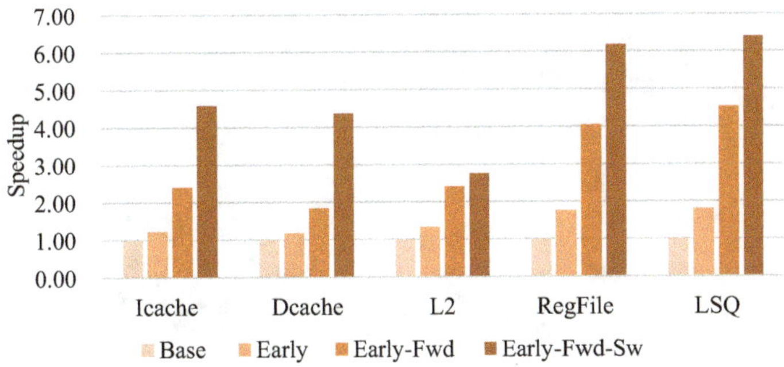


*Figure 8.7 Fault injection campaign speedups compared to baseline*

architecture-level detail for the PVF. The possible losses in the accuracy of the result reside in the possibility that the fault would cause further problems after its first visible effect. Switching will only capture the impact of this effect on the final output.

**Early limit and result prediction:** This feature is used to stop the simulation early after the first access. It will limit the manifestation period by a user-defined amount of time, in which the fault has the opportunity to get activated. This feature is intended to be used for predicting the result after a given time. For instance, the analysis presented in [12] shows how to effectively use this feature in order to achieve high acceleration in certain hardware components. Most pipeline structures can completely skip the Idle epochs, while most hardware resources that are related to the architectural state allow prediction of the result with minor accuracy losses.

Experimental evaluation of the referred techniques shows that significant acceleration of the assessment can be achieved without accuracy losses. The evaluation was performed on five microarchitectural components using three combinations of acceleration features. The "early" configuration has enabled early stop on overwrite, "Early-Fwd" is additionally using checkpointing feature, while "Early-Fwd-Sw" is additionally switching to emulation after 100 thousand clock cycles from the fault activation. The three configurations were compared against the baseline fault injection, and the results of the reliability assessment are the same. However, the assessment was significantly faster. Figure 8.7 shows how the campaign time was more than 6× shorter, compared to the baseline, while even the worst case component achieves a 3× speedup.

## *8.6.2 Fault list reduction*

Another effective methodology to reduce the reliability assessment time is by running less fault-injection simulations. But instead of reducing the number of faultmasks, this can also be achieved by running one simulation to classify multiple faults that share some common characteristics and are expected to have identical effect [9]. This requires two additional steps of *preprocessing* and *fault list reduction* before running the fault injection simulations and can be applied in a typical fault injection campaign configuration (Figure 8.8).

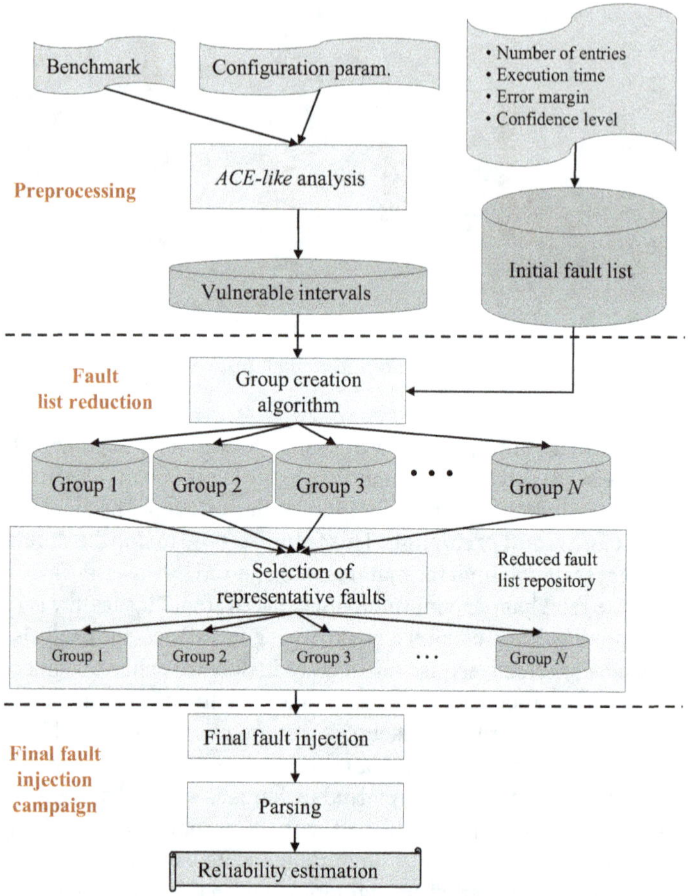


*Figure 8.8 Fault list reduction flow*

The first step of the method (*Preprocessing*) consists of a profiling stage that determines the data intervals that are being accessed by the hardware. This ACE-like analysis identifies all the vulnerable intervals (ACE-like intervals) of each entry of a hardware structure. A vulnerable interval for an entry of the structure being analyzed is an interval that:

- Starts with a write operation that is followed by a committed read of the same entry. The read marks the end of the interval.
- Starts with a committed read that is followed by another committed read of the same entry. The second read marks the end of the interval.

At the end of the Preprocessing step of the method, the vulnerable intervals repository contains all the vulnerable intervals for all the entries of the structure under study, accompanied by the information (instruction pointer and microprogram counter) of the committed read operation that accesses the entry at the end of the vulnerable interval.

At the beginning of the second step of the methodology, a Group Creation Algorithm uses as input the information that is stored in the vulnerable intervals repository and the initial fault list from the first phase of the method. This algorithm groups all the faults that are injected in different or the same vulnerable intervals in the same or different entries of the hardware structure. Consequently, all the groups created by the Group Creation Algorithm contain faults that will probably manifest the same fault effect. Next, we select the most representative faults from each group, selecting more than one fault that target the same entry in cases that these faults hit a different byte of the entry.

The last step of this method (final fault-injection campaign) is the injection simulation of all representative faults from each created group. The rest of the fault-injection campaign remains the same, but the classification is applied to all of the faults inside a group.

Figure 8.9 presents a simple example of the Preprocessing step and the Group Creation Algorithm that are used in our method for three entries of a hardware structure. During the lifetime of entryA, three vulnerable intervals were created (t1–t3, t11–t12, and t12–t15), because the read operation at t2 was squashed. In the same way, two vulnerable intervals (t5–t8 and t10–t14) were created during the lifetime of entryB and finally, two vulnerable intervals (t4–t6 and t9–t13) were created for entryC. After the execution of the Group Creation Algorithm, only four groups of faults were created (assuming that faults were injected in each vulnerable interval of the three entries). The faults that are injected during the intervals t5–t8, t12–t15, and t9–t13 consist the first group because the read operations that access these entries have the same instruction pointer and microprogram counter (rip C and uPC 3). In the same way, the second group contains all the faults that are injected in intervals t1–t3 and t10–t14 with rip A and uPC 0. Finally, the third and the fourth groups contain faults that are injected during the intervals t11–t12 and t4–t6, respectively, because there is no other ACE-like interval to be accessed by a read operation with the same instruction pointer and microprogram counter.

This methodology has been evaluated in three main structures (integer register file, store queue, and L1 data cache) of the x86-64 architecture [9]. In summary, the results revealed that the reliability estimation can be accelerated by

- 130.2×, 81.3×, and 60.9× on average for an integer register file with 256, 128, and 64 registers, respectively, with error less than 0.48 percentile points in all fault-effect categories;
- 298.0×, 252.8×, and 200.5× on average for a store queue with 64, 32, and 16 entries, respectively, with error less than 1.08 percentile points in all fault-effect categories; and
- 69.2×, 70.1×, and 69.5× on average for an L1 data cache with size of 64, 32, and 16 kB, respectively, with error less than $X$ percentile points in all fault-effect categories.

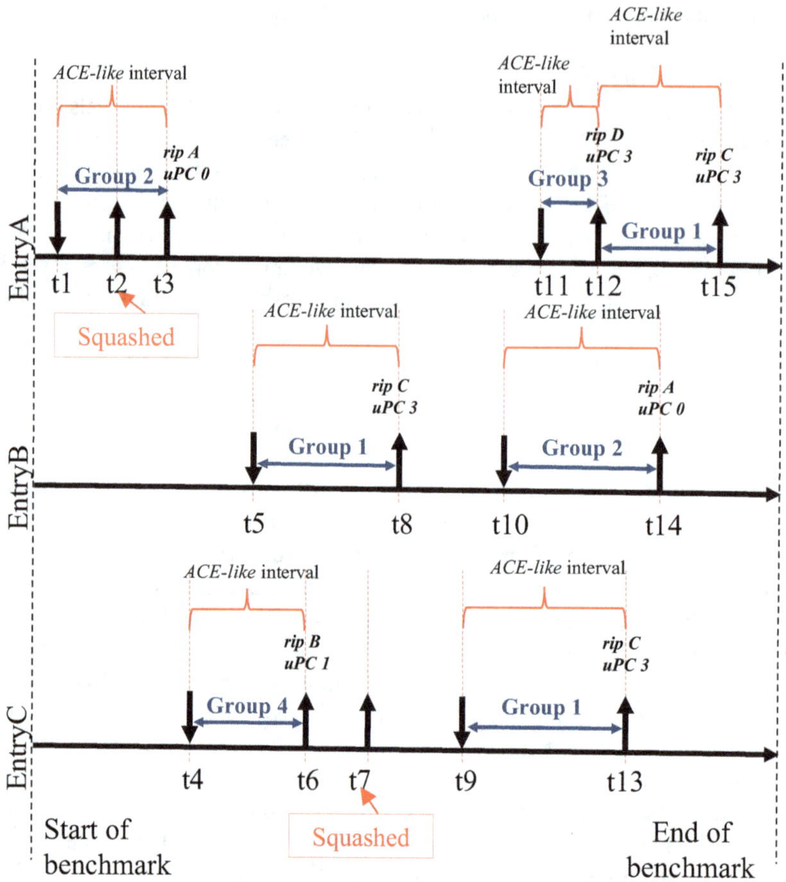


*Figure 8.9 Preprocessing and Group Creation Algorithm example*

## 8.7 Estimation accuracy

Microarchitecture-level reliability assessment targets early design stage reliability evaluation. The scope of the result is intended to be used for design decisions and initiate redesign cycles with low cost, compared to other alternatives. Given the absence of the actual design, microarchitecture level is the most detailed model capable of running full-system workloads on a clock-cycle accuracy. To answer how accurate microarchitecture-level reliability assessment is, a comparative assessment has been performed on an ARM Cortex-A9 platform using fault injection on a microarchitectural model of Gem5 against neutron beaming experiments on a Xilinx Zynq platform that uses ARM Cortex-A9 [18].

Beam experiments are the most realistic way to measure the error rate of a code or device in conditions that are as close as possible to the actual ones after system deployment. However, beam experiments are coarse grain, i.e., errors can be observed only when they manifest at the application output or when they compromise the system responsiveness. There is no information about the spatial/temporal location of the original fault and no information on its propagation pattern through the microarchitecture, the system software, or the application software layers. It is, therefore, hard to identify the most vulnerable hardware resources or software code portions and to extend/generalize results to other applications or devices when relying on beam experiments only.

The comparison was performed on an extended set of 13 benchmarks, using more than 250,000 fault injections at the microarchitecture level and an equivalent of 2.9 million years of terrestrial radiation exposure. The estimation resulted in FIT rates, and the results were categorized in three classes: SDC FIT, application crash FIT, and system crash FIT.

The reported SDC FIT rate estimations were very close between the two methodologies, in the scale of 1.1× in average, while application and system crashes differ an average of 4× and 10× accordingly, as shown in Figure 8.10. This interesting observation is attributed to the differences in the two setups. The neutron beaming was targeting the whole SoC, including peripheral devices, interconnects, memory controllers, and buses, while the fault injection was only targeting the CPU core. Among the different classes, only SDCs can be produced by the CPU core, which is

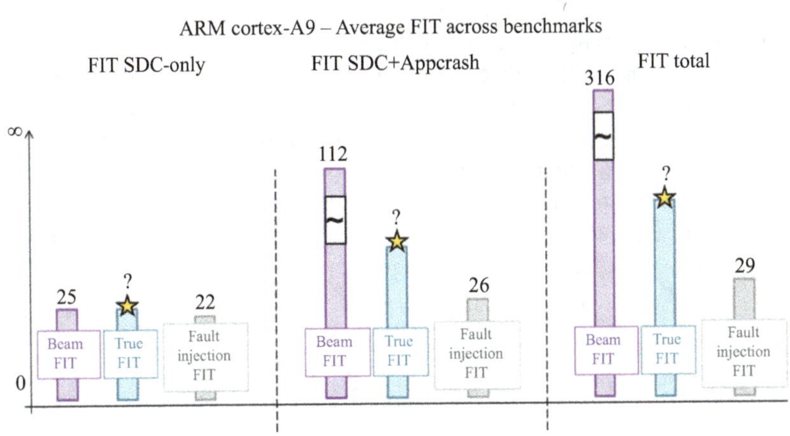


*Figure 8.10 Overview of the comparison between beam and fault injection FIT rates. Fault injection average FIT rate (gray bars) is dominated by the SDC FIT rate (leftmost bars) and is only slightly increased when the application crashes are added (middle bars) or the system crashes are also added (rightmost bars). On the contrary, the Beam FIT rates (purple bars) are increased when the two types of crashes are added. However, the SDC FIT rate of Beam is very close to the fault injection SDC FIT rate and also the differences when one or both types of crashes are added are close to only one order of magnitude.*

common in both setups, while crashes can source from multiple other components that are not targeted by the FI but are exposed to the neutron beaming. Additionally, the chip that was exposed to the neutron beam was a final, fully detailed implementation, which means that it also included the combinational and sequential logic that is absent in the microarchitecture-level model. Notice that while the FI is presented as underestimating the result, beam experiments appear to overestimate the result, and this is due to the instrumentation infrastructure that is required; the true FIT lies between the two.

This experiment revealed that the microarchitecture-level fault injection can be trusted for reliability estimation that concerns SDCs, while at the same time, it will miss most of the application and system crash sources, unless the same components outside the CPU are also targeted by fault injections.

## 8.8 Conclusions

In this chapter, we have presented the methodologies and challenges of microarchitecture-level fault injection. By being the earliest hardware model available in the design chain, microarchitectural models give the opportunity of significantly reducing redesign costs through the early level reliability assessment. We have described the fault-injection methodology and the purpose of each step in the evaluation. Next, we presented how microarchitecture-level models can be used to assess the intermediate layers of the system's vulnerability stack.

The two major challenges of microarchitecture-level fault injection are the throughput and the accuracy. We have gone through different techniques that have been developed to boost the performance of the assessment and have presented a comparison between neutron beaming assessments against the results of fault injection. Microarchitecture-level fault injection delivers a good balance between early availability, accuracy, and throughput that gives system designers a powerful design-decision aid at the early design stages.

## References

[1] R. Lucas, J. Ang, K. Bergman, *et al.*, "DOE Advanced Scientific Computing Advisory Subcommittee (ASCAC) Report: Top Ten Exascale Research Challenges," Office of Scientific and Technical Information (OSTI), 2014.

[2] M. Snir, R. W. Wisniewski, J. A. Abraham, *et al.*, "Addressing failures in exascale computing," *International Journal of High Performance Computing Applications*, pp. 1–45, 2014.

[3] R. Leveugle, A. Calvez, P. Maistri and P. Vanhauwaert, "Statistical fault injection: Quantified error and confidence," in *Design, Automation Test in Europe Conference Exhibition (DATE)*, Nice, France, 2009.

[4] H. Cho, S. Mirkhani, C.-Y. Cher, J. A. Abraham and S. Mitra, "Quantitative evaluation of soft error injection techniques for robust system design," in *Proceedings of the 50th Annual Design Automation Conference on – DAC 13*, Austin, TX, USA, 2013.

[5] S. Feng, S. Gupta, A. Ansari and S. Mahlke, "Shoestring: Probabilistic soft error reliability on the cheap," in *Architectural Support for Programming Languages and Operating Systems (ASPLOS)*, Pittsburgh, PE, USA, 2010.

[6] N. Foutris, M. Kaliorakis, S. Tselonis and D. Gizopoulos, "Versatile architecture-level fault injection framework for reliability evaluation: A first report," in *International On-Line Testing Symposium (IOLTS)*, Platja d'Aro, Girona, Spain, 2014.

[7] N. George, C. R. Elks, B. W. Johnson and J. Lach, "Transient fault models and AVF estimation revisited," in *2010 IEEE/IFIP International Conference on Dependable Systems & Networks (DSN)*, Chicago, IL, USA, 2010.

[8] S. K. S. Hari, S. V. Adve, H. Naeimi and P. Ramachandran, "Relyzer: Exploiting application-level fault equivalence to analyze application resiliency to transient faults," in *ACM International Conference on Architectural Support for Programming Languages and Operating Systems*, London, England, UK, 2012.

[9] M. Kaliorakis, D. Gizopoulos, R. Canal and A. Gonzalez, "MeRLiN: Exploiting dynamic instruction behavior for fast and accurate microarchitecture level reliability assessment," in *International Symposium on Computer Architecture (ISCA)*, Toronto, ON, Canada, 2017.

[10] S. S. Mukherjee, C. Weaver, J. Emer, S. K. Reinhardt and T. Austin, "A systematic methodology to compute the architectural vulnerability factors for a high-performance microprocessor," in *IEEE/ACM International Symposium on Microarchitecture (MICRO)*, Washington, DC, USA, 2003.

[11] M. Kaliorakis, S. Tselonis, A. Chatzidimitriou, N. Foutris and D. Gizopoulos, "Differential fault injection on microarchitectural simulators," in *2015 IEEE International Symposium on Workload Characterization*, Atlanta, GA, USA, 2015.

[12] A. Chatzidimitriou and D. Gizopoulos, "Anatomy of microarchitecture-level reliability assessment: Throughput and accuracy," in *IEEE International Symposium on Performance Analysis of Systems and Software (ISPASS)*, Uppsala, Sweden, 2016.

[13] A. Avizienis, J.-C. Laprie, B. Randell and C. Landwehr, "Basic concepts and taxonomy of dependable and secure computing," *IEEE Transactions on Dependable and Secure Computing*, vol. 1, no. 1, pp. 11–33, 2004.

[14] C. Constantinescu, "Intermittent faults and effects on reliability of integrated circuits," in *Reliability and Maintainability Symposium (RAMS)*, Las Vegas, NV, USA, Ischia, Italy, 2008.

[15] V. Sridharan and D. R. Kaeli, "Using hardware vulnerability factors to enhance AVF analysis," in *International Symposium on Computer Architecture (ISCA)*, New York, NY, USA, 2010.

[16] V. Sridharan and D. R. Kaeli, "Quantifying software vulnerability," in *Workshop on Radiation Effects and Fault Tolerance in Nanometer Technologies*, Ischia, Italy, 2008.

[17] V. Sridharan and D. R. Kaeli, "Eliminating microarchitectural dependency from architectural vulnerability," in *IEEE International Symposium on High Performance Computer Architecture (HPCA)*, Raleigh, NC, USA, 2009.

[18] A. Chatzidimitriou, P. Bodmann, G. Papadimitriou, D. Gizopoulos and P. Rech, "Demystifying soft error assessment strategies on ARM CPUs: Microarchitectural fault injection vs. neutron beam experiments," in *IEEE/IFIP International Conference on Dependable Systems and Networks (DSN)*, Portland, OR, USA, 2019.

## *Chapter 9*

# Fault injection at the instruction set architecture (ISA) level

*Karthik Pattabiraman*[1] *and Guanpeng Li*[1]

Fault Injection (FI) is a commonly used technique to evaluate the reliability of systems. As soft errors become more commonplace in computer systems, it is often necessary to involve the software in the overall system's resilience. Therefore, it is important to inject faults at the ISA level to emulate soft errors that are visible to the software, in order to test software resilience mechanisms. Consequently, there is a need to develop Instruction Set Architecture (ISA)-level FI tools and techniques.

We start by outlining the goals of ISA-level FI, followed by the main metrics that can be measured by the same. We then present a survey of techniques in the literature that attempt to inject faults at the ISA-level and up in the system stack. Finally, we present an overview of LLFI and PINFI, two fault injectors developed in our research group, that allow programers to inject faults at the LLVM compiler's Intermediate Representation (IR) level and x86 assembly code level, respectively. We conclude with a survey of the open challenges in the area.

## 9.1 Introduction

FI is the process of perturbing a target system to emulate the effects of faults that occur in it, in order to study its reaction. FI is a commonly used reliability evaluation technique for both hardware and software faults—in this chapter, we focus primarily on hardware faults caused due to soft errors. Typically, FI serves two purposes: (1) to evaluate the inherent resilience of the system to faults, and understand its weaknesses (2) to evaluate the coverage of various fault tolerance mechanisms that are deployed in it, and to choose appropriate mechanisms to obtain the desired coverage subject to performance and power overheads.

As soft errors have continued to become more and more prominent in commodity systems, there has been an increasing effort to deploy protection mechanisms further up in the system stack [1–4]. This is because researchers have found that only a small fraction of the errors that occur at the circuit or gate level propagate up the

[1]Electrical and Computer Engineering Department, University of British Columbia, Vancouver, Canada

system stack and manifest as impactful errors, i.e., errors that ultimately result in system failures [5–7]. This effect is due to significant masking and redundancy that is inherent in modern computer system stacks, and the fact that not all errors cause failures in applications. Further, many applications are able to tolerate some degree of inaccuracy or approximation in their outputs, e.g., video decoding can tolerate some errors in the pixels of individual frames. Therefore, traditional techniques that deploy protection at the hardware level that treat all errors are potentially impactful are often wasteful and incur significant power and performance overheads—this necessitates software techniques for soft error mitigation.

Examples of software protection techniques for soft errors are selective duplication of important code or data [2–4], control-flow checking [8,9], and the use of detectors or assertions in the program, also known as symptom-based detectors [10,11]. While useful, these techniques may not always provide 100% detection for hardware faults (due to performance and energy constraints). Further, deploying different combinations of these techniques can result in different coverage-performance-power consumption trade-offs. Hence, it is important to evaluate the coverage of these techniques using FI and enable software developers to choose the appropriate combination of techniques for their applications.

FI techniques need to be developed at the software ISA level to evaluate software fault-tolerance techniques and to enable rapid exploration of the design space of these techniques prior to system deployment. This is because detailed microarchitectural (or lower) FI is often too slow for evaluating software-based techniques, as it is important for the application to be executed to completion for such techniques. Further, it is important for the developer to be able to relate the results of the FI to the software, as developers are responsible for choosing the techniques to deploy (if any). This is often not possible with FI techniques that operate lower than the ISA level. Finally, ISA-level (or higher) FI has the advantage that it can be performed in a platform-neutral manner for the most part.

The rest of this chapter is organized as follows. Section 9.2 presents a brief summary of the terms and definitions, as well as the metrics and fault model in ISA-level FI. Section 9.3 classifies the ISA-level FI tools in the literature along three dimensions and highlights the advantages and disadvantages of each. Section 9.4 overviews the design of the LLFI and PINFI tools, which we have been developing in our research group at the University of British Columbia (UBC), and the design trade-offs we made in the tools. Finally, Section 9.5 presents the open challenges in ISA-level FI and concludes.

## 9.2 Background

We start by defining the terms and definitions used, followed by the failure outcomes and different metrics that can be measured by FI at the ISA level. We then provide an overview of the FI process. Finally, we present the fault model considered by techniques in this space.

### *9.2.1 Terms and definitions*

- **Fault:** A fault is a perturbation of a target system. In our case, we consider hardware faults due to soft errors.
- **Fault occurrence:** The event corresponding to the occurrence of a hardware fault in the processor.
- **Fault activation:** The event corresponding to the manifestation of the fault to the software, corrupting some portion of the software state (e.g., register, memory location).
- **Error:** Once a fault is activated, it becomes an error. The error may or may not result in a failure (see later).
- **Error propagation:** Error propagation means that the fault was activated and has affected some other portion of the program state, say "X." In this case, we say the fault has propagated to state X. We focus on the faults that affect the program state and therefore consider error propagation at the application level.

### *9.2.2 Failure outcomes*

An error may lead to one of the following three possible outcomes:

- **Crash:** The raising of a hardware trap or exception due to the error, because the program attempted to perform an action it should not have (e.g., read outside its memory segments). The Operating System terminates the program as a result.
- **Silent Data Corruption (SDC):** There is a mismatch between the output of a faulty program run and that of an error-free execution of the program, though it completed successfully.
- **Benign faults:** Program output matches that of the error-free execution even though a fault occurred during its execution. This means either the fault was masked or overwritten by the program.

Different failure outcomes have different severities. Benign faults, as the name suggests, are not considered to be harmful as the program produces the same result with its fault-free execution. The same holds for inactivated errors. However, SDCs and crashes affect programs' normal executions and need to be mitigated depending on the reliability targets of programs. For example, if an application requires strict correctness of program output, SDCs are considered to be more severe, as the program produces incorrect outputs without any warning [12]. On the other hand, if the availability of a system is critical, faults that lead to crashes need to be mitigated as crashes halt program execution and hence lead to the service becoming unavailable [13].

Figure 9.1 illustrates the typical progress of a fault into failures corresponding to the different severity levels. In the figure, SDCs are considered more severe than crashes.

### *9.2.3 Metrics*

Because FI performed at the ISA level or higher lacks visibility into the detailed hardware-level microarchitectural structures, it emulates the effect of hardware faults

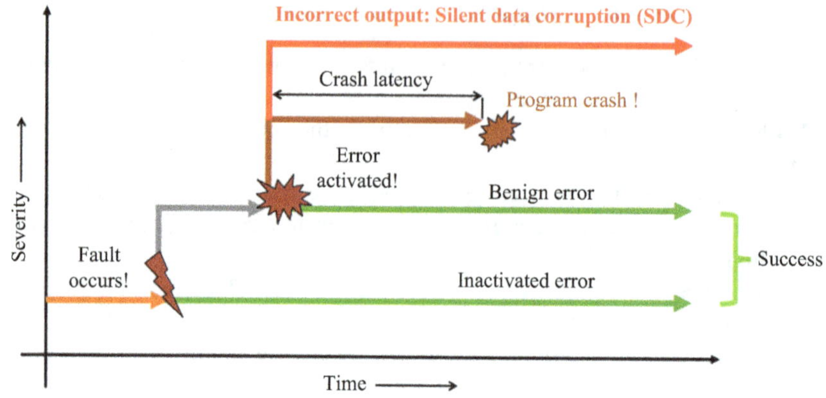


*Figure 9.1 Severity levels of a fault affecting the software*

by flipping bits in the program's data or instructions. An important question that arises in the design of FI tools at the ISA level is the metric being measured, as that can impact the accuracy of the tool. For example, an ISA-level FI tool would not be appropriate for studying masking at the hardware level, as it typically abstracts away such details in its design. On the other hand, studying masking due to the software characteristics is possible at the ISA level. We provide a brief overview of some of the key metrics that are often used in this area—the goal is not to be exhaustive but to provide an insight about what is possible to measure at this level.

- **SDC probability**: This is the conditional probability that a fault results in an SDC, given that it occurs and manifests at the software level.
- **Crash probability**: This is the conditional probability that a fault results in a program crash, given that it occurs and manifests at the software level.
- **Benign probability**: This is the conditional probability that a fault results in a benign outcome, given that it occurs and manifests at the software level.
- **Failure probability**: This is the conditional probability that a fault results in a failure (i.e., not benign), given that it occurs and manifests at the software level. It is the sum of the SDC and crash probabilities.
- **Crash latency**: The number of dynamic instructions executed between when the fault manifests at the software level and the crash of the program (assuming it crashes).
- **Propagation latency**: The number of dynamic instructions executed between when the fault manifests at the software level and the propagation of the error (assuming it propagates).

Note that the first four definitions previously described refer to *conditional probabilities*, as they are conditioned on the fault occurring and manifesting itself to the program. They do not measure the *a priori* probability of the fault occurring in the first place or manifesting at the software level, as these are a function of the hardware platform. More importantly, the overall probability of a fault resulting in a failure is a function of both the hardware and the software. ISA-level FI typically can measure the latter but not the former as it lacks detailed hardware or architectural models. This is the main difference between ISA-level FI and lower-level FI techniques that can measure the overall *vulnerability* of the program, e.g., the Architecture Vulnerability Factor [7].

### *9.2.4 Fault Injection process*

An FI experiment typically consists of numerous FI runs. As shown in Figure 9.2, each FI run artificially introduces a fault during a single-program execution and waits for the completion of the execution to observe the failure outcome (if any). The same program input is typically provided in each FI run. Note that only one fault is introduced in each run to ensure controllability of the process. An FI experiment typically consists of thousands of FI runs in order to achieve statistically significant results.

### *9.2.5 Fault model*

Another important question that arises in ISA-level FI tools is the fault model, or how to model hardware faults at the software level. As we are primarily concerned with soft errors, we can assume that the effect of the fault is confined to at most a few instructions or data. Further, most FI tools at the ISA level do not have visibility into the low-level hardware details and often assume a uniform distribution of the fault probability over the set of all program instructions or data. While it is possible to inject into both instructions and data using ISA-level FI, most existing tools consider the data as they assume that the instructions are protected through other means, for example, by encoding techniques [14,15]. Thus, most tools assume that faults can occur uniformly and randomly over the program's data.

There are three possible data sources to inject into in a program, namely (1) memory data, (2) architectural registers, and (3) instruction operands or results. Faults in memory data can occur either when the data is residing in memory or when it is read from or written to (e.g., due to bus errors). Faults in architectural registers can occur when the data is residing in the registers and/or when it read from or written to by the processor. Finally, faults in instruction operands and/or their result can occur when the

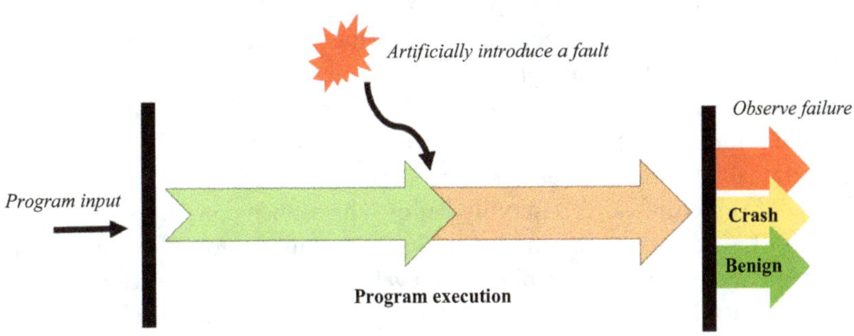


*Figure 9.2 Fault Injection run and failure outcomes*

data is being processed or computed in the processor's pipeline or microarchitectural structures. These include faults that occur both in the combinational elements of processors (e.g., the Arithmetic and Logic Unit), as well as those that occur in the pipeline latches and storage elements of the processor (e.g., Reordering buffer). The first two kinds of faults can be mitigated by incorporating Error Correcting Codes (ECC) in the memory and register file, respectively. However, the third kind of fault, namely, instruction operand and result faults, is very difficult to mitigate using those techniques as it is often prohibitively expensive to protect all the combinational elements and pipeline latches with ECC. Therefore, faults in instruction operands and results are often used to emulate software-visible faults that occur during the program's execution [2,4,16].

Finally, there is the question of how to model soft errors in software once we have identified the specific data element or instruction to inject. The most common fault model is that of a single bit-flip, where a single bit is uniformly chosen from the set of available bits in the data item or instruction, and its value is inverted (i.e., 1 if it was a 0 and vice versa). While this is a simple model to implement, there have been concerns about its lack of representativeness of how hardware faults can manifest in software [17]. Therefore, it has been suggested that we need to inject multiple bit-flips in software to accurately represent faults due to soft errors. However, recent studies have shown that for most of the metrics listed earlier, the precise choice of the bit-flip model (single or multiple) does not have much of an impact on the overall accuracy [18,19]. As multiple bit-flip FI significantly increases the FI space, it is not clear how significant are the benefits in terms of increased accuracy versus its cost. We return to this question later in this chapter (Section 9.5).

## 9.3 Classification of injection techniques

We classify FI tools that operate at the ISA-level along four dimensions:

- **Simulation-based or direct**: This dimension refers to whether we use a detailed simulation of the hardware platform, or whether we natively execute the program on the hardware platform with appropriate hooks for performing the injection. Simulation-based techniques typically require a detailed architectural model (but not necessarily microarchitectural) model of the hardware, while direct techniques require the actual hardware platform.
- **Intrusiveness**: This dimension refers to whether we modify the target program or system prior to injecting the faults (intrusive), or perform the injections through an external interface such as a debugger (nonintrusive).
- **Level of injection (e.g., assembly code)**: This dimension refers to the level at which FI is performed. Examples are injecting faults into the assembly code of the target program, or a compiler's IR level, or the program source code.
- **Target platform**: This dimension refers to the platform for which the FI experiments are targeted. Examples are CPU, GPU (i.e., Graphics Processing Unit), and FPGAs (i.e., Field Programmable Gate Arrays).

We briefly overview the advantages and disadvantages of each of the previously mentioned categories, before presenting the classification results.

### *9.3.1 Simulation versus direct*

Simulation techniques use a simulator to inject faults into the target system. Their main advantage is that they do not need the actual hardware platform to run the experiment, which is especially useful in the case of platforms that are not widely available or those that are under development. They also make it easy to experiment with different features in the hardware platform in simulation. In contrast, direct techniques need the actual platform to execute the program, and it is not easy to modify the platform in these techniques.

On the other hand, simulation techniques are often slower than direct techniques as they need to simulate the details of the platform (at least in sufficient detail so as to remain faithful to the original implementation). They also need to consider the effects of faults on the platform behavior, which is often not straightforward. Further, there is the question of accuracy, as these techniques are only as accurate as the accuracy of the simulation—this is because few simulators model all the behaviors of the target platform, especially when faults occur.

### *9.3.2 Intrusive versus nonintrusive*

The intrusiveness of an FI tool depends on whether it modifies the target program or system for inserting hooks for injecting faults. For example, an intrusive injector may instrument the target program with function calls to perturb the values of selected program variables or instruction operands at different points in the code. On the other hand, a nonintrusive injector may use a debugger or an external interface to the program (e.g., Ptrace) to inject faults without modifying the target program. The main advantage of intrusive injectors is that they can inject faults into any arbitrary location in the program, which is important for controllability. However, intrusive injectors may considerably modify the behavior of the system being studied, e.g., introducing timing delays, cache pollution, etc.

In contrast, nonintrusive injectors have a smaller footprint in terms of the modifications they make to the system but may not offer fine-grained controllability. This is because they are often limited by the granularity of the external interfaces. For example, a debugger may only allow breakpoints to be placed at certain locations in the program, and hence faults can only be injected at those locations.

### *9.3.3 Level of injection*

The third dimension for classifying ISA-level FI is the level at which they operate. On one extreme, we have tools that directly inject faults into the machine code or assembly code of a program. While they are straightforward to implement, it is often difficult to map back the results of the injection to a form that is usable by programmers who typically do not program directly in assembly or machine code. On the other hand, there are tools that operate directly at the program's source code, which is directly usable by programers. Unfortunately, such tools are not very accurate as there is not a one-to-one mapping between the source code and the assembly code of the program.

In recent years, there have been a wide variety of tools that inject faults at the compiler's IR level. For example, many tools inject faults at the LLVM compiler's IR level [20]. LLVM is a popular open-source compiler, and its IR level is low enough to model individual instructions, while at the same time, being expressive enough to model source code constructs such as loops, and program-type information. Therefore, it has emerged as a reasonable compromise between accuracy and usability. Experimental studies performed in our group [16,21] have found that LLVM IR-level FI is almost as accurate as assembly code level FI for SDC measurement.

### 9.3.4 Platform

As we consider FI techniques at the ISA level (and upwards), the FI technique is dependent on the ISA exposed by the target platform. Therefore, different techniques have been proposed for different platforms. Techniques that work at the LLVM IR level or higher are often oblivious of the target platform, provided the IR code can be translated to the platform's assembly code. However, techniques that work at the assembly code level depend on the semantics of the assembly language and the ISA exposed by the platform. Further, the applications that are written for each platform are often different in structure. For example, GPU applications typically constitute of computationally intensive kernels executed on the GPU, with the intervening data movement operations executed on the CPU. Therefore, the FI tool also has to take into account differences in application structure [22].

### 9.3.5 Classification results

Table 9.1 presents the classification of existing ISA-level FI tools. There are five broad categories of tools as shown in the table. LLFI [16], FlipIt [23], and KULFI [24] operate at the LLVM compiler's IR level and use direct technique to inject faults. All three tools extensively instrument the program with function calls for FI, and hence

*Table 9.1 Classification of ISA-level Fault Injection tools*

| **Fault injectors** | **Simulation versus direct** | **Intrusive versus nonintrusive** | **Level of injection** | **Platform** |
|---|---|---|---|---|
| LLFI [16], FlipIt [23], KULFI [24] | Direct | Intrusive | LLVM IR | CPU and GPU |
| PINFI [16], Xception [25], NFTAPE [26], F-SEFI [27] | Direct | Nonintrusive | Assembly | CPU |
| SASSIFI [28] | Direct | Intrusive | Assembly | GPU |
| GPU-Qin [29] | Direct | Nonintrusive | Assembly | GPU |
| GUFI [30], GemFI [31] | Simulation | Nonintrusive | Assembly | GPU and CPU |
| CAROL-FI [32] | Direct | Nonintrusive | Source code | CPU |
| Hauberk [33] | Direct | Intrusive | Source code | GPU |

they are intrusive. As they operate at the LLVM IR level, they are platform-neutral and can hence work on both CPUs and GPUs. With that said, only LLFI has been actually tested on GPU platforms, to the best of our knowledge [34].

PINFI [16], Xception [25], NFTAPE [26], and F-SEFI [27] all operate on the program's assembly code and are nonintrusive in the sense that they rely upon debuggers or emulators to modify the program's state. NFTAPE and Xception use the standard P-trace interface in UNIX systems, while PINFI and F-SEFI use the Pin tool from Intel [35] and QEMU hypervisor [36], respectively. All these systems are targeted for CPU programs.

SASSIFI [28] is a fault injector for GPUs that leverages compiler-based instrumentation for injecting faults into CUDA (Computer Unified Device Architecture) programs running on Nvidia GPUs. SASSIFI operates on the SAS assembly code (Standard Annotations and Semantics) executed by the Nvidia GPUs. In contrast, GPU-Qin [29] uses the CUDA–gdb (GNU Debugger) interface to inject faults into GPU programs also executing on Nvidia GPUs. Thus, SASSIFI is an intrusive injector, while GPU-Qin is nonintrusive.

GUFI [30] and GEMFI [31] use full system simulators to inject faults into GPU and CPU programs, respectively. GUFI uses the GPGPU-Sim [37] simulator for GPUs, while GEMFI uses the GEM5 simulator for CPUs. Both tools are nonintrusive and can operate directly at the assembly language level.

Finally, CAROL-FI [32] and Hauberk [33] inject faults into the program's source code for CPU and GPU programs, respectively. CAROL-FI uses the gdb to locate program variables of interest in the program's source code and perturbs them at run-time. So it is a nonintrusive injector. Hauberk injects faults into the source code of GPU programs by placing hooks in the code at compile-time using a source-to-source translator [33]. Thus, it is an intrusive injector.

## 9.4 LLFI and PINFI fault injectors

In this section, we describe two ISA-level FI tools that we have developed in our group, *LLFI** and *PINFI*.† LLFI operates at the LLVM compiler's IR level, while PINFI operates on the x86 assembly level. As such, they represent different design points in ISA-level FI tools. However, they also have many common features as explained later.

- **Fault model:** Both LLFI and PINFI primarily inject faults into program data that is either used as instruction operands or produced as the results of instructions. Thus, they both emulate the effects of computational faults in the processor.
- **Phases:** Both PINFI and LLFI operate in a two-phase mode. First, they profile the program by executing it with an input to obtain the frequency of the different dynamic instructions executed at runtime. In the second phase, they sample an instruction uniformly from the set of all executed instructions (corresponding to the fault criterion) and then inject the fault with the same input.

*Available at *https://github.com/DependableSystemsLab/LLFI.*
†Available at *https://github.com/DependableSystemsLab/PINFI.*

- **Activated faults:** Both PINFI and LLFI only consider activated faults, or those faults that are injected *and* read by another instruction in the program. Thus, they do not consider faults that are overwritten before they are read, or those faults that are not read.
- **Bit-flip model:** Both tools inject a *single bit-flip* fault into the chosen operand, from the set of all bits that are actually used by the instruction. For example, if an operand is used in a 32-bit instruction, only the first 32 bits of the operand are considered for injection, even if the operand is stored in a 64-bit register. This is again to satisfy the constraint of only considering activated faults.
- **Randomization:** Both tools allow the randomization of instruction and the bit to be chosen via a random seed that is specified by the user, for reproducibility purposes. It is also possible to vary the random distribution considered in both tools.
- **Statistics gathering:** Finally, both tools have extensive support for gathering statistics after the injection, such as the number of crash failures, hangs, and SDCs (these are obtained by comparing the output of the injected run to the golden output, or the output of the program when no faults are injected). They also support custom statistics gathering scripts that are able to parse the logs produced by the tools.

With the earlier-mentioned ones, there are also many important differences between the tools, due to the different levels of abstraction at which they operate. First, LLFI operates at the LLVM compiler's IR level and hence has limited visibility into architectural registers. Instead, it uses the notion of a virtual register file, which has an unlimited number of registers. Second, because LLFI operates at the level of LLVM instructions, which are generic, it does not know which instructions are actually used at the assembly language level to represent the operations. However, this is also an advantage, as it allows LLFI to be architecture neutral to a large extent. In fact, LLFI can work on any hardware platform that is supported by the LLVM compiler [16,34]. Finally, LLFI is invoked prior to the backend optimization passes of the compiler, and hence it may not have visibility into the code introduced by the compiler backend.

### *9.4.1 LLVM fault injector: LLFI*

We developed LLFI, a program level FI tool for performing the FI experiments [38, 39]. LLFI works at the LLVM compiler's intermediate level and enables tracing the propagation of the fault in the program by instrumenting the program at selected points. LLFI is closely integrated with the LLVM compiler and can hence support a wide variety of programs and programing languages [20].

LLVM uses the Static Single Assignment (SSA) form [40]. SSA requires a variable be assigned exactly once in the program, i.e., every variable in the program has a unique instruction that assigns to it. Further, its IR is a typed language, in which source-level constructs can be easily represented. In particular, the IR preserves the variable and function names, making source mapping feasible.

In each run, LLFI injects a fault, i.e., a single bit-flip into the destination register of exactly one dynamic instance of an instruction chosen at random (from the executed instructions) and classifies the outcome of the fault by comparing the final outcome with the fault-free outcome. The fault-free outcome is obtained by running the original executable with the same input, but without any injected faults.

**Workflow:** Figure 9.3 shows the workflow of LLFI. LLFI first takes the program's source code as input and converts it to the LLVM IR using the LLVM compiler. LLFI then instruments the IR code with FI functions at every program point where a fault can potentially be injected. The determination of these program points is based on the compile-time options provided by the user. Examples of compile-time options are the types of instructions and the register operands of the instructions. After this phase, the instrumented IR code is passed to the LLVM compiler's backend and the machine code is generated. We link the generated machine code with custom libraries to generate two executables. The first executable is the *profiling executable*, which profiles each FI site in the program to find the number of times it is executed. This information is passed to the second FI executable, which chooses an injection site at random from the total number of FI site executions (gathered by the profiling phase). The fault is then injected at this site by our custom FI library, and the program is executed to completion. The type of fault injected as well as the number of faults is decided based on a set of runtime options that are specified by the user. The FI results are then gathered and aggregate statistics are computed. Note that the compilation and instrumentation process is done only once for a given set of compile-time options, regardless of how many faults are injected at runtime.

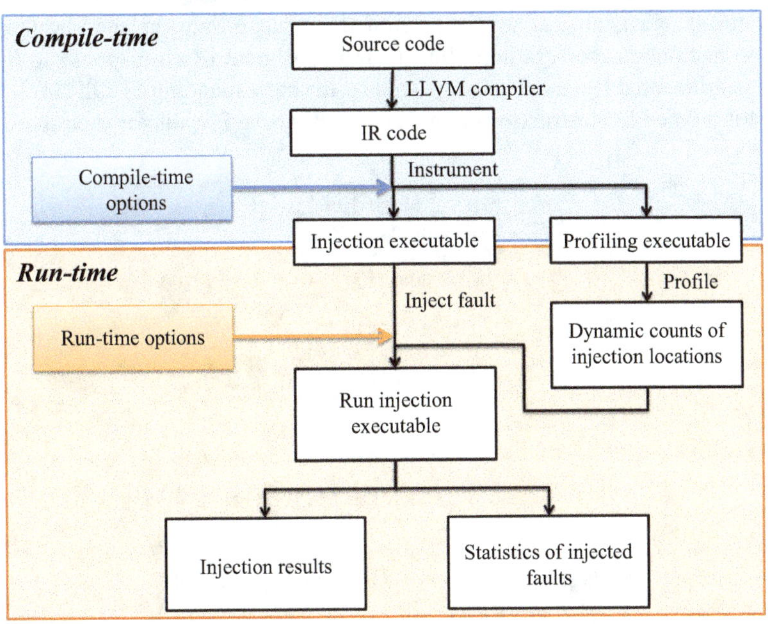


*Figure 9.3 LLFI workflow*

**Customizability and analysis**: LLFI has the following features for easy customization.

- *Instruction selection*: LLFI allows choosing a specific instruction or instruction type to inject into. This is done at compile-time.
- *Register selection*: LLFI allows choosing a specific register or operand of the instruction to inject into. This is also done at compile-time.
- *Fault type selection*: LLFI gives the user the ability to specify different kinds of faults to inject at selected points in the program. This is done at runtime.

**Example:** We explain the FI process using the factorial program in Figure 9.4 as an example. The original IR for the corresponding C code is shown in Figure 9.5. The value of `n` (the number whose factorial is calculated) is assigned in line 7 of the C code. This corresponds to %2 in line 6 of Figure 9.5. The basic block `bb1` in lines 15–18 in the IR corresponds to the loop header at line 9 in the C code. The `phi`[‡] node at line 15 gets the value of the iterator variable `i`, which is initialized to 1 (from basic block `entry`) or obtained from the incremented value within the loop at basic block `bb`. Similarly, `fact` is also assigned the `phi` node value where it is either 1 (initial value), or the value of `fact` from within the loop at line 10.

Figure 9.6 shows the original IR statically instrumented for FI. Each instruction with a return value has a call to the FI function `injectFault`. For example, lines 10 and 11 in Figure 9.5 have corresponding FI calls at lines 17 and 19 in the instrumented IR in Figure 9.6. The arguments of the `injectFault` function are the static fault ID, the type of call (FI call type has value 0), and the result of the instruction itself. This function flips a randomly chosen bit in the instruction's result and returns this faulty result. Next, all `uses` of the original instruction are replaced by the fault injected instruction. For example, line 15 in the original IR corresponds to line 23 in the instrumented IR, and %4 is replaced by the corresponding FI call %fi7. Since LLVM is a typed IR, instructions of different types have FI calls for each type.

```
 1 #include<stdio.h>
 2 main(argc, argv)
 3 int argc;
 4 char *argv[];
 5 {
 6   int i,fact, n;
 7   n = atoi(argv[1]);
 8   fact = 1;
 9   for(i=1;i<=n;i++)
10   {
11     fact = fact * i;
12   }
13   printf("%d\n",fact);
14 }
```

*Figure 9.4 C code for factorial program*

[‡] Phi is a construct used in SSA form.

```
define i32 @main(i32 %argc, i8** %argv) nounwind {
entry:
  %"alloca point" = bitcast i32 0 to i32
  %0 = getelementptr inbounds i8** %argv, i64 1
  %1 = load i8** %0, align 1
  %2 = call i32 (...)* @atoi(i8* %1) nounwind
  br label %bb1

bb:                                               ; preds = %bb1
  %3 = mul nsw i32 %fact.0, %i.0
  %4 = add nsw i32 %i.0, 1
  br label %bb1

bb1:                                              ; preds = %bb, %entry
  %i.0 = phi i32 [ 1, %entry ], [ %4, %bb ]
  %fact.0 = phi i32 [ 1, %entry ], [ %3, %bb ]
  %5 = icmp sle i32 %i.0, %2
  br i1 %5, label %bb, label %bb2

bb2:                                              ; preds = %bb1
  %6 = call i32 (i8*, ...)* @printf(i8* noalias getelementptr
        inbounds ([4 x i8]* @.str, i64 0, i64 0), i32 %fact.0) nounwind
  br label %return

return:                                           ; preds = %bb2
  ret i32 undef
}
```

*Figure 9.5 LLVM IR for the factorial program*

At runtime, we first obtain the total number of dynamic instructions by running the original factorial IR. Second, for each FI run, we choose a random instruction instance from the set of all dynamic instructions for FI. This is done by the function `initInjections` inserted at the beginning of the program (at line 3 in Figure 9.6). Third, for each instruction, the `injectFault` function checks if the particular instance is the dynamic instance to be fault injected, and if so, flips a single bit in the instruction result. This fault-injected result is propagated to all the `uses` of this instruction.

**Accuracy of LLFI:** While the LLVM intermediate code is close to the assembly code, it does not correspond one-to-one with the assembly language. We have quantified the impact of these differences between LLFI and PINFI in terms of the overall failure probability, as well as for specific instruction types [16]. Our results show that the difference is statistically indistinguishable from natural experimental variations as far as SDCs are concerned. However, this is not the case for program crashes, for which there is a statistically significant difference between the two injectors. We return to this issue in Section 9.5.

### *9.4.2 PINFI*

As mentioned earlier, PINFI works at the x86 assembly language level. PINFI is built with Intel Pin [35], an assembly-level instrumentation tool for x86-architecture

```
define i32 @main(i32 %argc, i8** %argv) nounwind {
entry:
  call void @initInjections(i8* getelementptr
        inbounds ([21 x i8]* @NameStr, i32 0, i32 0))
  %"alloca point" = bitcast i32 0 to i32
  %fi2 = call i32 @injectFault0(i32 2, i32 0, i32 %"alloca point")
  %0 = getelementptr inbounds i8** %argv, i64 1
  %fi3 = call i8** @injectFault1(i32 3, i32 0, i8** %0)
  %1 = load i8** %fi3, align 1
  %fi4 = call i8* @injectFault2(i32 4, i32 0, i8* %1)
  %2 = call i32 (...)* @atoi(i8* %fi4) nounwind
  %fi5 = call i32 @injectFault0(i32 5, i32 0, i32 %2)
  br label %bb1

bb:                                              ; preds = %bb1
  %3 = mul nsw i32 %fi8, %fi1
  %fi6 = call i32 @injectFault0(i32 6, i32 0, i32 %3)
  %4 = add nsw i32 %fi1, 1
  %fi7 = call i32 @injectFault0(i32 7, i32 0, i32 %4)
  br label %bb1

bb1:                                             ; preds = %bb, %entry
  %i.0 = phi i32 [ 1, %entry ], [ %fi7, %bb ]
  %fact.0 = phi i32 [ 1, %entry ], [ %fi6, %bb ]
  %fi8 = call i32 @injectFault0(i32 8, i32 0, i32 %fact.0)
  %fi1 = call i32 @injectFault0(i32 1, i32 0, i32 %i.0)
  %5 = icmp sle i32 %fi1, %fi5
  %fi9 = call i1 @injectFault3(i32 9, i32 0, i1 %5)
  br i1 %fi9, label %bb, label %bb2

bb2:                                             ; preds = %bb1
  %6 = call i32 (i8*, ...)* @printf(i8* noalias getelementptr
        inbounds ([4 x i8]* @.str, i64 0, i64 0), i32 %fi8) nounwind
  %fi10 = call i32 @injectFault0(i32 10, i32 0, i32 %6)
  br label %return

return:                                          ; preds = %bb2
  call void @postInjections()
  ret i32 undef
}
```

*Figure 9.6 LLVM IR of Figure 9.5 instrumented with Fault Injection calls*

processors. PINFI is built as a plugin in Pin and uses the API exposed by Pin to inject faults. PINFI works as follows: first, at runtime, PINFI loads the program executable and instruments the assembly instructions with FI functions that perturb their data. Second, it executes the instrumented executable and randomly selects a runtime instance of the instrumented instructions to trigger the FI function. The workflow of PINFI is similar to LLFI except: (1) LLFI performs instrumentation at compile time, while PINFI performs the instrumentation at runtime (when the program is loaded), and (2) LLFI performs the FI (instrumentation) at the IR code level, while PINFI performs the injection at the x86 assembly code level.

An additional challenge in PINFI is that we need to ensure that all injected faults are activated. In the case of LLFI, the LLVM compiler will automatically identify the "def-use" chain of an instruction, and hence we can avoid injecting faults into instructions whose value is not used. However, this analysis is much more complicated at the assembly language level where PINFI operates, and hence we devised the following heuristics for injecting faults using PINFI.

- In x86 assembly, branch condition instructions set the flag register, and different conditional jump instructions read different bits in the flag register to decide their branch target (Figure 9.7). To ensure fault activation, we first find the dependent flag register bit(s) of the conditional jumps and only inject faults into the dependent bit(s) before the conditional jump instructions. For example, in Figure 9.7, the *cmp* instruction sets the flag register, while the *jl* instruction only reads the *OF* (*bit* 11) of flag register to decide the branch target. So we only inject faults into this bit.
- Assume that we wish to inject into instructions of a certain category, e.g., load instructions. However, at the assembly code level, the instruction opcode is insufficient to decide the instruction category. For instance, in x86 assembly, *mov* instructions are used both to move data from memory to register (loads), and from one register to another. As shown in Figure 9.7, only the third *mov* instruction is a load instruction. Therefore, we need to disambiguate among different types of instructions by parsing the entire instruction in PINFI.
- For floating point operations, x86 instructions usually use XMM registers as the destination register, and hence all 128 bits of XMM register are FI candidate bits. However, double-precision floating-point operations only use the lower 64 bits for computation (Figure 9.7), and hence we prune the target injection space to the lower 64 bits for double-precision floating point operations.

**Advantages and disadvantages of PINFI**: As mentioned earlier, PINFI is closer to the x86 assembly code and hence injecting faults with it is arguably more accurate than injecting faults with LLFI. It is also capable of injecting faults into binary programs without going through the compilation process and hence does not require the program's source code unlike LLFI. With that said, there are also a couple of disadvantages in using PINFI. First, it only works on x86 platforms as the PIN tool on which it is built is only available on the x86 platform. This limits its portability to other platforms. Second, because PINFI operates directly on the assembly code, it lacks LLFI's deep analysis capabilities to inject faults in a targeted fashion and map back the results to the source code. Therefore, PINFI may not be as useful for software developers who want to understand and improve their program's resilience.

## 9.5 Open challenges and conclusion

In this section, we examine three open challenges in the design of ISA-level FI tools and then conclude the chapter.

```
cmp $0xa4, %eax
jl 4006e0
```

```
mov $0x4008a0, %edi
. . .
mov %eax, %esi
. . .
movsd 0x20d(%rip), %xmm1
. . .
movsd %xmm0, 0x8(%rsp)
```

```
addsd %xmm2, xmm0
```

*Figure 9.7 Examples of PINFI heuristics to increase fault activation: (a) flags, (b) loads, and (c) floating point*

### 9.5.1 Challenge 1: level of injection

As mentioned in Section 9.3, there are different FI tools implemented at different levels, from the LLVM IR level to the assembly code level. Unfortunately, it is not clear how accurate are the former compared to the latter. More generally, as one increases the level of abstraction in the tool, there is likely to be a loss in accuracy, which can affect the overall results. Therefore, there is a need to rigorously characterize and validate the accuracy of tools at different levels.

In our prior work [16], we have validated the accuracy of LLFI, a fault injector implemented at the LLVM IR level, versus the PINFI injector implemented at the assembly code level. However, we only considered overall failure probabilities in that experiment, and also the SDC and crash probabilities of individual instructions. It is an open question however as to how accurate are higher level tools for other measures as detailed in Section 9.2. Further, there may be differences among benchmarks, and compiler optimization levels, for each of the measures, and so these need to be rigorously quantified as well. We have studied the effect of compiler optimizations on the accuracy of high-level FI tools in a recent paper [21].

There are three challenges in rigorous quantification of FI methods at different levels. First, we need to standardize a set of benchmark applications to compare the techniques. While prior work has used typical benchmark suites such as SPEC [41], it is not clear how representative these suites are of programs that are interesting to the resilience and fault-tolerance community [42]. Second, we need to ensure that the fault models and injection methodology are standardized across the different techniques to enable an "apples-to-apples" comparison. Finally, there needs to be a standardized set of metrics to quantitatively compare the different tools with each other.

### 9.5.2 Challenge 2: target platform

Another open challenge in ISA-level FI tools is how to deal with emerging platforms such as Machine-Learning (ML) accelerators such as Tensor Processing Unit [43] and Diannao [44]. In the recent past, there have been a plethora of custom hardware accelerator platforms proposed primarily for ML applications. This is due to the phenomenon of Dark Silicon [45], and the emergence of ML as an important application in many domains. Unfortunately, these ML accelerators are very different from traditional platforms such as CPUs and GPUs, as they expose only a limited ISA to the software. Further, most of the operations that are critical to the overall correctness of the applications are hidden from the software layer, and hence traditional ISA-level FI will not be able to inject faults in them.

In recent work, we have investigated the error resilience of ML accelerators employing Deep Neural Network (DNN) applications in self-driving cars [10]. We performed FI experiments using a simulator and injecting the faults by modifying the simulator. We investigated the resilience of an open-source accelerator called Eyeriss [46] and found that its resilience did not meet the minimum requirements in the ISO26262 safety standard for road vehicles [47]. We also investigated different combinations of DNN parameters and hardware configurations and found that there was significant variability in the resilience based on the parameters considered. Finally, we found many outcomes that lead to safety violations in the system, e.g., the accelerator misclassified an image of a truck as a bird due to the error. Other studies have reported similar results [48,49].

While the earlier-mentioned tools are useful, they are often specific to the hardware platform and its implementation and hence do not consider the resilience of the ML application alone. Therefore, we believe that there is a need for FI tools to investigate the resilience of ML applications at the ISA level. However, the main challenge in doing so is to abstract the effects of low-level faults in the operations used in ML operations such as convolution to the high-level at which they are expressed. Toward this goal, we have built *TensorFI* [50] to inject faults in ML applications developed using the TensorFlow framework [43]. We have also built *BinFI*, an enhancement of *TensorFI*, for efficient FI in safety-critical ML applications [51].

### 9.5.3 Challenge 3: bit-flip model

A controversial choice in many FI tools is the use of single bit-flips to represent soft errors. As mentioned in Section 9.3, the use of the single bit-flip fault model has been criticized by some researchers as not being representative of how soft errors typically affect programs [17]. However, other work (including our own) has found that the differences between single and bit-flip fault models is negligible for aggregate SDC probability measurement [18,19]. While this is encouraging, it is an open question as to how applicable this result is for the other metrics in Section 9.3. Further, in certain safety-critical applications, worst case failure probability is a better measure of the system's resilience than the aggregate failure probability. Therefore, it may be necessary to inject multiple bit-flip faults—we found that even in this case, we need only a small number of bit-flips (three) [18].

The main challenge in injecting multiple bit-flips is the exponential increase in the FI space with the number of bits (to be flipped). Unfortunately, today there is no way to prune the space of multiple bit-flip faults, and this can lead to significant increase in the FI time.

### *9.5.4 Conclusion*

In this chapter, we examined the goals and current state-of-the-art ISA-level FI tools and techniques. We started by defining the terms and metrics used and then described the FI process, followed by the fault model at the ISA level for soft errors. We then examined the different FI techniques at the ISA level and classified them based on whether they are simulation-based, intrusive, the level at which faults are injected, and the target platform. Finally, we presented an overview of the LLFI and PINFI tools, which are two FI tools at the ISA level that were developed in our group.

The main take-away from this chapter is that ISA-level FI tools are both useful and accurate in many cases, especially for fault-tolerance techniques implemented at the software level. However, there are still open challenges such as their accuracy for different resilience metrics, their use for emerging hardware platforms, and fault models to efficiently represent multiple bit-flips. These challenges will need to be addressed before ISA-level FI techniques become widely applicable.

## Acknowledgments

This work was funded in part by the Natural Sciences and Engineering Research Council of Canada (NSERC), and industry gifts from AMD, Cisco, and Intel. We thank all past and current members of the Dependable Systems Lab group at UBC for contributing to the design of LLFI and PINFI.

## References

[1] Carter NP, Naeimi H, and Gardner DS. Design techniques for cross-layer resilience. In: Design, Automation & Test in Europe Conference & Exhibition (DATE). Dresden, Germany: IEEE; 2010. p. 1023–1028.

[2] Li G, Pattabiraman K, Hari SKS, *et al.* Modeling soft-error propagation in programs. In: IEEE/IFIP International Conference on Dependable Systems and Networks (DSN). Luxembourg: IEEE; 2018.

[3] Feng S, Gupta S, Ansari A, *et al.* Shoestring: Probabilistic soft error reliability on the cheap. In: ACM SIGARCH Computer Architecture News. vol. 38. Pittsburgh, PA, USA: ACM; 2010. p. 385–396.

[4] Hari SKS, Adve SV, and Naeimi H. Low-cost program-level detectors for reducing silent data corruptions. In: IEEE/IFIP International Conference on Dependable Systems and Networks (DSN). Boston, MA, USA: IEEE; 2012. p. 1–12.

[5] Narayanan S, Sartori J, Kumar R, *et al.* Scalable stochastic processors. In: Design, Automation & Test in Europe Conference & Exhibition (DATE). Dresden, Germany: IEEE; 2010. p. 335–338.

[6] Liu S, Pattabiraman K, Moscibroda T, *et al.* Flikker: Saving DRAM Refresh-Power Through Critical Data Partitioning. 2011. p. 213–224. Available from: http://doi.acm.org/10.1145/1950365.1950391.

[7] Mukherjee SS, Weaver C, Emer J, *et al.* A systematic methodology to compute the architectural vulnerability factors for a high-performance microprocessor. In: IEEE/ACM International Symposium on Microarchitecture (MICRO). San Diego, CA, USA: IEEE; 2003. p. 29–40.

[8] Oh N, Shirvani PP, and McCluskey EJ. Control-flow checking by software signatures. IEEE Transactions on Reliability. 2002;51(1):111–122.

[9] Saxena NR and McCluskey EJ. Control-flow checking using watchdog assists and extended-precision checksums. IEEE Transactions on Computers. 1990;39(4):554–559.

[10] Li G, Hari SKS, Sullivan M, *et al.* Understanding error propagation in deep learning neural network (DNN) accelerators and applications. In: Proceedings of the International Conference for High Performance Computing, Networking, Storage and Analysis. Denver, CO, USA: ACM; 2017. p. 8.

[11] Pattabiraman K, Saggese GP, Chen D, *et al.* Dynamic derivation of application-specific error detectors and their implementation in hardware. In: European Dependable Computing Conference (EDCC). Coimbra, Portugal: IEEE; 2006. p. 97–108.

[12] Lu Q, Pattabiraman K, Gupta MS, *et al.* SDCTune: A model for predicting the SDC proneness of an application for configurable protection. In: Proceedings of the 2014 International Conference on Compilers, Architecture and Synthesis for Embedded Systems. Jaypee Greens, India: ACM; 2014. p. 23.

[13] Li G, Lu Q, and Pattabiraman K. Fine-grained characterization of faults causing long latency crashes in programs. In: Dependable Systems and Networks (DSN), 2015 45th Annual IEEE/IFIP International Conference on. Rio de Janeiro, Brazil: IEEE; 2015. p. 450–461.

[14] Wang S and Duan G. On the characterization and optimization of system-level vulnerability for instruction caches in embedded processors. Microprocessors and Microsystems. 2015;39(8):686–692.

[15] Wang S, Hu J, and Ziavras SG. On the characterization and optimization of on-chip cache reliability against soft errors. IEEE Transactions on Computers. 2009;58(9):1171–1184.

[16] Wei J, Thomas A, Li G, *et al.* Quantifying the accuracy of high-level fault injection techniques for hardware faults. In: IEEE/IFIP International Conference on Dependable Systems and Networks (DSN). Atlanta, Georgia USA: IEEE; 2014. p. 375–382.

[17] Cho H, Mirkhani S, Cher CY, *et al.* Quantitative evaluation of soft error injection techniques for robust system design. In: Proceedings of the 50th Annual Design Automation Conference. Austin, Texas, USA: ACM; 2013. p. 101.

[18] Sangchoolie B, Pattabiraman K, and Karlsson J. One bit is (not) enough: An empirical study of the impact of single and multiple bit-flip errors. In: 2017 IEEE/IFIP International Conference on Dependable Systems and Networks (DSN). Denver, CO, USA: IEEE; 2017. p. 97–108.

[19] Chang CK, Lym S, Kelly N, *et al.* Evaluating and accelerating high-fidelity error injection for HPC. International Conference for High Performance Computing, Networking, Storage and Analysis. Dallas, TX, USA: IEEE; 2018. p. 577–589

[20] Lattner C and Adve V. LLVM: A compilation framework for lifelong program analysis & transformation. In: Proceedings of the International Symposium on Code Generation and Optimization: Feedback-Directed and Runtime Optimization. Washington, DC, USA: IEEE Computer Society; 2004. p. 75.

[21] Palazi L, Li G, Fang B, *et al.* A tale of two injectors: End-to-end comparison of IR-level and assembly-level fault injection. In: IEEE International Symposium on Software Reliability Engineering (ISSRE). Berlin, Germany: IEEE; 2019. p. 151–162.

[22] Gomez LB, Cappello F, Carro L, *et al.* GPGPUs: How to combine high computational power with high reliability. In: Proceedings of the Conference on Design, Automation & Test in Europe. European Design and Automation Association. Dresden, Germany: IEEE; 2014. p. 341.

[23] Calhoun J, Olson L, and Snir M. FlipIt: An LLVM based fault injector for HPC. In: European Conference on Parallel Processing. Porto, Portugal: Springer; 2014. p. 547–558.

[24] Sharma VC, Haran A, Rakamaric Z, *et al.* Towards formal approaches to system resilience. In: IEEE Pacific Rim International Symposium on Dependable Computing (PRDC). Vancouver, British Columbia, Canada: IEEE; 2013. p. 41–50.

[25] Carreira J, Madeira H, and Silva JG. Xception: A technique for the experimental evaluation of dependability in modern computers. IEEE Transactions on Software Engineering. 1998;24(2):125–136.

[26] Stott DT, Floering B, Burke D, *et al.* NFTAPE: A framework for assessing dependability in distributed systems with lightweight fault injectors. In: IEEE Computer Performance and Dependability Symposium. Chicago, IL, USA: IEEE; 2000. p. 91–100.

[27] Guan Q, Debardeleben N, Blanchard S, *et al.* FSefi: A fine-grained soft error fault injection tool for profiling application vulnerability. In: IEEE Parallel and Distributed Processing Symposium. Phoenix, AZ, USA: IEEE; 2014. p. 1245–1254.

[28] Hari SKS, Tsai T, Stephenson M, *et al.* SASSIFI: Evaluating resilience of GPU applications. In: Proceedings of the Workshop on Silicon Errors in Logic-System Effects (SELSE); 2015.

[29] Fang B, Pattabiraman K, Ripeanu M, *et al.* GPU-Qin: A methodology for evaluating the error resilience of GPGPU applications. In: IEEE International Symposium on Performance Analysis of Systems and Software (ISPASS). Monterey, CA, USA: IEEE; 2014. p. 221–230.

[30] Tselonis S and Gizopoulos D. GUFI: A framework for GPUs reliability assessment. In: IEEE International Symposium on Performance Analysis of Systems and Software (ISPASS). Uppsala, Sweden: IEEE; 2016. p. 90–100.

[31] Parasyris K, Tziantzoulis G, Antonopoulos CD, *et al.* GemFI: A fault injection tool for studying the behavior of applications on unreliable substrates. In: IEEE/IFIP International Conference on Dependable Systems and Networks (DSN). Atlanta, GA, USA: IEEE; 2014. p. 622–629.

[32] Oliveira D, Frattin V, Navaux P, *et al.* CAROL-FI: An efficient fault-injection tool for vulnerability evaluation of modern HPC parallel accelerators. In: Proceedings of the Computing Frontiers Conference. Siena, Italy: ACM; 2017. p. 295–298.

[33] Yim KS, Pham C, Saleheen M, *et al.* Hauberk: Lightweight silent data corruption error detector for GPGPU. In: IEEE Parallel & Distributed Processing Symposium (IPDPS). Anchorage, Alaska, USA: IEEE; 2011. p. 287–300.

[34] Li G, Pattabiraman K, Cher CY, *et al.* Understanding error propagation in GPGPU applications. In: International Conference for High Performance Computing, Networking, Storage and Analysis (SC). Salt Lake City, USA: IEEE; 2016. p. 240–251.

[35] Luk CK, Cohn R, Muth R, *et al.* Pin: Building customized program analysis tools with dynamic instrumentation. In: ACM SIGPLAN Notices. vol. 40. Chicago, IL, USA: ACM; 2005. p. 190–200.

[36] Bellard F. QEMU, a fast and portable dynamic translator. In: USENIX Annual Technical Conference, FREENIX Track. vol. 41. California, USA: USENIX; 2005. p. 46.

[37] Bakhoda A, Yuan GL, Fung WW, *et al.* Analyzing CUDA workloads using a detailed GPU simulator. In: IEEE International Symposium on Performance Analysis of Systems and Software (ISPASS). Boston, MA, USA: IEEE; 2009. p. 163–174.

[38] Thomas A and Pattabiraman K. LLFI: An intermediate code level fault injector for soft computing applications. In: Workshop on Silicon Errors in Logic System Effects (SELSE). Boston, MA, USA: IEEE; 2013.

[39] Lu Q, Farahani M, Wei J, *et al.* LLFI: An intermediate code-level fault injection tool for hardware faults. In: 2015 IEEE International Conference on Software Quality, Reliability and Security. Vancouver, BC, Canada: IEEE; 2015. p. 11–16.

[40] Cytron R, Ferrante J, Rosen BK, *et al.* Efficiently computing static single assignment form and the control dependence graph. ACM Transactions on Programming Languages and Systems (TOPLAS). 1991;13(4):451–490.

[41] Henning JL. SPEC CPU2006 benchmark descriptions. SIGARCH Computer Architecture News. 2006;34(4):1–17. Available from: http://doi.acm.org/10.1145/1186736.1186737.

[42] Kanoun K and Spainhower L. Dependability Benchmarking for Computer Systems. Washington, DC, USA: Wiley-IEEE Computer Society Press; 2008.

[43] Abadi M, Barham P, Chen J, *et al.* TensorFlow: A system for large-scale machine learning. In: 13th USENIX Symposium on Operating Systems Design

and Implementation (OSDI). vol. 16. Savannah, GA, USA: USENIX; 2016. p. 265–283.

[44] Chen T, Du Z, Sun N, *et al.* Diannao: A small-footprint high-throughput accelerator for ubiquitous machine-learning. ACM SIGPLAN Notices. 2014;49(4):269–284.

[45] Esmaeilzadeh H, Blem E, Amant RS, *et al.* Dark silicon and the end of multicore scaling. In: IEEE International Symposium on Computer Architecture (ISCA). San Jose, CA, USA: IEEE; 2011. p. 365–376.

[46] Chen YH, Krishna T, Emer JS, *et al.* Eyeriss: An energy-efficient reconfigurable accelerator for deep convolutional neural networks. IEEE Journal of Solid-State Circuits. 2017;52(1):127–138.

[47] ISO 26262. Road Vehicles—Functional Safety. International Standard ISO/FDIS; 2011.

[48] Fernandes F, Weigel L, Jung C, *et al.* Evaluation of histogram of oriented gradients soft errors criticality for automotive applications. ACM Transactions on Architecture and Code Optimization (TACO). 2016;13(4):38.

[49] Reagen B, Gupta U, Pentecost L, *et al.* Ares: A framework for quantifying the resilience of deep neural networks. In: Proceedings of the 55th Annual Design Automation Conference. DAC '18. New York, NY, USA: ACM; 2018. p. 17:1–17:6. Available from: http://doi.acm.org/10.1145/3195970.3195997.

[50] Li G, Pattabiraman K, and DeBardeleben N. TensorFI: A configurable fault injector for TensorFlow applications. In: IEEE International Symposium on Software Reliability Engineering Workshops (ISSREW). Memphis, TN, USA: IEEE; 2018. p. 313–320.

[51] Chen Z, Li G, Pattabiraman K, *et al.* BinFI: An efficient fault injector for safety-critical machine learning systems. In: International Conference for High Performance Computing, Networking, Storage, and Analysis (SC). Denver, CO, USA: ACM; 2019. p. 1–23.

# Chapter 10
# Analytical modeling for cross-layer resiliency

*Arijit Biswas*[1]

Analytical models, techniques and methods are mathematical models that have a closed form solution. In other words, the solution to the equations used to describe the changes in the system being modeled can be expressed as a mathematical analytic function. They are used widely in the design of complex systems. In microprocessor design, they are most often employed to analyze performance, power and reliability.

In reliability, analytical models are typically employed in the computation of vulnerability metrics, such as Architectural Vulnerability Factor (AVF) [1] or Program Vulnerability Factor (PVF) [2]. Vulnerability factors quantify the probability that a bit corruption or fault results in an actual error by determining whether the fault propagates to an observable error state.

Analytical models can compute vulnerability factors of a system orders of magnitude faster than brute force methods such as statistical fault injection [3], at the cost of some accuracy [4]. This speed/accuracy trade-off is at the heart of a well-balanced analytical model. Analytical models may be used at any level of modeling abstraction, from Register Transfer Level (RTL) models, through emulation models, all the way up to architectural or functional models [5–9]. Most of the analytical modeling techniques described in this chapter are used either at the functional model level, the architectural/micro-architectural level or at the RTL. Some are hybrid techniques that use multiple levels of abstraction to achieve high levels of accuracy [10]. These hybrid techniques achieve high levels of accuracy from lower level models such as RTL while retaining the speed benefits of higher level architectural models.

## 10.1 Introduction

Analytical models can compute the vulnerability factors of multiple components of a system simultaneously in a single run, thus resulting in statistically significant results in a very short amount of time, compared to brute force methods [11–13]. The detail level provided in the underlying model is critical to ensuring that the analytical model results accurately portray the behavior of the modeled system.

[1] Intel, Hudson, MA, USA

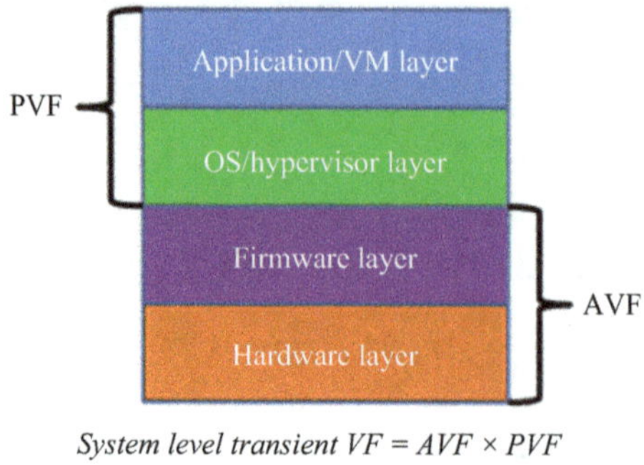


*Figure 10.1 System stack showing vulnerability factors at different layers*

The different analytical methods for reliability vulnerability modeling covered in this chapter include Architecturally Correct Execution (ACE) lifetime analysis [1]—including enhancements to ACE lifetime analysis such as Hamming Distance Analysis (HDA) and Bit Field Analysis (BFA) [14], sequential AVF analysis [10], and PVF Analysis or PVF [15]. The latter sections also cover modeling artifacts and how to deal with them, reduction of unknowns as well as directions and trends for future work.

All these techniques model vulnerability factors for various layers of the system and together can provide a vulnerability factor for the complete system. Cross-layer resiliency focuses on targeting different techniques at different layers of the system stack [16]. As such, different techniques must be used to quantify the effectiveness of each technique based on the layer of the system stack it targets.

Figure 10.1 shows how vulnerability at different levels can be used to compute the overall system-level vulnerability.

## 10.2 ACE lifetime analysis

ACE lifetime analysis is an analytical modeling technique used to compute the AVF of system components [1]. It has been effectively used to compute AVFs for key microprocessor structures such as caches, Translation Lookaside Buffers (TLBs), register files, instruction queues, data paths, arithmetic units and so forth. ACE lifetime analysis is a technique that is taught broadly in universities and used widely in industry.

ACE lifetime analysis is usually deployed in cycle-accurate micro-architectural performance models to achieve a level of detail that provides a good balance between accuracy and speed. A model instrumented to employ this method can compute the AVFs of multiple components in a single run. ACE lifetime analysis utilizes the concept that only a subset of a program's data flow graph needs to be correct. As shown in Figure 10.2, this path is known as ACE or necessary for ACE. Everything else is considered un-ACE (Unnecessary for ACE) and can be derated away.

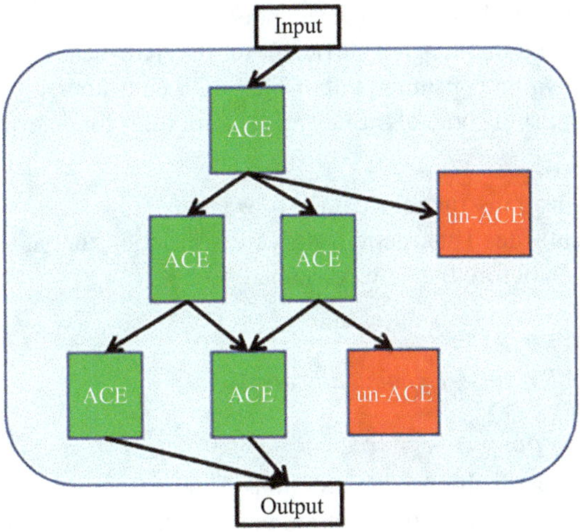


*Figure 10.2 Data flow graph showing ACE and un-ACE paths through a system*

This technique breaks the lifetime of a component down into segments with each access of the structure denoting the end of the previous segment and the start of the next segment. Each segment is then tagged as being ACE or un-ACE. The AVF of the component is the sum of the durations of all the ACE segments for that component divided by the total lifetime duration.

$$AVF = \frac{\sum ACE\ lifetimes}{Total\ lifetimes}$$

### *10.2.1 Un-ACE and ACE*

The determination of ACE-ness for each lifetime segment of a component is the key to accurate AVF modeling using ACE lifetime analysis. At a high level, a segment is ACE from the time it is produced to when it is consumed. For example, consider a component such as a register file with read and write access events, any segment that starts with a write (producer) and ends with a read (consumer) may be considered as ACE. By extension, any segment that ends with a write (producer) can be considered un-ACE regardless of the access event that started the segment.

- Producer to consumer=ACE
- Consumer to consumer=ACE
- Producer to producer=un-ACE
- Consumer to producer=un-ACE (this is a subcategory of un-ACE also called ex-ACE)

This concept of ACE/un-ACE, however, is simplistic. The above-mentioned criteria are necessary but not sufficient to accurately determine AVF. Without further refinement, it will provide a conservative upper bound for AVF. To refine this notion of ACE, we determine that it is not sufficient for the ACE-ness of a lifetime segment to be determined by *any* consumer, but only by *ACE* consumers.

Following is a list of some of the key refinement rules for determining ACE-ness.

- Dynamically dead instructions/data
  - Dynamically dead instructions/data are instructions or data whose results are overwritten before they are consumed.
  - Example:
    *R8=ADD R0, R1*
    *R9=ADD R1, R8←Dynamically dead*
    *R9=SUB R8, R2*
    *R10=ADD R9, R3*
- Transitively dynamically dead instructions/data
  - Transitively dynamically dead instructions/data are similar to dynamically dead except that these are instructions/data that only feed instructions/data that are dynamically dead. In this case, the entire chain of instructions/data that feeds the dynamically dead one is considered dynamically dead as well.
  - Example:
    *R8=ADD R0, R1*
    *R9=ADD R1, R8←Transitively dynamically dead*
    *R10=ADD R9, R3←Dynamically dead*
    *R10=SUB R8, R2*

AVF and error rates are attributes of bits in hardware. Hence, the notions of ACE and un-ACE can also be associated with hardware bits. Figure 10.3 shows the entries of an instruction queue tagged as ACE or un-ACE with un-ACE broken into a few different subcategories.

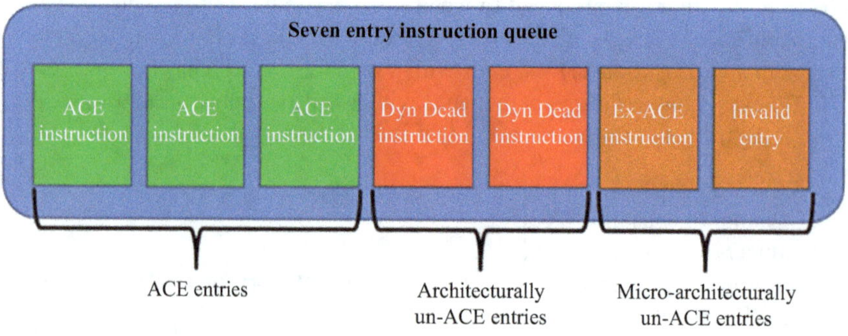


*Figure 10.3 Example instruction queue showing entries holding ACE or un-ACE values*

## *10.2.2 Little's law*

Little's law comes from queuing theory. It states that the average number of items in a system is the product of the average arrival rate and the average duration of items in the system. It is given by the following formula in which $L$ is the average number of items in the system, $\lambda$ is the average arrival rate and $W$ is the average duration.

$$L = \lambda \times W$$

It can be used to compute the average utilization of a system given by the following equation:

$$N = \frac{Tput \times Lat}{\#Bits \times total\ duration}$$

In the previous equation, $N$ is the average utilization, $Tput$ is the average throughput and $Lat$ is the average latency. Utilization is an upper bound for the AVF of any component. This equation can be further enhanced to compute the AVF by qualifying it with the concept of ACE. In this case, the AVF of a component is the product of its ACE throughput and ACE latency. This is given by the following equation:

$$AVF = \frac{Tput_{ACE} \times Lat_{ACE}}{\#Bits \times total\ duration}$$

Figure 10.4 illustrates the use of Little's law to compute utilization and AVF and also shows the difference between the two forms. Note that for any system $N$ (utilization) will always be greater than or equal to AVF.

## *10.2.3 Example of ACE lifetime analysis*

ACE lifetime analysis essentially captures the $Tput_{ACE}$ and $Lat_{ACE}$ terms of the AVF equation. It accumulates the throughput and latency information over time as per the analytical model simulation then applies the ACE/un-ACE concepts discussed previously in this section. The resulting AVF is computed as per the equation given in Little's law subsection as a product of the $Tput_{ACE}$ and $Lat_{ACE}$ averaged over the course

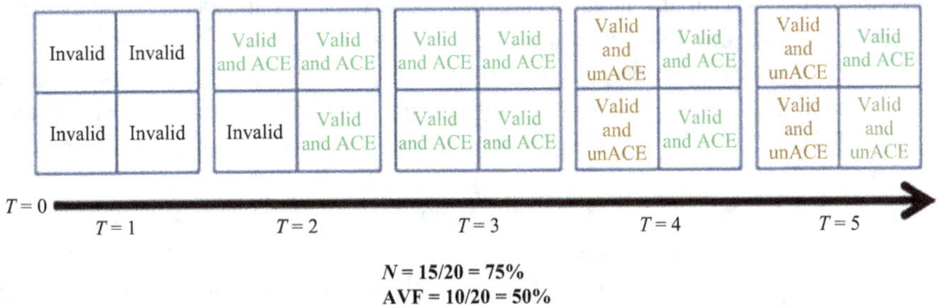


*Figure 10.4 Example of utilization and AVF computed for a four-entry buffer over five cycles*

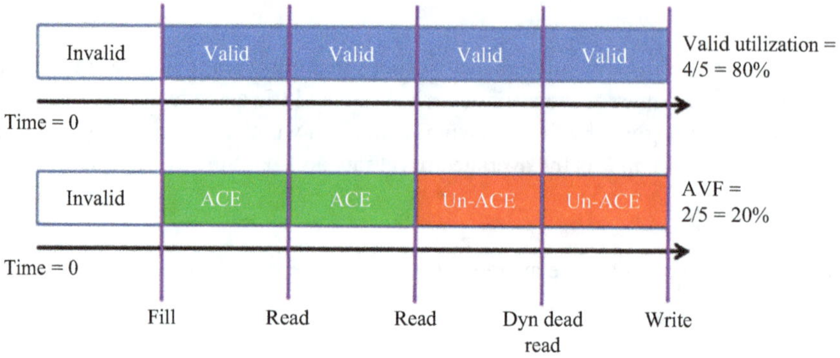


*Figure 10.5 Example of ACE lifetime analysis for a simple cache line with fill, read and write events occurring at equal intervals*

of the simulation model run. Figure 10.5 is an example of ACE lifetime analysis used to compute the AVF of a cache line.

### *10.2.4 AVFs of various structures and workloads using ACE lifetime analysis*

AVFs can vary significantly based on the component and the workload since it is a factor of the architectural and micro-architectural behavior of the underlying system. Also, architectural error mitigation mechanisms (such as parity and Error Correction Code) can be evaluated by observing their impact on the component's AVF. AVF can be particularly useful to compute the effectiveness of partial mitigation schemes, since these only protect a component part of the time based usually on the frequency of certain scenarios or events. Simulation models can compute the occurrence frequency of these scenarios during which, the component is partially protected. Hence, the AVF can be used to measure their effectiveness.

There are certain components that tend to have fairly static AVF values. Purely speculative structures such as branch prediction tables will have AVFs of 0, since errors in these structures typically never cause functional failures. Similarly, structures that contain only committed architectural state, such as program counters, will tend to have AVF near 100%, since an error in these bits almost always results in some sort of functional error. However, many components of modern microprocessors contain a mix of speculative and architectural states. In these cases, it is important to understand the AVF behavior as it can vary widely in terms of both components on a given workload as well as a single component across a variety of workloads.

In cases where timing matters from a functional perspective, such as in cases of real-time operation, functional safety or cycle-accurate lock-step—even faulting behavior that would only result in a performance glitch (typically un-ACE) could be considered to be ACE and faults in hardware such as branch predictors should in fact be modeled as ACE.

The AVF can be used to determine the relative significance of a structure from a reliability perspective based on its contribution to the overall error rate. This enables cost-effective targeting of design resources and effort by quantifying which structures or components are bigger contributors to the overall system error rate than others.

### *10.2.5 Hamming Distance Analysis and bit field analysis*

ACE lifetime analysis provides the foundation for computing AVF in an analytical manner. However, it can produce overly conservative results in certain cases unless further refinement is done. This is akin to adding more detail into the model regarding the behavior of certain structures or components.

Basic ACE lifetime analysis tends to be employed on an *entry* level where an *entry* is the largest "word" that can be read or written for a specific component. For example, a byte addressable cache would have a "word" size of a single byte, and AVF analysis would be performed at that level. However, there are structures where this can lead to conservatism in the AVF computation.

Tag arrays, such as those whose values are consumed by Content Addressable Memory structures (where array values are compared against a given value, and action is taken if the values either match or do not match), are one such type of component. AVF analysis is typically done for tag arrays at a "word" size equal to the size of the tag. However, this can lead to overly conservative AVFs. In order to address this, AVF models may employ a technique called HDA, typically Hamming-Distance-1 (HD-1) Analysis for single-bit faults [7]. HD-1 analysis can further refine the AVF values computed by ACE lifetime analysis to better account for the way in which tag values are actually consumed and used by the system.

An example of two values that are HD-1 apart is as follows:

- Value1: 10001
- Value2: 10000

In the previous example, the least significant bit is the only bit that is different between these two values. Hence, they are HD-1 apart.

HD-1 analysis is used to better account for the way a bit corruption in a tag entry is consumed and propagated through the system. Tag reads are essentially lookups where a given tag value is compared against a set of tag values from a tag array. The AVF of the tag array, if ACE lifetime analysis only is used, will usually indicate that the AVF of such a component is very high. However, in reality, this is not true for many types of tag arrays. Write-Through Caches and TLBs, for example, result in one of two results based on a tag lookup. The tag can either *hit* (the tag values match) or *miss* (the tag values do not match). In most cases, if the tag values miss, then the system will simply go re-fetch the data from a higher level of the memory subsystem. In the case of a hit, we need to look at what kind of bit corruption would result if a tag value hits when it should have missed. In a single-bit fault model, this is only the case if the tag value in the array is HD-1 away from the tag value being looked up. As such, only the single bit that is HD-1 away is ACE, the other bits of the tag entry are effectively un-ACE. As a result, HD-1 analysis can greatly change the AVFs for tag arrays by singling out only the *bit* of an entry that is ACE at each lookup access.

Another type of refinement is called BFA. Similar to HD-1 analysis, for which only a single bit of an entry can be ACE, BFA is used on arrays whose entries contain different bit fields that have differing behavior and can be read/written at different times. These bit fields can vary in size from a single bit to many bits. BFA breaks these entries down into the individual bit fields and simply computes the AVF of each bit field independently from the rest. Examples of components that can have multiple embedded bit fields are Retirement Order Buffers, Instruction Queues, Reservation Stations and so forth. These structures tend to contain a mix of data, address and meta-data values contained in specific bit ranges within a single entry.

### *10.2.6 Hamming Distance Analysis and multi-bit fault modeling*

HD-1 analysis is used to refine the AVF values computed using ACE lifetime analysis for tag structures. It specifically identifies whether any entry in the match set is HD-1 away from the lookup value. The underlying assumption is that one and only one bit will be corrupt for any tag in the set. While this works fine for single-fault models, there is an increasing need to understand how these techniques can be modified in order to comprehend multi-bit faults.

As technologies improve and geometries get ever smaller and threshold voltages ever lower, multi-bit faults become a significant contributor to system error rates. Single-bit faults have dominated the overall system error rate in the past but that can no longer be assumed to be the case in the future.

For HDA, we can address multi-bit faults using HD-*X* analysis where *X* is the largest hamming distance for which we wish to analyze our tags. HD-*X* can range from HD-1 to HD-<size of tag entry>. Of course, if we set *X* to the size of the entire tag, then we can simply employ ACE lifetime analysis at the tag entry granularity and no HD analysis is needed. However, if we set *X* to something more reasonable such as 2 or 3, we can cover 1-bit, 2-bit and 3-bit faults. An ACE model that could cover all 1-bit, 2-bit and 3-bit faults in tag arrays would employ HD-1, Hamming-Distance-2 (HD-2) and HD-3 analyses in order to identify all tag values that are 1, 2 and 3 hamming distance away from the lookup value.

HD-2 example:

- Value1: 10011
- Value2: 00010

The previous example shows two values that are only different in the most and least significant bits; hence, they are HD-2 apart. Based on the type of error however, it may be useful to only look at values that are hamming distance apart for *adjacent* bits.

Example of Adjacent HD-2:

- Value1: 10001
- Value2: 10010

In the first HD-2 example, the values were HD-2 apart but **not** Adjacent HD-2 apart. In the previous example, only the least significant and least significant+1 bits are different; hence, these values are Adjacent HD-2 apart.

In practice it is useful to compute the AVFs of each multi-bit scenario separately rather than combining them together, since the actual error rate usually varies (depending on the error type) for each. In this way, the HD-1 AVF can be multiplied with the *raw* single-bit error rate to compute the *effective* single-bit error rate, the HD-2 AVF can be multiplied with the *raw* 2-bit error rate to compute the *effective* 2-bit error rate and so forth. The final *effective* error rates (in Failures in Time) can be added together to provide a final *total* error rate for the associated component.

## 10.3 Sequential AVF analysis

One of the major criticisms leveled against ACE lifetime analysis is a lack of sufficient detail based on the level of abstraction used in the analytical modeling. While more detail can be added to the ACE model to alleviate this issue, there will always be a detail gap between that model and the lower level RTL models. Otherwise, the higher level model would embody all the detail of the RTL models and they would be the same, thus the higher level abstraction would lose its benefits in speed.

This detail gap becomes apparent in the absence of sequentials in the higher level models. Flops and latches used for staging, pipelining and so forth are a level of detail found in the lower level RTL models but not in the higher level models. Performance models tend to gloss over the detailed flops and latches and simply model their impact by way of tagging signals with bandwidth and latency limits which dictate how much data can be transmitted and received between modeled components in a specific number of cycles.

This detail gap has become more pronounced as these unmodeled sequentials have become increasing contributors to error rates in modern devices. The ability to compute the AVFs of flops and latches in a design is an important capability that is missing from the standard ACE lifetime analysis analytical model.

Sequential AVF analysis is a hybrid technique that leverages the speed and analytical abstractions of the ACE lifetime analysis techniques while leveraging the sequential detail contained in the lower level RTL models. It does this *without* needing to run the lower level model—instead leveraging set theory rules to propagate AVF values from the ACE model into a data graph extracted from the RTL model and thus providing AVF values for every flop and latch in the design *analytically*.

### *10.3.1 port AVF (pAVF) and structure AVF*

The sequential AVF computation technique starts with the computation of port AVF (pAVF) for each component in the ACE lifetime model [10]. The pAVF of a bit in a structure's interface (or port) is the probability of ACE data being transmitted to or from the component through that bit, as shown in Figure 10.6. For a read port, $\text{pAVF}_R$ is calculated by dividing the number of ACE reads from the component by the total

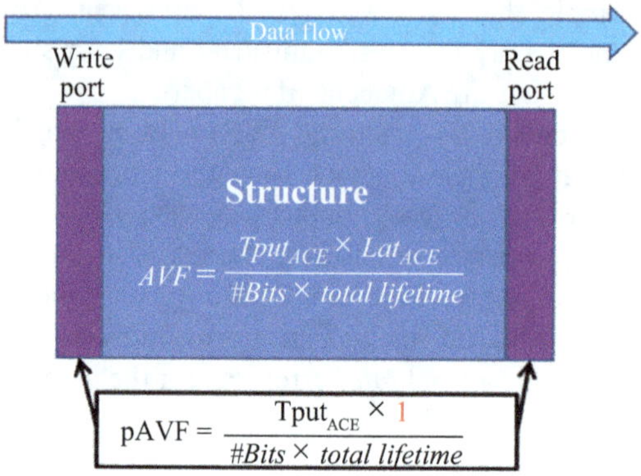


*Figure 10.6 Illustration of structure vs read/write ports and how the AVF equation changes for structure AVF vs pAVF*

number of cycles simulated. For a write port, $pAVF_W$ is computed by dividing the number of ACE writes to the component by the number of simulated cycles.

The ACE-read and ACE-write count values for each structure are reported by the ACE lifetime model. These values represent the ACE data rate of the interface circuits of the associated component (the logic that represents the read/write ports of the component). While both component or structure AVF and pAVF measure the same characteristic of a bit, pAVF is used specifically for sequential AVF to avoid confusion with structural AVF, as well as to describe how the value is calculated. A pAVF value can be calculated for all nodes (both combinational and sequential) in a design.

### *10.3.2 Sequential AVF computation*

Sequential AVF computation is done by extracting the sequential and node connectivity detail from the RTL model into a node graph representation that can be traversed. This effectively creates a detailed netlist that includes all flops and latches.

The specific nodes and sequentials that represent the port bits of components modeled in the ACE lifetime model are identified in the netlist. These primal nodes are then tagged with the pAVF values for those bits calculated from the ACE lifetime model. At this point, a subset of nodes and states in the netlist is tagged with pAVF values. Those values are then propagated throughout the entire netlist using specific set theory rules depending on the type of connection. Read pAVF values are propagated forward from inputs to outputs, while write pAVF values are propagated backward from outputs to inputs. Once the values are fully propagated, every connected state and node in the design netlist is tagged with an AVF value. Propagation of pAVF values ends when an ACE lifetime model component, a primal node or a primary input/output for the design netlist are reached. In this manner, an AVF value is computed for every flop and latch in the entire design without ever needing to run the RTL model. The

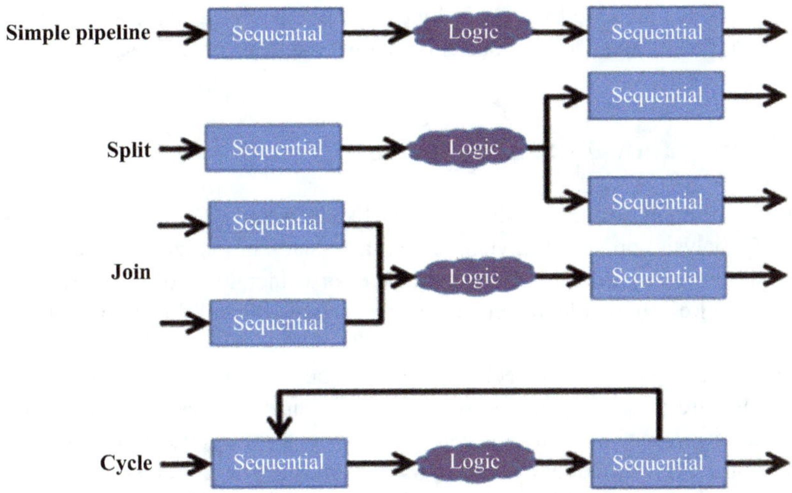


*Figure 10.7 Example illustrations of pipeline, split, join and cycle-type path*

pAVF values for the primal nodes themselves come from the analytical ACE model and can be based on an average of workloads or on a specific workload.

pAVF propagation rules are based on the connectivity between state nodes in the netlist. These consist of simple pipelines (straight point-to-point connect), splits (fan-outs), joins (fan-ins) and cycles (paths that begin and end at the same node), as illustrated in Figure 10.7.

- Simple pipeline: pAVF at input of pipeline=pAVF at output of pipeline.
- Splits: pAVF of each path after split=pAVF of the path before the split.
- Joins: pAVF after join=union of the pAVFs of all the inputs into the join. If there is no overlap between the inputs, then this is simply the sum of the pAVFs of all the inputs into the join (capped at 1.0 since this is a probability).
- Cycles: Special case usually identified by propagation reaching a node that has already been reached multiple times. In this case, we can either keep updating the node pAVF until it stops changing (or changes negligibly) or by assigning cycles some fixed value based on the design.

Since multiple paths through the design can affect each node in the netlist, we need a final rule to establish how we update the AVF of a node when the node already has a value from a previous propagation. In this case, the update rule is given as:

$$New\ node\ AVF = MIN(old\ node\ AVF,\ new\ propagated\ AVF)$$

This technique utilizes the ACE lifetime analysis to initialize a netlist generated from the lower level RTL model. The rules established by the sequential AVF technique then provide a means of propagating those values analytically throughout that entire netlist. Thus, we are able to analytically compute AVFs for every component in the design, be it memories, queues, buffers or individual flops or latches.

## 10.4 Program vulnerability factor

ACE lifetime analysis and its enhancements, like BFA and HDA, along with sequential AVF analysis, provide analytical techniques that can model error vulnerability probabilities in the hardware. This leaves the error vulnerability of the software layer to still be addressed in order to get a complete view of vulnerability across the system stack.

PVF is another analytical technique that can be used to address the vulnerability of the software layer [2]. PVF is similar to AVF but computes the vulnerability of program state instead of hardware state. PVF computes the vulnerability of program values (data and instructions) stored in main memory. Where AVF is the probability that errors propagate to *observable* or *architectural state*, PVF is the probability that errors in architectural state propagate to *program output*.

Program impact can be defined as changes in program output values or changes in program behavior, including hangs (livelocks or deadlocks) and crashes due to faulting behavior. Program impact may also be defined based on user expectations. For instance, a user may decide that they only care about values written to a specific log file. Therefore, only the instruction streams and data values associated with values written to that specific log file are vulnerable, and the PVF computation can be defined to take this into account.

PVF is usually computed using a functional or architectural model such as Simics that is capable of modeling, at a functional level, the target instruction set architecture and associated architectural state. The program state needed for correct execution is identified as it moves through memory and architectural state, and its vulnerability is computed. This computation is based on the fraction of time the program state is vulnerable to changes in architectural state between that state being written and the state being output by the program. Program output takes the form of instructions executed or data written to permanent IO such as disk, network, display or other IO.

### *10.4.1 Cross-layer modeling using AVF and PVF*

Once AVF and PVF are computed for a system, the *system-level vulnerability factor* can be computed as the product of AVF×PVF. AVF is the probability of a fault in hardware propagating to architectural state, while PVF is the probability that faults in architectural state propagate to program output. Thus, the product of the two vulnerabilities is the probability that a fault in hardware propagates to program output which is effectively the system-level vulnerability.

The following example shows five hardware faults that occur over the course of a program's run. Assume a system with five faults and only one propagates all the way through the hardware and firmware stacks, up through the OS and application layers to manifest as a visible error at the system level. Hence, in this example, the

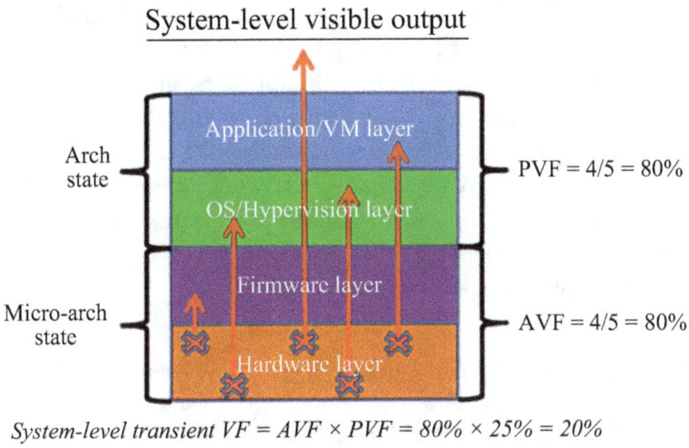


*Figure 10.8 Example of system level VF broken down into AVF and PVF*

system-level vulnerability is 1/5=20%. We can also see from the example that we can break this down to the various components—AVF and PVF. In this example, four out of the five faults escape from micro-architectural state (encompassed by the hardware and firmware layers) into architectural state (encompassed by the OS and application layers). As a result, the AVF=4/5=80%. However, of the four faults that reach architectural state, only one actually impacts the correct operation of the program. Thus, the PVF=1/4=25%. We can then apply the equation for cross-layer modeling as AVF×PVF=80%×25%=20%, which is the result we expected.

Usually, we do not have the luxury of already knowing the system-level impact as we do in this example. Rather, we have analytical models for each portion of the system stack as discussed throughout this chapter. By combining them as shown in Figure 10.8, we can model the behavior at the system level. AVF can be computed for every hardware component in the system, and PVF can be computed for all layers of the software stack. Together, these analytical techniques can provide vulnerability modeling for all layers of the system.

## 10.5 Artifacts of analytical vulnerability modeling and mitigations

Modeling techniques in general all have inherent limitations depending on the type of modeling used. These are usually the result of limitations in program execution and/or underlying assumptions on which the modeling is based. For analytical modeling, some of these limitations can result in inaccurate or incomplete results due to certain artifacts of the underlying model.

For many of the vulnerability modeling techniques discussed in this section, these artifacts can result in inaccurate results or large unknowns in resolving ACE lifetimes. One way these techniques help one to deal with these artifacts is to ensure that underlying assumptions are inherently conservative such that these limitations will result in higher, rather than lower, AVF values. Sometimes, this can result in overly conservative results or results that are not actionable due to large relative inaccuracies. A number of ways have been identified to help one to reduce modeling artifacts and unknowns to improve accuracy and reduce conservatism. These methods include modeling the actual impact of masking on data values, reducing unknowns using cooldown and warm-up techniques, and using iterative annealing techniques to break down large models for speed and complexity.

### *10.5.1 Significance of data values in analytical modeling*

The analytical modeling techniques discussed in this chapter are usually agnostic of the actual data values flowing through the system. It assumes that any bit corruption will alter the data values in an unacceptable manner. However, in practice, this is a conservative assumption and is not always true. Here, we discuss two specific cases where this assumption can result in conservative AVFs—and how we can use this knowledge to account for these cases.

The first case is lucky loads. A lucky load is a load whose address has become corrupt but still manages to return the same value as if the load had accessed the correct address. Computer programs often store the same value in many different locations in memory. One popular value is zero that occurs frequently in computer programs, and techniques have been developed to improve performance or power efficiency by taking advantage of this phenomenon. Lucky loads is one case where taking into account the actual data value being returned can help one to determine whether an address corruption is in fact ACE. In this case, an address bit would be considered ACE not only if it is HD-1 away from the lookup address but *also* if the resulting address returns a data value *different* from the correct address.

Another broad category of cases where the data values can determine ACE-ness is logical masking. Logical masking can impact any data or control path where logical or arithmetic operations or transforms occur, but the usual target is the main execution data paths. Logical masking is where two or more values are logically or mathematically transformed (for instance, by an ALU) such that the resulting value is the *same* even if one or more of the input values are corrupt. In this case, the actual operation and the actual data values must be considered in order to determine final ACE-ness. Following are some examples of logical masking.

- Value A×Value 0: Any value multiplied by 0 results in 0 so any bit corruption in value A would always be un-ACE.
- Value A AND Value 0: Similar to multiply by 0—any value ANDed with 0 is 0 so any bit corruption in value A would always be un-ACE.
- Value A AND Value B; Value A OR Value B: A bit corruption in *either* value A or B is not necessarily ACE. We need to know both values to determine whether a bit corruption in either value would alter the correct outcome of the operation and therefore its ACE-ness.

- Shift operations: Any bit corruption in the value being shifted, if the corrupt bit is shifted out then it is un-ACE so ACE-ness is determined by the shift value.
- Compares: If the compare operation still resolves correctly then the corruption is un-ACE. For example, if value A<0 then the actual value of A does not matter except that it be less than zero.

These are just a few examples of logical masking where the actual data values and the logical operation being performed will determine the ACE-ness of the bit(s).

Since ACE lifetime analysis tends to treat all bits as ACE regardless of value, the AVF computed will always be conservative without taking data values into account. How conservative those values are, depends on the workloads and the data values.

### *10.5.2 Reducing unknowns—warmup and cooldown*

Other artifacts of modeling are the effects of averaging and the resolution of unknowns. Both arise from the fact that models are typically limited to running snippets of workloads in isolation and not full programs in a full system environment. This results in two problems. The first is that the model tends to start cold—with empty caches, buffers and queues. This results in very different model behavior when the workload starts than in the middle or when the program ends. In an actual system, this rarely happens as workloads are rarely run in isolation on a cold system—usually there are at least parts of OS/kernel in the system and parts of other workloads or background processes in the system. Even if that is not the case, systems usually tend to run a workload multiple times so that the caches and other components are "Warmed up" before taking actual measurements of performance. The second problem is that models do not usually run programs to completion, so at the end of the simulation the lifetimes of much of the state is left "unknown," meaning that it is not known whether the next operation would be a producer or a consumer of any given bit.

The method to mitigate the first problem, "warming up" the system to average out any cold start behavior, is very similar in the model as on a system. In a system, the workload is usually run multiple times before measurements are taken. In the model, we usually run several thousands to a couple of million instructions of the program trace before we start computing the AVFs. This allows the cache and TLB component models to "warm up," eliminating any cold start artifacts from affecting the simulated measurements. This is a common practice in performance modeling and is used in reliability modeling as well. The exact length of warm-up that is appropriate for a given architecture/design may vary and should be determined by studying the impact of several different warm-up lengths to find the shortest duration of warm-up needed to not perturb the steady-state operation results.

The method to mitigate the second problem, resolving unknowns, is called "cooldown." In cooldown, the simulation is run for a few thousands to a couple of million instructions past the point at which we stop taking simulation measurements. The information from the extra simulation cycles is then used to resolve the final lifetimes from the end of the main part of the simulation. The exact length of cooldown needed is determined by the workload and the architecture and may require some studies to determine the optimal duration of cooldown to reduce/eliminate the unknowns significantly while not extending the simulation time unreasonably.

### 10.5.3 Dealing with large and complex models

One final artifact of modeling is the simple fact that modeling modern computing systems is a hugely complex task. It can be expensive in terms of model size and capacity. This can result in simulation slowdown, memory capacity problems and so forth. Many times, a divide and conquer type of approach may provide excellent results and even make modeling very large systems feasible. However, such breaking up of a system to improve modeling throughput and capacity may result in additional complexity in simply managing the pieces of the model.

One technique discussed in this chapter that benefits from this modular approach is sequential AVF, since the netlist resulting from a modern processor can be extremely large and unwieldy if done as a single monolithic data structure. In this case, the approach used for sequential AVF modeling is to break the netlist into functional blocks and propagate the pAVFs within each block first. Then an iterative approach is used similar to annealing, in which the pAVFs are then propagated between blocks. The newly propagated pAVFs are then propagated through each block again and values updated as per the rules established by that technique. This iterative process continues until the block-level AVFs are either no longer changing or are only changing within some finite range, in which case we determine that it has reached steady-state equilibrium. In this way, we can model extremely large and complex systems while still keeping the compute and memory requirements within reason.

## 10.6 Future directions for analytical technique

This chapter has covered a few of the key high-level analytical modeling techniques that have been used in academia and industry for modeling reliability and vulnerability of hardware and software systems. New work is also in progress to discover new techniques and continue to improve on established ones.

One area that is fast becoming of greater importance is Multi-Bit Upset AVF (MBU AVF). The modeling techniques described here are all essentially single-fault models and thus compute Single-Bit Upset (SBU) AVF. MBUs are orders of magnitude less frequent than SBUs. However, as our understanding of SBU has steadily improved and our ability to mitigate SBU has improved, MBU is becoming significant, especially as geometries become denser and voltages continue to decrease.

MBU AVF is an area that could benefit from further research, in both the hardware and software spaces. MBU can occur across both space (single fault causes multiple bit corruptions) and time (multiple faults cause multiple bit corruptions). There can also be a cross-layer dimension where faults in hardware can cause convergent or divergent faults in software.

- multiple HW faults cause multiple SW faults across space/time
- single HW fault causes multiple SW faults across space/time

- multiple HW faults cause single SW fault across space/time
- combinations of the above

A good MBU AVF model should attempt to encompass all of the previous vectors in order to provide a complete picture of MBU vulnerability across all layers of the system stack.

## 10.7 Summary of analytical modeling for vulnerability

Analytical models are powerful tools that can be used to determine various critical aspects of system behavior, including performance, power and reliability. Analytical models can provide statistically significant results across a broad range of workloads and system behaviors in much less time and with fewer compute resources than brute force techniques while still maintaining a higher level of accuracy than purely statistical models.

ACE lifetime analysis is the analytical modeling technique employed on performance models that is used to compute the AVF of bits in hardware. It can be enhanced and improved considerably through the use of techniques such as hamming distance and BFA. Complementary techniques such as sequential AVF analysis can be used to extend ACE lifetime analysis to also cover details found only in lower level models such as individual flops and latches.

PVF analysis utilizes analytical methods to compute the vulnerability of the software layer. The product of AVF and PVF can provide a vulnerability factor for the system by effectively computing the probability that a fault in hardware will adversely impact system behavior or output.

By employing a range of analytical reliability modeling techniques, we can effectively compute the vulnerability factor for different layers (Figure 10.9) of the system stack—hardware, firmware and software. By taking the weighted average of the

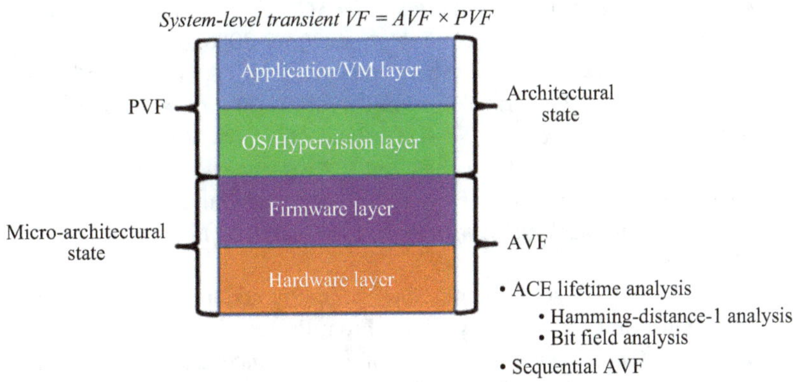


*Figure 10.9 Detailed system stack diagram associating AVF and PVF modeling with each layer of the stack*

vulnerability factors at each layer of the system stack (weighted by bits/area) and then taking the product of the weighted average vulnerabilities of each layer, we can compute an effective vulnerability for the entire system.

## References


[1] S. S. Mukherjee, C. Weaver, J. Emer, S. K. Reinhardt and T. Austin, "A Systematic Methodology to Compute the Architectural Vulnerability Factors for a High-Performance Microprocessor," in *IEEE/ACM International Symposium on Microarchitecture (MICRO)*, Washington, DC, USA, 2003.

[2] V. Sridharan and D. R. Kaeli, "Quantifying Software Vulnerability," in *Workshop on Radiation Effects and Fault Tolerance in Nanometer Technologies* (WREFT), Ischia, Italy, 2008.

[3] R. Leveugle, A. Calvez, P. Maistri and P. Vanhauwaert, "Statistical Fault Injection: Quantified Error and Confidence," in *Design, Automation Test in Europe Conference Exhibition (DATE)*, Nice, France, 2009.

[4] N. Wang, J. Quek, T. Rafacz and S. Patel, "Characterizing the Effects of Transient Faults on a High-Performance Processor Pipeline," in *International Conference on Dependable Systems and Networks (DSN)*, Florence, Italy, 2004.

[5] N. George, C. R. Elks, B. W. Johnson and J. Lach, "Transient Fault Models and AVF Estimation Revisited," in *2010 IEEE/IFIP International Conference on Dependable Systems & Networks (DSN)*, Chicago, IL, USA, 2010.

[6] X. Li, S. V. Adve, P. Bose and J. A. Rivers, "Architecture-Level Soft Error Analysis: Examining the Limits of Common Assumptions," in *EEE/IFIP International Conference on Dependable Systems and Networks (DSN)*, Edinburgh, UK, 2007.

[7] A. Biswas, P. Racunas, J. Emer and S. S. Mukherjee, "Computing Accurate AVFs using ACE Analysis on Performance Models: A Rebuttal," *IEEE Computer Architecture Letters*, vol. 7, pp. 21–24, 2008.

[8] X. Fu, T. Li and J. A. B. Fortes, *Sim-SODA: A Unified Framework for Architectural Level Software Reliability Analysis*, 2006.

[9] A. A. Nair, S. Eyerman, L. Eeckhout and L. K. John, "A First-Order Mechanistic Model for Architectural Vulnerability Factor," in *International Symposium on Computer Architecture (ISCA)*, Portland, OR, USA, 2012.

[10] S. Raasch, A. Biswas, J. Stephan and P. E. Racunas, "A Fast and Accurate Analytical Technique to Compute the AVF of Sequential Bits in a Processor," in *IEEE/ACM International Symposium on Microarchitecture (MICRO)*, Waikiki, HI, USA, 2015.

[11] L. Duan, B. Li and L. Peng, "Versatile Prediction and Fast Estimation of Architectural Vulnerability Factor From Processor Performance Metrics," in *International Symposium on High Performance Computer Architecture (HPCA)*, Raleigh, NC, USA, 2009.

[12] G.-H. Asadi, V. Sridharan, M.D. Tahoori, and D. Kaeli, "Balancing Performance and Reliability in the Memory Hierarchy," in *Proceedings of the IEEE*

*International Symposium on Performance Analysis of Systems and Software, 2005*, Washington, DC, USA, 2005.

[13] A. Vallero, A. Savino, G. Politano, *et al.*, "Cross-Layer System Reliability Assessment Framework for Hardware Faults," in *2016 IEEE International Test Conference (ITC)*, Fort Worth, TX, USA, 2016.

[14] A. Biswas, P. Racunas, R. Cheveresan, J. Emer, S. Mukherjee and R. Mukherjee, "Computing Architectural Vulnerability Factors for Address-Based Structures," in *International Symposium on Computer Architecture (ISCA)*, Madison, WI, USA, 2005.

[15] V. Sridharan and D. R. Kaeli, "Eliminating Microarchitectural Dependency From Architectural Vulnerability," in *IEEE International Symposium on High Performance Computer Architecture (HPCA)*, Raleigh, NC, USA, 2009.

[16] V. Sridharan and D. R. Kaeli, "Using Hardware Vulnerability Factors to Enhance AVF Analysis," in *International Symposium on Computer Architecture (ISCA)*, New York, NY, USA, 2010.

# Chapter 11
# Stochastic methods

*Alessandro Savino[1], Alessandro Vallero[1], and Stefano Di Carlo[1]*

It is clear from the previous chapters of this book that both fault-injection techniques and analytical approaches for cross-layer reliability analysis have both positive and negative aspects that must be carefully analyzed whenever choosing the best approach to evaluate the reliability of a computing system, and none of them alone represents an optimal solution.

With the increasing complexity of future computing systems, analyzing the impact on system reliability of any change in the technology, circuit, microarchitecture and software is a critical and complex design task that requires proper tools and models. The adoption of cross-layer reliability techniques makes this analysis even more complex and challenging.

Therefore, there is an increasing interest into stochastic reliability models that are able to combine the benefit of fault-injection techniques at different abstraction levels and analytical approaches such as Register Data Lifetime [1] or Architectural Correct Execution [2–4] analysis into a unified stochastic model that is able to cope with the complexity of the target design together. This is favored by an increasing discussion on the use of these models in the framework of relevant reliability and safety-related standards such as the IEC 61508 [5] and ISO 26262 [6].

## 11.1 Introduction

A stochastic model is a representation of a process or system where uncertainty is present. In other words, it is a model for a process that has some kind of randomness. The word stochastic comes from the Greek word *stokhazesthai* meaning to aim or guess. In the real word, uncertainty is a part of everyday life, so a stochastic model could literally represent anything, including a complex computing system and the occurrence of faults in its hardware.

The opposite is a deterministic model that predicts outcomes with 100% certainty. Deterministic models are often analytical techniques and methods based on

[1]Department of Control and Computer Engineering, Politecnico di Torino, Torino, Italy

mathematical formulations that have a closed form solution. In other words, the solution to the equations used to describe the changes in the system being modeled can be expressed as a mathematical function. Analytical models are used widely in the design of complex systems. In microprocessor design, they are often employed to model performance, power and reliability. However, they poorly scale with the complexity of the modeled systems, thus requiring strong simplifications.

Stochastic models are usually efficient when dealing with very complex systems but, on the other hand, they likely produce different results every time the model is executed or simulated. Fault-injection experiments by means of statistical fault sampling can be considered a very simple stochastic approach to the reliability analysis. Nevertheless, in this chapter, rather than limiting the stochasticity to the fault sampling we want to focus on its use to model complex characteristics of the system under analysis.

Modern Systems-on-Chip and Cyber Physical Systems show a tight integration and interaction between hardware and software built upon an integration capacity of billions of transistors. On top of the hardware, a multilayered software stack, including an operating system and complex applications, is usually executed and must be taken into account when performing the reliability analysis. In the cross-layer reliability context, random variables can be associated, for instance, with different hardware/software components of the system or to different states of the system. In fact, the overall system reliability assessment can usually be simplified by breaking up the system into components, which may or may not be statistically independent. In component-based reliability modeling, system reliability is estimated through reliability parameters of individual system components and the way they interact in the system. Component-based reliability modeling can be easily integrated with typical component-based system design workflows [7].

In general, all stochastic reliability models that will be considered in this chapter have the following in common:

- the target system is modeled associating its parts with a set of random variables;
- probabilities are assigned to events related to the identified variables;
- these probabilities are used to make predictions or supply other relevant information about the analyzed process.

In general, a stochastic approach to cross-layer reliability analysis starts from the characterization in terms of failure probability of individual portions of the system. The next step is the selection of an appropriate dependability modeling technique. Some of the widely used dependability modeling techniques include Reliability Block Diagrams (RBDs) [8], Fault Trees [9], Markov Chains (MCs) [10] and Bayesian Models (BMs) [11]. The selection among these modeling techniques depends upon numerous factors that include the level of available details, size and complexity of the system, etc. Once the modeling technique is selected, the third and the last step is the analysis of the generated model in order to calculate system level reliability metrics such as Mean-Time To Failure (MTTF) [12], Mean-Time Between Failure (MTBF) [12] and Mean-Time To Repair [12]. These metrics play a particularly useful role at the design stage of a system for scrutinizing the design alternatives without building the actual system.

The purpose of this chapter is to provide a generic overview of the stochastic models that are being utilized for cross-layer reliability analysis. These formal methods primarily include (i) RBDs, (ii) Markov models and (iii) BMs.

## 11.2 Methodologies

### *11.2.1 Reliability Block Diagrams*

RBDs are a simple formalism that has been used to asses various failure-related characteristics (e.g., reliability [13], availability [14] and dependability [15]) of a wide range of engineering systems.

Using RBDs, a computing system is represented in terms of *blocks* and *connector lines* [16–18]. In particular, in cross-layer reliability of computing systems, blocks may represent either the computational hardware cores (e.g., a CPU/GPU or a subunit of a GPU/CPU) or software functions, each characterized with some given failure rate. The connectors between them may be used to describe various alternative paths required for a successful computation.

By using an RBD, the failure characteristics of a system can be judged based on the failure rates of individual components, whereas the system's failure happens if all paths for successful execution fail. RBD-based analysis enables us to evaluate the impact of component failures on the overall system safety and reliability; thus, it is widely used for assessing the trade-offs of various possible system configurations at the system design stage.

In an RBD, the state of each block $i$ is represented by a state variable denoted with $x_i$ that can assume one of two states (i.e., *faulty* or *fault free*):

$$x_i = \begin{cases} 1 & \text{if block i is fault free} \\ 0 & \text{otherwise} \end{cases} \tag{11.1}$$

The vector $\mathbf{x} = (x_1, x_2, \ldots, x_n)$ is called the *state vector*.

The state of the full system can then be described by means of a binary function called *structure function* defined as

$$\phi(\mathbf{x}) = \begin{cases} 1 & \text{if the system is fault free} \\ 0 & \text{otherwise} \end{cases} \tag{11.2}$$

At a high level, the construction of the RBD model of a system is a composition of three basic architectures named here:

- series,
- active redundancy and
- standby redundancy.

In a series architecture, shown in Figure 11.1, the system is fault free if and only if all of its $n$ components are fault free. The structure function for this architecture is defined as

$$\phi_{series}(\mathbf{x}) = x_1 \cdot x_2 \cdots x_n = \prod_{i=1}^{n} x_i \tag{11.3}$$

In active redundancy architecture, the system is fault free if all components in at least one of the redundant stages are fault free. In this architecture, components may be connected in a parallel architecture or in a combination of series and parallel architecture as shown in Figure 11.2.

In parallel architecture, the structure function of the system can be defined as

$$\phi_{parallel}(\mathbf{x}) = 1 - (1 - x_1)(1 - x_2) \cdots (1 - x_m) = 1 - \prod_{i=1}^{m} (1 - x_i) \tag{11.4}$$

In a standby redundancy architecture, at least $k$ out of $n$ redundant blocks are required to be fault free in order to have a fault-free system, as represented in Figure 11.3.

In this architecture, the structure function of the system can be defined as

$$\phi_{standby}(\mathbf{x}) = \begin{cases} 1 & \text{if } \sum_{i=1}^{n} x_i \geq k \\ 0 & \text{otherwise} \end{cases} \tag{11.5}$$

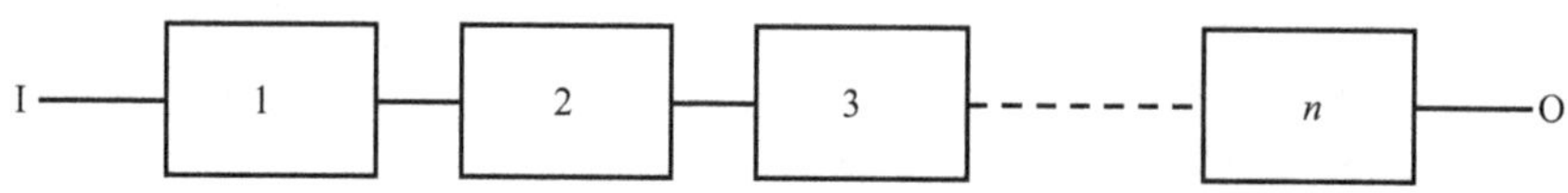


*Figure 11.1 Reliability Block Diagram of a series architecture*

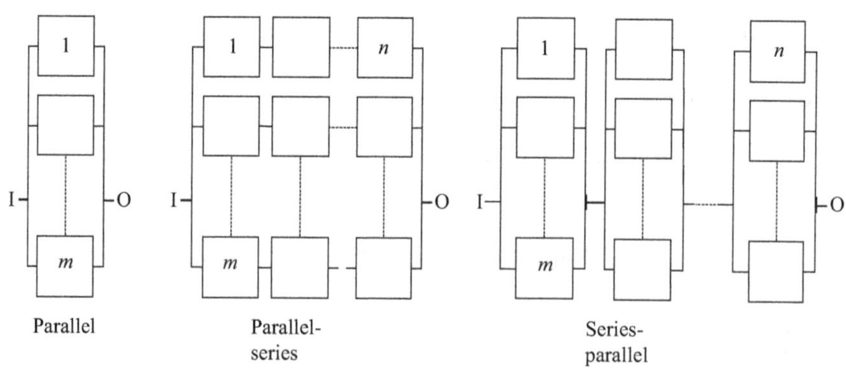


*Figure 11.2 Reliability Block Diagram of an active redundancy architecture*

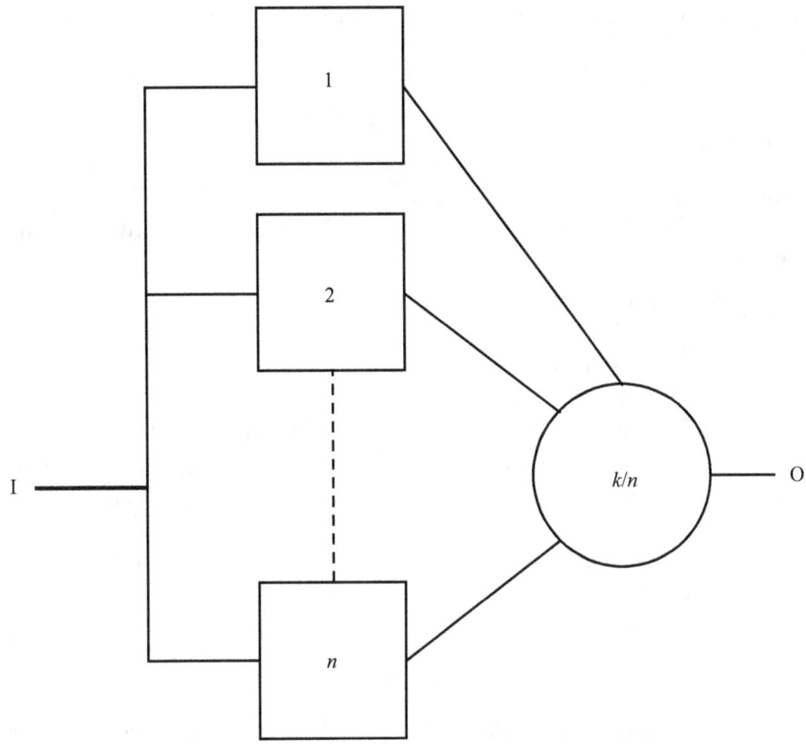


*Figure 11.3 Reliability Block Diagram of a standby redundancy architecture*

In order to define the structure function for a complex system with $i$ different blocks, the designer must first determine the *minimal path* sets of the system and then consider that the system is fault free if the blocks of at least one path set are fault free.

**Definition 11.1.** *A path set is a set of items that, if fault free, secure that the system is fault free. A path set is minimal if it cannot be reduced without losing its status as a path set.*

Alternatively, it is possible to determine the *minimal cut* sets and consider that the system fails if the items contained in one or more of the cut sets fail.

**Definition 11.2.** *A minimal cut is a set of items that, by failing, secure that the system fails. A cut set is said to be minimal if it cannot be reduced without losing its status as a path set.*

In the structure function, the state variables $x_i$ are deterministic quantities (faulty or fault free). However, in system reliability analysis, it is important to consider state variables as stochastic variables dependent on time. We denote the stochastic state variables as $X_i(t)$. Given the stochastic nature of the system failure, in reliability analysis it is important to quantify the probability of being in a specific state (i.e., faulty or fault free). This in turns requires to define the *survival function* $R_i(t)$ for block $i$:

$$\begin{aligned} Pr(X_i(t) = 1) &= R_i(t) \\ Pr(X_i(t) = 0) &= 1 - R_i(t) \end{aligned} \tag{11.6}$$

The survival function for a block, under the assumption of exponential time to failure (e.g., in the case of radiation-induced soft-errors in hardware), can be defined as

$$R_i(t) = e^{-\lambda_i t} \tag{11.7}$$

where $\lambda_i$ is the constant failure rate of block $i$, and $t$ is the time at which the survival probability is calculated.

Similarly to the structure function, the survival function of the full system can be computed by composing the functions of the single blocks as reported in Table 11.1. Note that the standby redundancy here is composed of identical components $R_1(t)$, $R_2(t), \ldots, p_R(t)$ all with equal survival function $R(t)$.

Given the survival function, the MTTF of the system can be calculated as

$$MTTF = \int_{t=0}^{\infty} R_s(t)dt \tag{11.8}$$

Traditionally, RBD-based analysis has been done using paper-and-pencil proof methods and computer simulations. However, these methods cannot ascertain absolute correctness due to their inherent limitations [19]. To overcome the earlier-mentioned inaccuracy problems, formal methods have been proposed for the RBD-based analysis as well. However, these methods have a limited scope and thus cannot be used to analyze all kinds of complex engineering systems. Moreover, RBDs assume the statistical independence of failures of each block. This is a strong assumption that does not always hold in reality.

### *11.2.2 Markov Chains*

MCs are a class of stochastic models that have received considerable attention in the reliability and safety domain [20].

*Table 11.1 Survival function composition at system level*

| System | Reliability function |
|---|---|
| Series architecture | $\prod_{i=1}^{n} R_i(t)$ |
| Parallel architecture | $1 - \prod_{i=1}^{m} (1 - R_i(t))$ |
| Standby redundancy | $\sum_{j=k}^{n} \binom{n}{j} R(t)^j (1 - R(t))^{n-j}$ |

MCs are stochastic processes whose futures are conditionally independent of their pasts given their present values. More formally, an MC is a stochastic process $X(t), t \geq 0$ defined as

$$\begin{aligned} Pr(X(t+1) &= x_{t+1}|X(t) = x_t, X(t-1) = x_{t-1}, \ldots, X(0) = x_0) \\ &= P(X(t+1) = x_{t+1}|X(t) = x_t) \end{aligned} \tag{11.9}$$

where $S = \{x_0, \ldots, x_{n-1}\}$ defines the *state space* of the stochastic process, i.e., a countable set of spaces the system may assume in time. Equation (11.9) defines the main property of an MC, also known as the *Markov property*:

**Definition 11.3.** *If the present state of the system is known, then the future of the system is dependent only on the present.*

The way time is considered, which makes it possible to distinguish two types of Markov models:

1. *Discrete Time MC* in which time is a discrete variable (and it is defined as a countable set) and
2. *Continuous Time MC (CTMC)* in which time is considered as a continuous variable.

In order to model a computing system using an MC, designers have to enumerate all possible states the system may assume. In a system described by means of different hardware and software components, with each component assuming different *internal* states, the state space $S$ enumerates all possible combinations of components' internal states.

Let us consider the simple example reported in Table 11.2. A simple system is split into two components, each characterized by two possible operational states (i.e., faulty or fault free). Globally the state space of the system is composed of four possible states obtained as a combination of the states of its components.

Once the state space of the system has been identified, it is possible to *create* a transition diagram representing how the system evolves from one state to the other

*Table 11.2 State space representation in a system composed of two components, each of them being in two possible states: faulty and fault free*

| **State** | **Component 1** | **Component 2** |
|---|---|---|
| 1 | Fault free | Fault free |
| 2 | Fault free | Faulty |
| 3 | Faulty | Fault free |
| 4 | Faulty | Faulty |

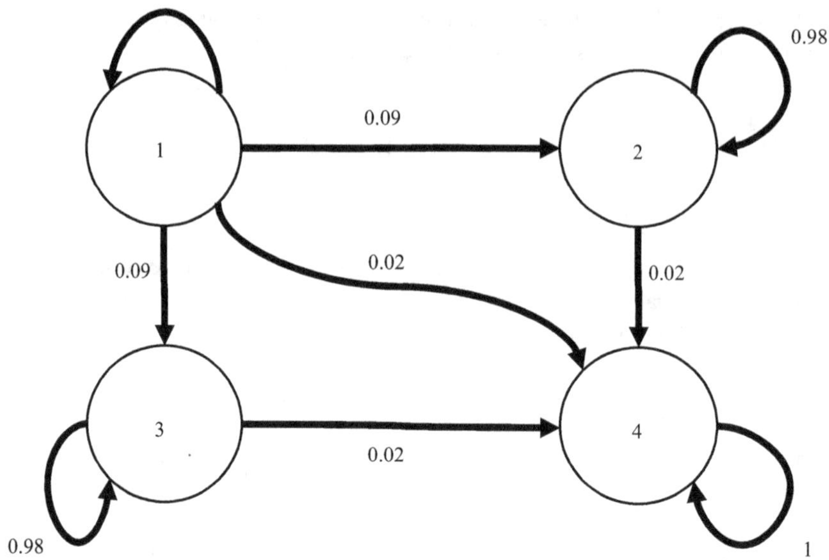


*Figure 11.4 Example of transition diagram based on the two-component example from Table 11.2*

ones. The transition graph is a weighted directed graph where nodes represent the system's states and edges represent the state transitions. Weights on the edges can represent either a probability or a *rate* at which the Markov process changes from one state to another (with the direction indicated by the arrow).

Figure 11.4 shows an example of such a diagram for the simple system introduced in Table 11.2. From a fully fault-free state (state 1), the system may enter different states in which one of its components (or both of them) becomes faulty (states 2–4). Supposing the system is without any repairing mechanism, there is no transition to go back from states 2–4 back to state 1, as well as it is not possible to move from state 2 to 3 (and vice versa).

Formally, an MC can be represented through its *transition matrix*:

$$\mathbb{TM}(t) = \begin{pmatrix} P_{1,1}(t) & P_{1,2}(t) & \cdots & P_{1,n}(t) \\ P_{2,1}(t) & P_{2,2}(t) & \cdots & P_{2,n}(t) \\ \vdots & \vdots & \ddots & \vdots \\ P_{n,1}(t) & P_{n,2}(t) & \cdots & P_{n,n}(t) \end{pmatrix} \tag{11.10}$$

Each element $P_{i,j}(t)$ of the matrix represents the probability of moving from state $i$ to state $j$. The entries in row $i$ represent the transitions out of state $i$, and the entries in column $j$ represent the transitions into state $j$.

Since all values in $\mathbb{TM}(t)$ are probabilities, the following property must be always respected:

$$\sum_{j=1}^{n} P_{i,j} = 1, \ 1 \leq j \leq n \ \wedge \ 0 \leq P_{i,j} \leq 1 \ i,j \in S \tag{11.11}$$

In order to properly define a system using MCs, we also need to define the initial state $P_x(0)$ of the system. Since MCs are stochastic models, this in fact means defining for each state the probability of that state to be an initial condition for the system. This set of probabilities can either be 1 for a single state (when the problem starts in a fixed and know state) or can contain uncertainty (by defining different probabilities with sum still equal to 1).

To solve the model and being able to determine the probability of moving from one state to another, the Kolmogorov forward equation can be applied:

$$\dot{P}_{x_i}(t) = \sum_{k=0}^{n-1} P_{k,i}(t) \cdot P_{x_k}(t) \tag{11.12}$$

which can be summarized as

$$\dot{P}_x(t) = \mathbb{TM}(t) \cdot P_x(t) \tag{11.13}$$

Equation (11.13) corresponds to a vector to matrix multiplication being able to evaluate all states at once and it is also note as *state equation*. Please note that both $P_x$ and $\dot{P}_x$ vectors add up to 1.

Most of the time CTMC are used to model reliability problems, and the Markov process is modeled as a process having two properties:

- the amount of time spent into each state $x_i$ is exponential with rate $\alpha_i$ and
- when the process leaves the state $x_i$, it enters a state $x_j$ with probability $P_{ij}$ (where $\sum_{j \neq i} P_{ij} = 1$).

Looking at each probability of transition, it is possible to define a *transition rate* from state $i \rightarrow j$ as

$$a_{ij} = \alpha_i \times P_{i,j} \Rightarrow \alpha_i = \sum_{j \neq i} a_{ij} \tag{11.14}$$

Since it is easy to move from the transition rate to the actual values of $\alpha_i$ and $P_{i,j}$, most of the works prefer to simplify the $\mathbb{TM}$ of a CTMC by storing only the different rates of the distribution, having a *transition rate matrix*:

$$\mathbb{TRM}(t) = \begin{pmatrix} a_{1,1} & a_{1,2} & \cdots & a_{1,n} \\ a_{2,1} & a_{2,2} & \cdots & a_{2,n} \\ \vdots & \vdots & \ddots & \vdots \\ a_{n,1} & a_{n,2} & \cdots & a_{n,n} \end{pmatrix} \tag{11.15}$$

In this case, all diagonal elements ($a_{i,i}$) must be filled with the $\alpha_i$ rates, such that the sum of all entries in each row is now equal to zero.

CTMCs are among the most common Markov models used for reliability purposes. Two characteristics make them suitable for the task:

1. The failure rate of the system is easily mapped with the rates defined for CTMC.
2. They can model a finite multistate system with different assumptions of repair [21].

When dealing with cross-layer analysis, the modeling challenge is to properly translate the impact of errors coming from the bottom layers of the system, i.e., the hardware layer, to the upper layers, i.e., the software layer. We can cite one study trying to capture hardware-related software failures using CTMC [22]. This chapter is a good example of what is required to model reliability into Markov processes but also of the issues involving this activity. The method defines three kinds of possible failures in a system:

1. *pure HW*: failures that do not directly affect the SW (e.g., direct shutdown of the system);
2. *pure SW*;
3. *HW/SW*: failures in the SW caused by HW modification.

In this scenario, the reliability of the system can be defined as

$$R(t) = R_{HW}(t) \cdot R_{SW}(t) \cdot R_{HW/SW}(t) \tag{11.16}$$

In a scenario where software failures are not considered, the $R_{SW}(t)$ reliability function is constant and equal to 1. Let us focus on the use of an MC to model the HW/SW interaction, which represents the cross-layer connection. For this reason, the reliabilities of HW and SW alone are defined by a single formula:

$$\begin{aligned} R_{HW}(t) &= e^{-\lambda t^{\beta}} \\ R_{SW}(t) &= e^{-(m(t+T)-m(T))} \end{aligned} \tag{11.17}$$

where $R_{HW}(t)$ represents the Weibull model function [23] and $R_{SW}(t)$ the NHPP model function [24].

Figure 11.5 describes a possible state diagram of HW/SW interaction. This example assumes that the HW failure causing the SW to misbehave comes at a specific rate $\lambda_1$ and first moves the system into a degradation stage, where three possible degradation effects are considered. Therefore, we need to weight the failure rate $\lambda_1$ by the probability of immediately detecting the failure ($p_1$ and the related probability of not detecting it $q_1 = 1 - p_1$) and, once detected, by the probability of recovering it with software methods ($p_2$ and its opposite $q_2 = 1 - p_2$). This stage shows the ability to model, assigning some specific rate ($\mu_{1a}$ and $\mu_{1b}$ in Figure 11.5), the capability of the system to gracefully recover and go back to a normal working state. This is where usually Markov models are mostly applied, when the model must include recovery mechanisms. Still looking at the example, other rates characterize the system reaching a failure state. Here three different cases are considered: (i) a complete abort of the execution, (ii) a total hardware failure or (iii) a software failure caused by the undetected hardware failure. Each specific rate can be estimated by performing ad-hoc experiments and simulations. As a general guideline, the transition rates can be purely theoretical or may be inferred from experimental analyses [25].

Once all rates are evaluated and the transition rate matrix is built, $R_{HW/SW}(t)$ of (11.16) can be computed by solving the obtained model. This starts from the definition of *steady state*.

**Definition 11.4.** *The steady state is the set of asymptotic probabilities ($P_i$) for each state of the model, i.e., when $t \rightarrow \infty$.*

All $P_i$ representing the steady state can be computed by solving the following system of equations (from (11.13)):

$$\begin{cases} P \cdot \mathbb{TMR}(t) = 0 \\ \sum_{i=0}^{n-1} P_i(t) = 1 \end{cases} \tag{11.18}$$

The steady state provides the probability of the system to be in every possible state. To compute the reliability of the system, it is enough to restrict the analysis to the fault-free states usually referred to as *functioning states*.

Considering the system described in Figure 11.5, we can define $W = \{0, 1a, 1b, 1c\}$ as the set of functioning states and $F = \{2, 3, 4\}$ as the set of failing states. Based on this, $R_{HW/SW}(t)$ can be computed as

$$R_{HW/SW}(t) = P_0(t) + P_{1a}(t) + P_{1b}(t) + P_{1c}(t) \tag{11.19}$$

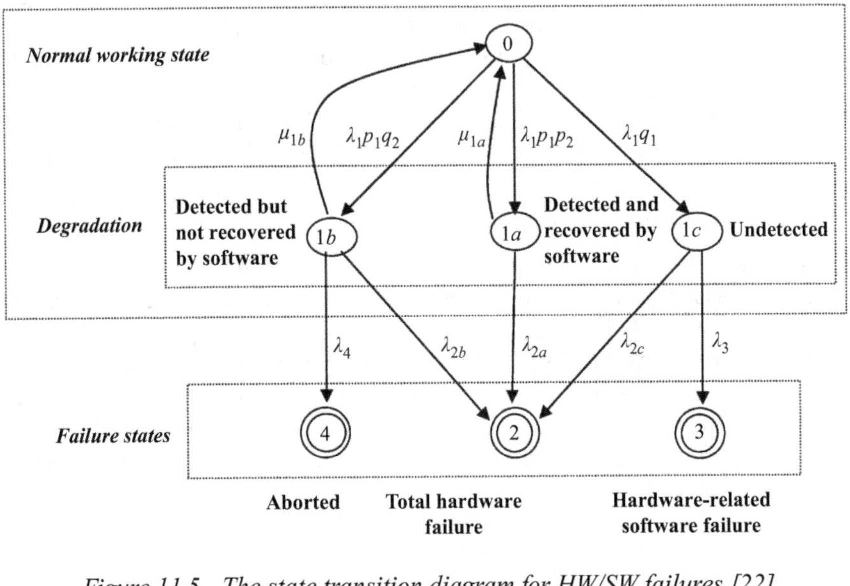


*Figure 11.5 The state transition diagram for HW/SW failures [22]*

where every $P_i(t)$ is computed according to (11.18).

In the considered example, the number of states is rather small, so the analysis of all states is quite simple. In real cases, in order to simplify the model evaluation in the presence of a larger set of states, all functional states are usually clustered into a single one [26].

The opposite of (11.19) is generally known as *Frequency of System Failures* ($\omega_F$), since it accounts for all failure states of the system. Using $\omega_F$, it is straightforward to evaluate the *MTBF* of the system.

$$MTBF = \frac{1}{\omega_F} \tag{11.20}$$

Similarly it is possible to evaluate several other reliability metrics [20].

While interesting, this approach lacks a real model of the propagation of errors. In Figure 11.5, states represent the macro-states of the whole system w.r.t. the failure. This representation is limited to *general* states and lacks of spatiality: there is no way to distinguish where the failure takes place and how it affects specific portions of the software. This limitation is particularly critical when the model is required to support space exploration tasks, where the goal is to choose an effective subset of components of the system to be protected.

In general, the reason for the lack of details in Markov models is a direct consequence of their poor scalability. As soon as the state space grows, the model becomes complex to analyze. Several examples in the literature proved that the Markov models are effective on the limited ones [27–29], usually for the rapid assessment of real-time scenarios or when reliability is considered from a safety perspectives [30].

To cope with the Markov model's complexity and to produce a rapid reliability assessment, Takaragi *et al.* [31] suggested to find and remove noneffective edges, i.e., edges that contribute less in the final steady-state calculation. This is an iterative process that produces a smaller graph, hence transition matrix, but it also introduces errors in the calculation.

Eventually, most of the presented approaches lack a complete toolset that is able to derive the different parameters required to build the model from fast simulations.

### *11.2.3 Bayesian Networks*

Bayesian Networks (BNs) are stochastic models that can be used to represent and analyze complex systems from a quantitative and a qualitative point of view through multivariate statistical distribution functions.

Weber *et al.* in a comprehensive review reported more than 200 papers published between 1998 and 2008 in international journals on applications of BNs in different fields of dependability, risk analysis and maintenance [32]. Differently from state-space-based models such as Markov models, BMs are better suited when component-based modeling is required.

A BN is a Directed Acyclic Graph formally defined as

$$S = (N, E, \Theta) \tag{11.21}$$

where

- $N = \{n_1, n_2, \ldots, n_m\}$ is the set of network nodes. Each node models a random variable. Random variables can be associated with different components of the system, thus modeling the set of states that the component may assume.
- $E = \{(n_i, n_j) \in N \times N\}$ is the set of arcs. Each arc defines a relationship between two nodes, indicating that the state of a node depends on the state of another node.
- $\Theta = \{\theta_1, \theta_2, \ldots, \theta_m\}$ is a set of Conditional Probability Tables (CPTs), one table for each node. The CPT $\theta_i$ of node $n_i$ defines the conditional probabilities of $n_i$ to be in a state $s_i$, given that its parent nodes are in a defined combination of states. Each such combination is called an *instantiation* of the parent set. The CPT maps each instantiation to a probability. Nodes without parents, called root nodes, are described according to their marginal probability distributions.

BNs are an interesting model for cross-layer reliability. Similarly to RBDs, they follow a component-based modeling approach in which the system is described in terms of components and their interaction. Component-based modeling facilitates the integration of the reliability analysis in the early phases of the design workflow. BNs are suitable to model systems where the information about what is going on is incomplete, as probability is one of the best ways to understand the modeled system. There are four relationship that BNs can model and give additional information about as follows:

- **Likelihood**: It determines if an event or a status of the system is more likely than another (e.g., the system is more likely to work properly than to fail).
- **Conditioning**: It is the representation of the Bayes conditionalization $P(A|B) = p$ (e.g., if a component of the system fails, the whole system can fail).
- **Causation**: It describes the relationship between two different propositions, whether one provokes the other (e.g., being the processor faulty caused the system to fail).
- **Relevance**: It describes the relationship between two different propositions in case there is a change of the belief. It is very similar to causation but, it also includes likelihood: two events are not relevant to each other when a common cause is predicted, but when a consequence is actually observed (e.g., whether the system fails depends on whether the processor is faulty).

An example of BN is shown in Figure 11.6.

The evaluation of a BN is the computation of the probabilities for each node $n_i$ of $S$ of being in every possible state:

$$P(n_1, n_2, \ldots, n_m) = \prod_{i=1}^{m} P(n_i|parents(n_i)) \tag{11.22}$$

Setting an *evidence* on a node of the network means modify our belief that the node is in a given state. In case no evidence is observed, solving the network is the belief propagation from the root nodes up to the leaf nodes.

Equation (11.22) is recursive. Its solution can be obtained resorting to two different approaches: one gives the exact solution while the other generates an approximate

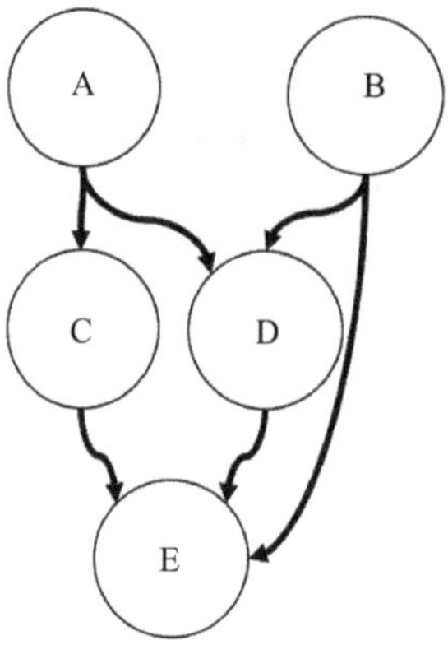


| $\theta_D$ | $A-1$ | | $A-2$ | | $A-3$ | |
|---|---|---|---|---|---|---|
| | $B-1$ | $B-2$ | $B-1$ | $B-2$ | $B-1$ | $B-2$ |
| $D-1$ | 0.5 | 0.2 | 0 | 0 | 0.3 | 0 |
| $D-2$ | 0 | 0.6 | 0.7 | 0 | 0.3 | 0.5 |
| $D-3$ | 0.25 | 0.1 | 0 | 1 | 0.3 | 0.5 |
| $D-4$ | 0.25 | 0.1 | 0.3 | 0 | 0.1 | 0 |

$P(D{=}D{=}2 \mid A{=}A{-}2,\, B = B{-}1)$

*Figure 11.6 These Bayesian Networks have several nodes and some of them are connected by directed arcs. The CPT of node D is composed of 24 entries. The total number of entries is given by the product of the number of states of node A, i.e., 3, the number of states of node B, i.e., 2 and the number of states of node D, i.e., 4. The sum of the entries of each column must be 1*

one. Evaluating the exact solution is in general NP hard so that the computation time may explode and its application is restricted to small BNs. Differently, the approximate approach enables shorter computation time in spite of inaccuracy of the results. In the literature, several algorithms were proposed concerning the computation of approximate solutions. The interested reader can refer to [33] for further information about the theory of BNs and the approaches commonly used to solve them.

The fields of application of BNs are many, including artificial intelligence, medical diagnosis and reliability. The remaining of this chapter presents two attempts of using BNs in the field of cross-layer reliability assessment we recently published. These are not exhaustive but provide an idea of the power of this model.

Vallero *et al.* [34,35] proposed to exploit BNs to model program traces, i.e., sequences of instructions issued by a microprocessor when a program is executed with a specific workload. The analysis of program traces is a powerful instrument to study the behavior of the program when hardware faults affect the microprocessor.

In this model, both the hardware and the software layers of the system are taken into account. The microprocessor Instruction Set Architecture (ISA) represents the direct link between the two layers. In particular, each instruction is considered with regard to the hardware resources required for its execution. To build a BN model of a program trace running in a system, it is essential to take into account the hardware resources, the error sources and the instructions of the analyzed trace.

Hardware resources, *Res*, are divided into two subsets: the storage resources subset ($S_{res}$), which includes registers and memories, and the computational resources subset ($C_{res}$), including functional units required for computation. Resources can be affected by errors during the program execution. In the case of soft errors, the reliability function of a resource ($R_{res}(t)$), i.e., the probability of an error-free resource in a period of time $t$, can be modeled as

$$R_{res}(t) = e^{-\lambda_{res} t} \tag{11.23}$$

$$\lambda_{res} = \lambda_{comp} \frac{A_{res}}{A_{comp}} \tag{11.24}$$

where $\lambda_{res}$ is the Soft Error Rate of the resource computed as a fraction of the error rate of the full hardware block assuming equal spatial distribution of soft-errors occurrence in the considered component.

The occurrence of errors in hardware resources can be masked at the hardware level due to several masking effects. A masking probability for each instruction of the ISA, representing the probability that an instruction prevents an error to propagate can therefore be defined. Masking of hardware resources can be obtained by error protection mechanisms implemented in the design (e.g., Error Correcting Codes for memories [36], Triple Module Redundancy (TMR) for registers [37], etc.) or by data transformations they perform (e.g., an AND functional block can mask a bit-lip of an operand if the not-faulty operand is equal to 0). This information can be characterized for every resource and every instruction by means of fault-injection campaign and then reused every time the execution of an application must be analyzed.

Figure 11.7 shows an example of a BN model for a simple program trace.

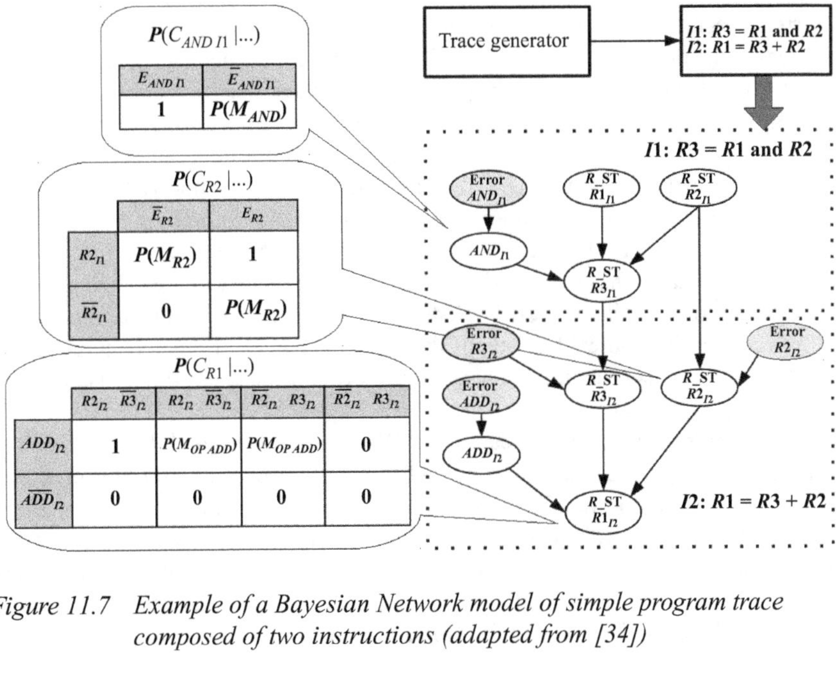


*Figure 11.7 Example of a Bayesian Network model of simple program trace composed of two instructions (adapted from [34])*

There are four types of nodes:

- **Error nodes modeling the source of errors occurring in each resource of each instruction**. The probability of an error-free error node connected to a computational resource is computed according to (11.23) with a proper $\lambda_{res}$ and $t$ equal to the execution time of the instruction. Instead, for storage resources, the probability of an error-free node is calculated by means of (11.23) considering a proper $\lambda_{res}$ but choosing $t$ equal to the period of time between the read and the write of the resource.
- **Nodes identifying input computational resources of an instruction** (e.g., $ANDI1$ in Figure 11.7). They are characterized by the masking probability of the hardware resource (e.g., $P(M_{AND})$).
- **Nodes identifying input storage resources of an instruction** (e.g., $R2_{I2}$ in Figure 11.7). They are characterized by the masking probability of the hardware resource (e.g., $P(M_{R2})$). In the proposed example, the CPT of these node models a condition in which if more than one node is faulty, the output is faulty as well (no masking is possible).
- **Nodes identifying output storage resources of an instruction** (e.g., $R1_{I2}$ in Figure 11.7). They model the probability that an instruction performed by the computational resource masks the error in an input storage resource (e.g., $P(M_{OP\,ADD})$). Also in this case, in our example, errors can be masked only when the input computational resource is error free and at most a single-input storage resource is faulty.

Once all nodes, arcs and CPTs of the BN are properly identified, the probability of correct execution of a trace can be estimated as the probability of all output storage resources of the program to be correct. After setting the evidence for these nodes to be correct, the BN can be solved and the probability of the evidence computed according to Equation (11.22). For more accurate results, several traces with different workloads can be obtained, each one producing a different trace and results can be averaged.

To give an example of the potential of this BM, Figure 11.8 shows results obtained considering three benchmark applications form MiBench [38] running on the x86-64 architecture. Traces have been extracted resorting to the MARSSx86 simulator [39] and the same simulator has been used to perform microarchitecture level fault injection to compare results obtained using the BMs with those obtained using fault injections.

As reported in Figure 11.8, one of the best advantages of this technique is that it significantly reduces the time required to perform the analysis (100× faster) while providing high accuracy.

A more comprehensive application of BN to cross-layer reliability analysis has been presented in [40–43]. Figure 11.9 provides a high-level overview of the proposed cross-layer reliability design framework. It provides two main functionalities: a fast and accurate model to evaluate the reliability of complex systems early in the design time and an efficient multiobjective Design Space Exploration (DSE) engine system supporting the designers during the initial design choices. The framework focuses on Radiation-Induced Failures leading to soft errors in memory structures.

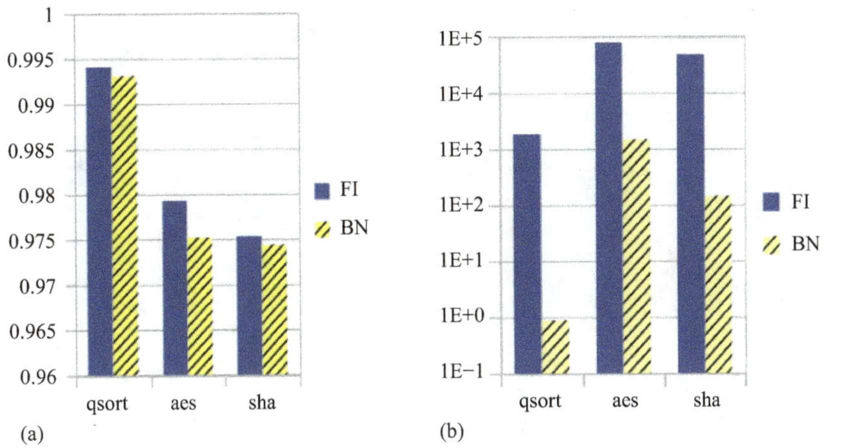


*Figure 11.8* *It shows the application of the proposed BM to a set of benchmarks running on the x86-64 architecture: (a) estimations of system level, masking probability for both the Bayesian Network (BN) and the fault injection (FI); (b) time required for the analysis expressed in second (adapted from [34])*

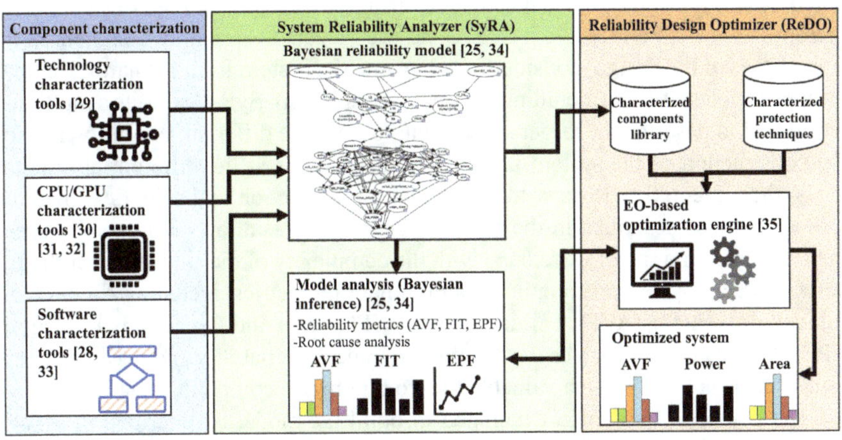


*Figure 11.9* *Overview of our complete cross-layer reliability framework. The component characterization toolset integrates a set of characterization tools for technologies [44], CPUs [45], GPUs [46,47] and software routines [40,48]. The tools are used to build the Bayesian reliability model that is at the core of our System Reliability Analyzer (SyRA) [41,43]. Eventually, the reliability model is exploited by our Reliability Design Optimizer (ReDO) [42] to evaluate several combinations of cross-layer mechanism and to trade off them with other design parameters such as Power and Area (adapted from [51])*

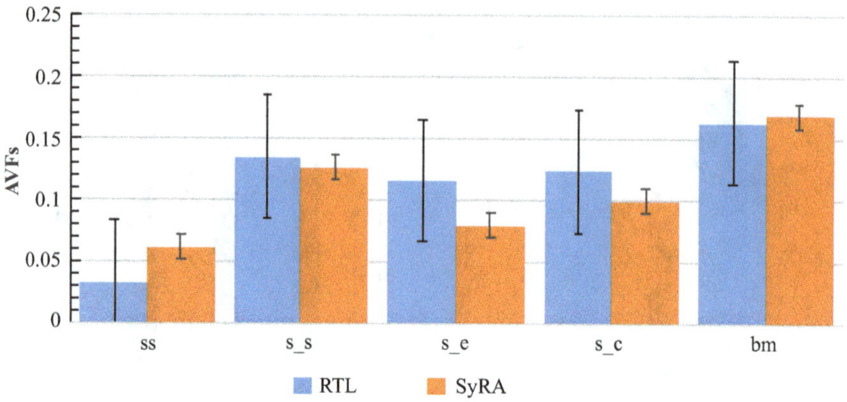


*Figure 11.10 Results obtained using SyRA to estimate the AVF of five different applications executed an ARM Cortex A9. The figure compares estimation provided by SyRA with those obtained using precise RTL fault injection. The full experimental setup is described in [43] (adapted from [51])*

BNs are used here to build a full SyRA being able to analyze the impact of radiation-induced soft errors in the memory arrays of a complex computing cores (i.e., microprocessor, GPUs) [41,43]. SyRA has been created to support designers in the early phases of the design, considering all layers of a system from the hardware up to the application software (including the operating system). SyRA exploits a multilevel hybrid BM to describe the target system and to estimate different reliability metrics. The construction of the system is based on simulations at the different abstraction levels, thus integrating BMs with traditional fault injection and analytical methods such as the ones described in the previous chapters. This allows designers to speed up the analysis and therefore to cope with the complexity of the simulation of the full software stack. SyRA can compute several reliability metrics, including Architecture Vulnerability Factor (AVF), Failures In Time (FIT) rate and Executions Per Failure (EPF). The last metric enables the designer to trade off reliability and performance in a single measure providing a valuable tool to optimize a computing system.

The complete tool-chain developed to build the model is described in [41,43] and an example of the accuracy of the analysis performed by SyRA is reported in Figure 11.10. The proposed framework scales efficiently with the complexity of the system. On average, it is 68% faster than full micro-architecture level fault injection and two orders of magnitude faster than RTL fault injection while maintaining a comparable accuracy [49].

Bayesian inference supported by SyRA enables speculation on the effects that different protection mechanisms have on the system. This feature has been used to build ReDO, a DSE framework to build soft-error resilient computing systems [42]. ReDO is designed to support the early phase of the design of a computing system. It evaluates the application of selected classes of cross-layer soft-error protection

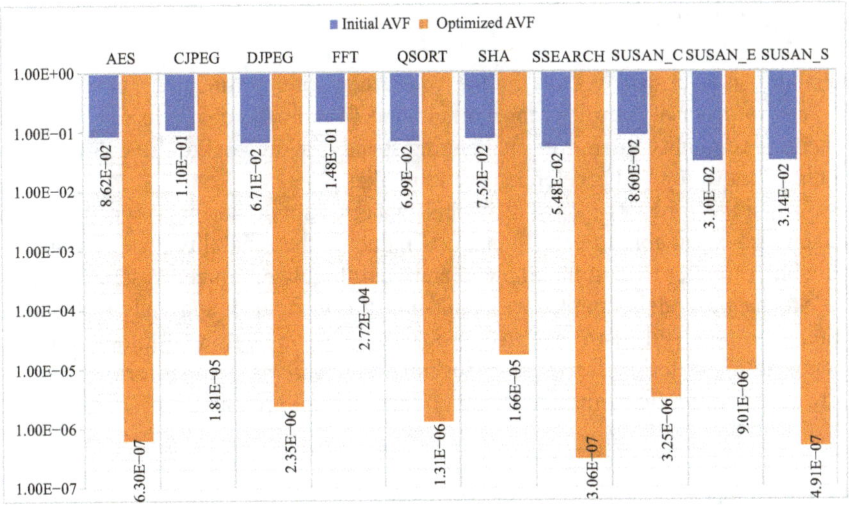


*Figure 11.11* *Results obtained using ReDO to optimize different microprocessor-based systems. The full experimental setup is described in (adapted from [51])*

techniques taking into account multiple design objectives. Exploiting this framework during the design phase, reliability can be traded off with other design constraints, i.e., hardware area, software size, performance and power consumption.

ReDO internally models the target system resorting to our proposed reliability BM [41,43]. This model provides a very compact component-based representation of the system stack (from the fabrication technology up to the software layer).

On top of the reliability model of the system, ReDO builds a new exploration heuristic, based on the Extremal Optimization (EO) theory [50]. The heuristic is designed to efficiently explore the design space composed of different combinations of cross-layer protection mechanisms applied to the components of the system. The goal of the EO is to optimize a global variable by improving local variables that involve coevolutionary avalanches. This is important in a cross-layer reliability scenario in which we want to evaluate how the application of different combinations of local protection mechanisms in selected components of the system (improvements of local variables) affect the global characteristics of the system in terms of reliability combined with power, area and performance (global variable).

The combination of the proposed reliability model with the DSE heuristic supports the analysis of a complex system in a limited computation time. This makes ReDO an interesting option to support designers in the early phases of the design cycle, when strategic decisions must be taken to design highly optimized systems.

The full framework is described in [42]. Figure 11.11 shows an example of the AVF improvement that ReDO can achieve for different software benchmarks when given the freedom to play with different microprocessor architectures and different HW/SW fault-tolerance mechanisms.

## 11.3 Conclusions

Reliability is a critical vector for the whole compute continuum. Understanding the requirements of the different markets and finding the cost-effective solutions to address reliability in an ever-challenging space is a key goal to drive down costs, increase innovation and accelerate time to market.

One of the primary needs to reach this goal is to support designers with dedicated frameworks being able to ease the task of reducing the reliability "tax" driving the available resources toward the implementation of highly optimized systems.

Stochastic models, coupled with fault-injection techniques and analytical models, represent a powerful instrument to cope with the complexity of future computing system, allowing designers to explore several design options taking informed decision early in the design flow when strong optimization is still possible.

## References

[1] Lee J and Shrivastava A. Static analysis of register file vulnerability. IEEE Transactions on Computer-Aided Design of Integrated Circuits and Systems. 2011;30(4):607–616.

[2] Mukherjee SS, Weaver C, Emer J, *et al.* A systematic methodology to compute the architectural vulnerability factors for a high-performance microprocessor. In: Proceedings of the 36th Annual IEEE/ACM International Symposium on Microarchitecture. MICRO 36. Washington, DC, USA: IEEE Computer Society; 2003. p. 29. Available from: http://dl.acm.org/citation.cfm?id=956417.956570.

[3] Li X, Adve SV, Bose P, *et al.* SoftArch: an architecture-level tool for modeling and analyzing soft errors. In: 2005 International Conference on Dependable Systems and Networks (DSN'05). Yokohama, Japan: IEEE; 2005. p. 496–505.

[4] Wang NJ, Mahesri A, and Patel SJ. Examining ACE analysis reliability estimates using fault-injection. In: Proceedings of the 34th Annual International Symposium on Computer Architecture. ISCA '07. New York, NY, USA: ACM; 2007. p. 460–469.

[5] Bell R and Bennet P. IEC 61508. Functional safety of electrical/electronic/programmable electronic safety-related systems. Computing & Control Engineering Journal. 2000;11(2):3–5. Available from: https://digital-library.theiet.org/content/journals/10.1049/cce_20000108.

[6] British Standards Institute Staff. Road vehicles. Functional safety. Guideline on ISO 26262. B S I Standards; 1912. Available from: https://books.google.it/books?id=xANfnAEACAAJ.

[7] Filieri A, Ghezzi C, Grassi V, *et al.* Reliability analysis of component-based systems with multiple failure modes. In: Grunske L, Reussner R, and Plasil F, editors. Component-Based Software Engineering. Berlin, Heidelberg: Springer Berlin Heidelberg; 2010. p. 1–20.

[8] Čepin M. Reliability block diagram. In: Assessment of Power System Reliability. Springer; 2011. p. 119–123. Available from: https://www.springer.com/gp/book/9780857296870

[9] Vesely WE, Goldberg FF, Roberts NH, *et al.* Fault tree handbook. Washington, DC: Nuclear Regulatory Commission; 1981.

[10] Gilks W. Markov chain Monte Carlo. Wiley Online Library; 2005. Available from: https://onlinelibrary.wiley.com/doi/abs/10.1002/0470011815.b2a14021

[11] Ben-Gal I. Bayesian networks. In: Encyclopedia of Statistics in Quality and Reliability. vol. 1. John Wiley & Sons, Ltd; 2008.

[12] Avizienis A, Laprie JC, Randell B, *et al.* Fundamental concepts of dependability. University of Newcastle upon Tyne, Computing Science; 2001.

[13] Lin CM, Teng HK, Yang CC, *et al.* A mesh network reliability analysis using reliability block diagram. In: 2010 8th IEEE International Conference on Industrial Informatics. Osaka, Japan: IEEE; 2010. p. 975–979.

[14] Huffman DL and Antelme F. Availability analysis of a solar power system with graceful degradation. In: 2009 Annual Reliability and Maintainability Symposium; 2009. p. 348–352.

[15] Wei B, Lin C, and Kong X. Dependability modeling and analysis for the virtual data center of cloud computing. In: 2011 IEEE International Conference on High Performance Computing and Communications; 2011. p. 784–789.

[16] Forche RF. Analysis of reliability block diagrams with multiple blocks per component. In: Annual Proceedings on Reliability and Maintainability Symposium. Los Angeles, California, USA: IEEE; 1990. p. 145–148.

[17] Guo H and Yang X. A simple reliability block diagram method for safety integrity verification. Reliability Engineering & System Safety. 2007;92(9):1267–1273. Critical Infrastructures. Available from: http://www.sciencedirect.com/science/article/pii/S0951832006001682.

[18] Trivedi KS. Probability & statistics with reliability, queuing and computer science applications. John Wiley & Sons; 2008. Available from: https://onlinelibrary.wiley.com/doi/book/10.1002/9781119285441.

[19] Hasan O, Ahmed W, Tahar S, *et al.* Reliability block diagrams based analysis: A survey. In: AIP Conference Proceedings. vol. 1648. Rodhes, Greece: AIP Publishing; 2015. p. 850129.

[20] Rausand M and Høyland A. System Reliability Theory: Models, Statistical Methods, and Applications. Wiley Series in Probability and Statistics. Wiley; 2004. Available from: https://books.google.it/books?id=gkUWz9AA-QEC.

[21] Lim TJ, and Lie CH. Analysis of system reliability with dependent repair modes. IEEE Transactions on Reliability. 2000;49(2):153–162.

[22] Teng X, Pham H, and Jeske DR. Reliability modeling of hardware and software interactions, and its applications. IEEE Transactions on Reliability. 2006;55(4):571–577.

[23] Bain L. Statistical analysis of reliability and life-testing models: theory and methods. Routledge; 2017. Available from: https://onlinelibrary.wiley.com/doi/abs/10.1002/zamm.19800601021.

[24] Pham H. Software reliability. Springer Science & Business Media; 2000. Available from: https://www.springer.com/gp/book/9781852339500
[25] Gupta G, Mishra RP, and Jain P. Reliability analysis and identification of critical components using Markov model. In: 2015 IEEE International Conference on Industrial Engineering and Engineering Management (IEEM). Singapore: IEEE; 2015. p. 777–781.
[26] Ghaderi A, Haghifam MR, and Abedi SM. Application of Monte Carlo simulation in Markov process for reliability analysis. In: 2010 IEEE 11th International Conference on Probabilistic Methods Applied to Power Systems. Singapore: IEEE; 2010. p. 293–298.
[27] Huang Y, Pan X, and Hu L. Rapid assessment of system-of-systems (SoS) mission reliability based on Markov chains. In: 2015 First International Conference on Reliability Systems Engineering (ICRSE). Beijing, China: IEEE; 2015. p. 1–6.
[28] Kolek L, Ibrahim MY, Gunawan I, *et al.* Evaluation of control system reliability using combined dynamic fault trees and Markov models. In: 2015 IEEE 13th International Conference on Industrial Informatics (INDIN). London, UK: IEEE; 2015. p. 536–543.
[29] Ammar M, Hamad GB, Mohamed OA, *et al.* System-level analysis of the vulnerability of processors exposed to single-event upsets via probabilistic model checking. IEEE Transactions on Nuclear Science. 2017;64(9): 2523–2530.
[30] Chen H and Qian Y. Reliability and safety analysis of cross-redundant structure based on Markov process. In: 2012 Fifth International Symposium on Computational Intelligence and Design. vol. 2. Hangzhou, China: IEEE; 2012. p. 406–409.
[31] Takaragi K, Sasaki R, and Shingai S. A method of rapid Markov reliability calculation. IEEE Transactions on Reliability. 1985;R-34(3):262–268.
[32] Weber P, Medina-Oliva G, Simon C, *et al.* Overview on Bayesian networks applications for dependability, risk analysis and maintenance areas. Engineering Applications of Artificial Intelligence. 2012;25(4): 671–682.
[33] Pearl J. Probabilistic reasoning in intelligent systems: networks of plausible inference. San Francisco, CA, USA: Morgan Kaufmann Publishers Inc.; 1988.
[34] Vallero A, Savino A, Tselonis S, *et al.* Bayesian network early reliability evaluation analysis for both permanent and transient faults. In: 2015 IEEE 21st International On-Line Testing Symposium (IOLTS). Halkidiki, Greece: IEEE; 2015. p. 7–12.
[35] Vallero A, Savino A, Tselonis S, *et al.* A Bayesian model for system level reliability estimation. In: 2015 20th IEEE European Test Symposium (ETS). Cluj-Napoca, Romania: IEEE; 2015. p. 1–2.
[36] Fabiano M, Indaco M, Di Carlo S, *et al.* Design and optimization of adaptable BCH codecs for NAND flash memories. Microprocessors and Microsystems. 2013;37(4-5):407–419.

[37] Hudson S, Sundar RS, and Koppu S. Fault control using triple modular redundancy (TMR). In: Progress in Computing, Analytics and Networking. Singapore: Springer; 2018. p. 471–480.

[38] University of Michigan at Ann Arbor. MiBench Version 1.0. Available from: http://vhosts.eecs.umich.edu/mibench//.

[39] Patel A, Afram F, Chen S, *et al.* MARSS: a full system simulator for multicore x86 CPUs. In: Proceedings of the 48th Design Automation Conference. New York, NY, USA: ACM; 2011. p. 1050–1055.

[40] Vallero A, Tselonis S, Foutris N, *et al.* Cross-layer reliability evaluation, moving from the hardware architecture to the system level: a CLERECO EU project overview. Microprocessors and Microsystems. 2015;39(8):1204–1214.

[41] Vallero A, Savino A, Politano G, *et al.* Cross-layer system reliability assessment framework for hardware faults. In: 2016 IEEE International Test Conference (ITC). Fort Worth, TX, USA: IEEE; 2016. p. 1–10.

[42] Savino A, Vallero A, and Di Carlo S. ReDO: cross-layer multi-objective design-exploration framework for efficient soft error resilient systems. IEEE Transactions on Computers. 2018;67(10):1462–1477.

[43] Vallero A, Savino A, Chatzidimitriou A, *et al.* SyRA: early system reliability analysis for cross-layer soft errors resilience in memory arrays of microprocessor systems. IEEE Transactions on Computers. 2019;68(5):765–783.

[44] Riera M, Canal R, Abella J, *et al.* A detailed methodology to compute soft error rates in advanced technologies. In: Proceedings of the 2016 Conference on Design, Automation & Test in Europe. EDA Consortium. Dresden, Germany: IEEE; 2016. p. 217–222.

[45] Kaliorakis M, Tselonis S, Chatzidimitriou A, *et al.* Differential fault injection on microarchitectural simulators. In: 2015 IEEE International Symposium on Workload Characterization. Atlanta, GA, USA: IEEE; 2015. p. 172–182.

[46] Vallero A, Gizopoulos D, and Di Carlo S. SIFI: AMD southern islands GPU microarchitectural level fault injector. In: 2017 IEEE 23rd International Symposium on On-Line Testing and Robust System Design (IOLTS). Thessaloniki, Greece: IEEE; 2017. p. 138–144.

[47] Tselonis S and Gizopoulos D. GUFI: a framework for GPUs reliability assessment. In: 2016 IEEE International Symposium on Performance Analysis of Systems and Software (ISPASS). Uppsala, Sweden: IEEE; 2016. p. 90–100.

[48] Kooli M, Di Natale G, and Bosio A. Cache-aware reliability evaluation through LLVM-based analysis and fault injection. In: 2016 IEEE 22nd International Symposium on On-Line Testing and Robust System Design (IOLTS). Sant Feliu de Guixols, Spain: IEEE; 2016. p. 19–22.

[49] Chatzidimitriou A, Kaliorakis M, Gizopoulos D, *et al.* RT level vs. microarchitecture-level reliability assessment: case study on ARM(R) Cortex(R)-A9 CPU. In: 2017 47th Annual IEEE/IFIP International Conference on Dependable Systems and Networks Workshops (DSN-W). Denver, CO, USA: IEEE; 2017. p. 117–120.

[50] Boettcher S. Extremal optimization: heuristics via coevolutionary avalanches. Computing in Science Engineering. 2000;2(6):75–82.

[51] Vallero A, Savino A, Carelli A, and Di Carlo S. Bayesian models for early cross-layer reliability analysis and design space exploration. In: 2019 IEEE 25th International Symposium on On-Line Testing and Robust System Design (IOLTS). Rhodes, Greece; 2019. p. 143–146, Available from: http://10.1109/IOLTS.2019.8854452.

# Index

www.ingramcontent.com/pod-product-compliance
Lightning Source LLC
LaVergne TN
LVHW011846200926
848484LV00002B/407